Springer Mineralogy

More information about this series at http://www.springer.com/series/13488

Victor Garanin · Konstantin Garanin ·
Galina Kriulina · George Samosorov

Diamonds from the Arkhangelsk Province, NW Russia

Springer

Victor Garanin
A. E. Fersman Mineralogical Museum
of RAS
Moscow, Russia

Galina Kriulina
Mineralogy Department, Geological
Faculty
M. V. Lomonosov Moscow State University
Moscow, Russia

Konstantin Garanin
OJSC "ALROSA"
Mirny, Russia

George Samosorov
Mineralogy Department, Geological
Faculty
M. V. Lomonosov Moscow State University
Moscow, Russia

ISSN 2366-1585 ISSN 2366-1593 (electronic)
Springer Mineralogy
ISBN 978-3-030-35716-0 ISBN 978-3-030-35717-7 (eBook)
https://doi.org/10.1007/978-3-030-35717-7

This Springer imprint is published by the registered company Springer Nature Switzerland AG
The registered company address is: Gewerbestrasse 11, 6330 Cham, Switzerland

Dedicated to the pioneers of Russia's diamond deposits, real field geologists, brilliant and gifted, who are among the best people of Russia

Introduction

The contemporary notion of typomorphism introduced into mineralogy by A. E. Fersman means "genetic causation of attributes and signs of minerals," i.e. typomorphic signs of minerals explicitly characterize the conditions of their origination. The typomorphic peculiarities of diamond as such that enable to predict the presence of diamondiferous kimberlites in the study area, identify diamond halos, reveal links to the known pipes or unidentified ore bodies, and evaluate the quality and cost of diamonds, comprise a key element in the system of predicting, prospecting and evaluating diamond fields (Vaganov 2000; Garanin et al. 2009; Kriulina et al. 2011, Kriulina 2012). A wide array of diamond typomorphic properties have been revealed to date—morphology, distribution of nitrogen and hydrogen impurity centers, internal structure, carbon isotope composition and others, specific to crystals of each non-placer. All those parameters make an indispensable set of tools for quality evaluation of rough diamonds and, ultimately, their cost property. This is particularly crucial in evaluation of diamond gemological characteristics, which, as polished diamonds, manifest the beauty and play of color that thrill and excite the girls and women as the fair sex, and the men as the other half of mankind which is captivated by the beauty of women in sumptuous diamond jewelry.

Opening the first diamond deposit (named after M. V. Lomonosov) in Arkhangelsk region in early 1980s, and afterwards the V. Grib deposit in the same region in 1996 allowed viewing the area as a new industrial source of diamonds dubbed Arkhangelsk Diamondiferous Province. At the time this name was likely prompted by the hope of the pioneers that a vast kimberlite province, similar to the one of Yakutia, would be opened in this European part of Russia. It is clear now that the industrial potential of the region is far from being assessed to the full extent. The area yet has two deposits, while Yakutia can boast of more than 25, so detailed survey of the properties of diamond as such, the paramount indicator of predicting and prospecting for new fields, is a focus of prospecting and exploration in the European part of Russia. We have been searching for diamond deposits, rather than its indicator and paragenetic minerals (olivine, pyrope, chromite, picroilmenite, chrome-diopside). But, as we know, indicator minerals play a crucial role in searching for diamond deposits. Researchers have been displaying keen interest for the area—the northwest of European Russia—over the past 40–45 years. Apart from seven medium- and high-diamondiferous kimberlite pipes (the M. V. Lomonosov and V. Grib deposits), the province has about 100 more bodies largely comprised by alkaline ultrabasic kimberlite and related rocks, which form a series of individual fields (clusters), with proven content of diamonds in some of them, though in negligent amounts. The most essential research efforts of investigating the deposits of diamonds, kimberlites and related rocks, and the diamond mineral as such in the Arkhangelsk Diamondiferous Province were held by Makhin et al. 1990; Makhin 1991; Bartoshinsky et al. 1992; Galimov et al. 1994; Poberezhskaya 1995; Sablukov 1995; Mineeva et al. 1996; Sobolev et al. 1997; Bogatikov et al. 1999; Sablukov et al. 2000; Garanin et al. 2001; Verzhak 2001; Zakharchenko et al. 2002; Verichev 2002; Golovin 2003; Garanin 2004; Kudryavtseva et al. 2004; Kudryavtseva et al. 2005; Verichev et al. 2005; Garanin 2006; Garanin et al. 2006; Khachatryan et al. 2006; Palazhchenko et al. 2006; Kononova et al. 2007; Khachatryan et al. 2008; Palazhchenko 2008; Tretyachenko 2008; Kopchikov 2009; Kriulina 2012; Khachatryan 2017. A variety of articles, monographs and atlases is primarily dedicated to studying the typomorphic properties of diamonds from the V. Grib and M. V. Lomonosov deposits, whereas the data on diamonds from pipes and even whole fields are unavailable in some cases. Besides, eight new bodies have been discovered in the province over the past decades, some of them with diamond crystals (Palazhchenko 2008; Gunin et al. 2010). In general, the time is right to generalize all possible materials on diamonds from that region in a single major monographic paper and pay tribute to all those people (geologists, drillers, technicians and so on) who have contributed to evolution of the diamond geology in the European Russia, which led to discovery of new diamondiferous pipes in Russia's Karelia and in Finland, and development of known diamond deposits in Arkhangelsk region. All those discoveries have helped understand clearly that minerагenic zoning must take place in view of all those new discoveries in a wider taxon of the Karelian-Kola subprovince in the East European province (as reported by NIGP (Geo-Scientific Research Enterprise) ALROSA (PJSC)). "*Archangelsk diamondiferous kimberlite-picrite region can be deemed as an element of the Early Hercynian Kola-Dvina subprovince of carbonatites and kimberlites (Frolov 2005, Belov 2008). Or else, the Arkhangelsk diamondiferous kimberlite-picrite region as an element of the global (regional) Baltic—White Sea subprovince of carbonatites and kimberlites,*" says Vladimir Tretyachenko (verbal statement). That is why now we need to consider the typochemistry of diamonds from bodies of the Zimny Bereg area in view of all those new data related to the entire north of the East European Craton. To this effect, the term "Archangelsk Diamondiferous Province" should be solely regarded from the historical side, meaning that at the initial discovery stage the name fueled confidence of local geologists that a new province comparable with the Yakutian one would soon originate in the area. Nowadays, this is more likely to be defined by the following taxon—the Arkhangelsk Diamondiferous Kimberlite-Picrite Region (ADR). We will stick to those definitions further when viewing the geology and diamond potential of the region.

This summary paper primarily focused on identifying the typomorphism of diamonds from deposits and other explored pipes and bodies of ADR's Zimny Bereg area to evaluate the conditions of diamond genesis in the entire province, whereby we paid attention to the use of data on typomorphic properties of diamond to resolve the exploration, prediction and evaluation tasks.

The paper relies on original results of studying some 50,000 diamond crystals from pipes of ADR's Zimny Bereg area that vary by their diamond potential, inter alia from seven prospecting blocks of the said area; those results were submitted by Arkhangelskgeoldobycha PJSC. Additional studies covered 20 diamond crystals from bodies of new discoveries in the ADP—the Kepina field pipes (Galina, Rozhdestvenskaya and sill 746b) received for study from ALROSA-Pomorye division of ALROSA.

The diamond crystals were submitted by ALROSA-Pomorye division of ALROSA, Severalmaz PJSC and Arkhangelskgeoldobycha. The diamond collection was compiled to address the crucial mineralogical, genetic, practical, prediction and prospecting tasks based on representative (quarter) samples from commercial production lots for Arkhangelskaya and Karpinskogo-1 pipes; geo-prospecting samples from Pionerskaya, Karpinskogo-2, Lomonosovskaya, and Pomorskaya pipes. For the comparison utilized diamonds of similar size groups: −12+9, −9+5, −4+3 (average weight 0.15–0.45, −0.05–0.15 and up to 0.05 ct, respectively).

In terms of diamond quantity, crystals from pipes of the Zolotitsa field (Snegurochka, Pervomayskaya, Koltsovskaya) prevail in the studied collection from pipes with low diamond potential. The total quantity of diamonds from low diamond potential pipes of other ADR fields does not exceed 150 crystals: the biggest number of crystals—136—comes from bodies of the alkaline ultrabasic magmatism in the Kepina field. The share of diamonds in individual pipes and bodies varies from single to several dozens of crystals (maximum 30). Since none of the crystal samples from individual bodies in the Kepina, Verkhotina and Izhmozero fields appear to be statistically representative to characterize the diamond properties, it would be proper to examine the data of the studied bodies as a whole.

Optical and scanning microscopy methods, including color cathodoluminescence imaging, mass spectroscopy and optical spectroscopy in the infrared, ultraviolet and visible spectra, photoluminescence spectroscopy and electron paramagnetic resonance are applied to explore the constitutional peculiarities of diamond from pipes and bodies of the ADR: external morphology and internal structure, chemistry, spectroscopic and other physical properties of diamond, external effects to physical properties of crystals, such as color, grade, toughness.

This entire abundant multi-year material on examining the properties of ADR diamonds has been accumulated at the Diamond Deposits Laboratory (Faculty of Geology at Moscow State University), with data on 45,000 diamond crystals, including from the M. V. Lomonosov and V. Grib deposits. Other reported materials were also used (data on 5000 diamond crystals), previously obtained by employees of the Central Scientific Research Geoprospecting Institute of Nonferrous and Precious Metals at the Ministry of Natural Resources (TsNIGRI at MNR of Russia), NIGP ALROSA, ALROSA-Pomorye division, and others.

Here, in the introduction, our immensely warm and sincere gratitude goes to the geologists of Arkhangelsk, whom the authors of the paper teamed and befriended for many years, who were giving the materials, remained close and committed partners. First and foremost, to pioneers and discoverers of diamond deposits in the region, V.

P. Grib, I. P. Dobeyko, S. V. Bortnik, E. M. Verichev, V. V. Verzhak, A. S. Galkin, N. N. Golovin, P. V. Grib, A. P. Gunin, N. A. Diakova, T. A. Pavlenko, V. V. Tretyachenko, N. V. Shirobokov and many other geologists, true professionals, transparent and decent persons.

We would also like to give acknowledgements to our mentors, associates, friends, students of the Faculty of Geology at the Lomonosov Moscow State University—academicians V. S. Urusov, A. A. Marakushev, professor G. A. Krutov, assistant professor G. I. Bocharova, senior research fellow A. V. Bovkun, Ph.D.s in geology and mineralogy O. V. Palazhchenko, Dr. M. B. Kopchikov, professors A. S. Marfunin, D. G. Koshchug.

We extend our deepest gratitude and acknowledgements to academician Nikolay Pavlovich Laverov, renowned ore geologist, scientist and statesman of Russia, who prompted us to head for Arkhangelsk region in early 1980s to prospect for diamond-bearing bodies and contributed greatly in establishing business and friendly relations with NPO Arkhangelskgeologiya which discovered the first diamondiferous pipes in Arkhangelsk region in the 1980s.

Authors of the paper would like to express their distinct attitude to and unfailing love for their mentor and companion, distinguished mineralogist, D.Sc. in geology and mineralogy, academic advisor for diamond-related topics in the Faculty of Geology at the Lomonosov Moscow State University, winner of the Prize of the Government of Russia—Galina Petrovna Kudryavtseva, who prematurely departed in 2006. We shall cherish the memories about her to carry them over to the existing and future diamond professionals in the Faculty of Geology at the M. V. Lomonosov Moscow State University, and to all young diamond-focused geologists of Russia and World.

References

Bartoshinsky, Z.V., Bekesha, S.N., Vinnichenko, T.G., Makhin, A.I.: Crystal morphology of diamond from kimberlites of Arkhangelsk diamondiferous province. Miner. Sb. Lvov University. **46**(2), 64–73 (1992)

Belov, S.V., Lapin, A.V., Tolstov, A.V., Frolov, A.A.: Metallogeny of platform magmatism (traps, carbonatites, kimberlites), p. 537. ND RAS, Novosibirsk (2008)

Bogatikov, O.A., Garanin, V.K., Kononova, V.A., et al.: Arkhangelsk diamondiferous province, p. 522. MSU Publishing, M. (1999)

Frolov, A.A., Lapin, A.V., Tolstov, A.V., et al.: Kimberlites and carbonatites (relationships, minerageny, and forecasting), p. 540. NIA-Priroda, M. (2005)

Galimov, E.M., Zakharchenko, O.D., Maltsev, K.A., Makhin, A.I.: Isotopic composition of carbon in diamonds from kimberlite pipes of Arkhangelsk region. Geochemistry **1**, 74–76 (1994)

Garanin, K.V.: Alkaline ultramafic magmatites of Zimny Bereg: their potential diamond-bearing capacity and commercial development prospects. Ph.D. in geology and mineralogy thesis, p. 50. MSU, M. (2004)

Garanin, K.V., Garanin, V.K., Kudryavtseva, G.P., Palazhchenko, O.V.: Morphological and spectroscopic peculiarities of diamonds from V. Grib deposit from Arkhangelsk diamondiferous province. Article 2. Spectroscopic characteristics and their relation to morphology of crystals. Proceedings of IHLs. Series Geology and Exploration, vol. 3. pp. 20–25 (2006)

Garanin, V.K., Bovkun, A.V., Garanin, K.V., Rotman, A.Ya., Serov, I.V.: Microcrystalline oxides from kimberlites and related rocks of Russia, p. 498. MSU publ., M. (2009)

Garanin, V.K.: Mineralogy of kimberlites and related rocks in diamondiferous provinces of Russia due to their genesis and prospecting. D.Sc. in geology and mineralogy thesis, p. 50. MSU, M. (2006)

Garanin, V.K., Kudryavtseva, G.P., Posukhova, T.V., Verzhak, V.V., Verichev, E.M., Garanin, K.V.: Two types of the diamondiferous kimberlites from the Arkhangelsk province. Geol. Explor. **4**, 36–49 (2001)

Golovin, N.N.: Geological structure, mineral composition and formation conditions of alkaline ultramafic rocks in the Kepina area. Ph.D. in geology and mineralogy thesis, p. 45. MSU, M. (2003).

Gunin, A.P., Verzhak, V.V., Minchenko, G.V., Sotnikov, V.I.: Geology, morphology and matter composition of new kimberlites in Arkhangelsk Diamondiferous Province. III–IV G. P. Kudryavtseva Scientific Readings Volume. Moscow, Institute of Applied Mineralogy, pp. 50–62 (2010)

Khachatryan, G.K.: Nitrogen and hydrogen in world diamonds as indicators of their genesis and tool for prospecting of primary diamond deposits. Abstract of D.Sc. in geology and mineralogy thesis, p. 48. TsNIGRI, Moscow (2017)

Khachatryan, G.K., Palazhchenko, O.V., Garanin, V.K., Ivannikov, P.V., Verichev, E.M.: Genesis of "non-equilibrated" diamonds crystals from Karpinskogo-1 pipe according to cathode luminescence and IR spectroscopy data. Newsletter of Moscow University, Series 4. Geology **2**, 38–45 (2008)

Khachatryan, G.K., Verichev, E.M., Garanin, V.K., Garanin, K.V., Kudryavtseva, G.P., Palazhchenko, O.V.: Distribution of structural defects in diamonds from the V. P. Grib pipe (Arkhangelsk diamondiferous province). Newsletter of Moscow University, Series 4. Geology **6**, 29–37 (2006)

Kononova, V.A., Golubeva, Y.Y., Bogatikov, O.A., Kargin, A.V.: Diamond content of Zimneberezhnoe Field kimberlites (Arkhangelsk region). Geology of Ore Deposits **6**, 483–505 (2007)

Kopchikov, M.B.: Typomorphic peculiarities of diamond from Arkhangelsk diamondiferous province. Abstract of Ph.D. in geology and mineralogy thesis, p.17. MSU, M. (2009)

Kriulina, G.Yu.: Constitutional characteristics of diamond fields of the Arkhangelsk and Yakutian diamondiferous provinces. Abstract from Ph.D. in geology and mineralogy thesis, p. 24. MSU publ., M. (2012)

Kriulina, G.Yu., Garanin, V.K., Rotman, A.Ya., Koval'chuk, O.E.: Peculiarities of diamonds from the commercial deposits of Russia. Newsletter of Moscow University, Series 4, Geology **3**, 23–34 (2011)

Kudryavtseva, G.P., Posukhova, T.V., Verzhak, V.V., Verichev, E.M., Garanin, V.K., Golovin, N.N., Zuev, V.M.: Atlas: morphogenesis of diamonds and their mineral-satellites from the kimberlites and other relative rocks from the Arkhangelsk diamondiferous province, 1st edn., p. 624. Polyarny Krug publ., M. (2005)

Kudryavtseva, G.P., Verichev, E.M., Garanin, V.K., Golovin, N.N., Palazhchenko, O.V., Posukhova, T.V.: Microcrystals of diamonds from kimberlite deposits of the Arkhangelsk diamondiferous province. News of Higher Educational Institutions: Geology and Exploration, RSGE Univ., Moscow **3**, 32–36 (2004)

Makhin, A.I.: Crystallomorphology and physical characteristics of the M. V. Lomonosov deposit diamonds (Arkhangelsk diamondiferous province). Abstract from Ph.D. in geology and mineralogy thesis, p. 20. Lvov (1991)

Makhin, A.I., Bartoshinsky, Z.V., Bekesha, S.N., Vinnichenko, T.G., Voloshinovsky, A.S., Vasiliv, V.V.: Technical report: to study of main typomorphic features of diamonds from Zolotitsa field kimberlites for the purposes of classification, to develop and implement a methodology for integrated study of diamonds in terms of the expedition. Archangelsk (1990)

Mineeva, R.M., Bershov, L.V., et al.: First announcement on peculiarities of paramagnetic centers in diamond crystals from kimberlites of Arkhangelsk province. Report RAS **348**(5), 668–670 (1996)

Palazhchenko, O.V. Diamonds from the deposits of Arkhangelsk diamondiferous province. Abstract of Ph.D. in geology and mineralogy thesis, p. 24. MSU, M. (2008)

Palazhchenko, O.V., Verichev, E.M., Garanin, V.K., Kudryavtseva, G.P.: Morphological and spectroscopic peculiarities of diamonds from V. Grib deposit from Arkhangelsk diamondiferous province. Article 1. Morphology of diamond crystals. Proceedings of IHLs. Series Geology and Exploration, vol. 2, pp. 14–22 (2006)

Poberezhskaya, I.V.: Typomorphism of diamonds from kimberlite pipes and placers of some ore fields from Arkhangelsk diamondiferous province. Abstract of Ph.D. in geology and mineralogy thesis, p. 24. LSU, Lvov (1995)

Sablukov, S.A., Sablukova, L.I., Shavyrina, M.V.: Mantle xenoliths from the Zimny Bereg kimberlite deposits of rounded diamonds (Arkhangelsk diamondiferous province). Petrology **8**(5), 518–548 (2000)

Sablukov, S.M.: Volcanism of Zimny Bereg and petrological criteria for the diamond potential of kimberlites. Ph.D. in geology and mineralogy thesis, p. 24. TsNIGRI, M. (1995)

Sobolev, N.V., Yefimova, E.S., Reimers, L.F., Zakharchenko, O.D., Makhin, A.I., Usova, L.V.: Mineral inclusions in diamonds of the Arkhangelsk kimberlite province. Russ. Geol. Geophy. **38**, 379–393 (1997)

Tretyachenko, V.V.: Mineragenic zoning of kimberlite area of South Eastern White Sea region. Ph.D. in geology and mineralogy thesis, p. 28. M., (2008)

Vaganov, V.I.: The diamond deposits of Russia and World: basics of forecasting, p. 369. Geoinformmark (2000)

Verichev, E.M.: Geological conditions of origination and exploration of the V. Grib diamond deposit. Ph.D. in geology and mineralogy thesis, p. 36. MSU, M. (2002)

Verichev, E.M., Verzhak, V.V., Posukhova, T.V., Palazhchenko, O.V., Kriulina, G.Yu., et al.: Mineralogy of diamonds from deposits of the Arkhangelsk kimberlite province. Geology of diamonds—the present and the future (geologists to the 50th anniversary of Mirny and the diamond-extraction industry of Russia), pp. 965–981. VSU publ., Voronezh (2005)

Verzhak, V.V.: Geological structure, matter composition, formation conditions and methods for exploration of the M. V. Lomonosov diamond deposit. Ph.D. in geology and mineralogy thesis, p. 36. MSU, M. (2001)

Zakharchenko O.D., Makhin A.I., Khachatryan G.K:. Atlas of typomorphic properties of diamond from the East European Craton (Lomonosovskaya deposit), p. 104. TsNIGRI, M. (2002)

Contents

Abbreviations

ADR	Archangelsk Diamondiferous Region
AKB	Autolithic Kimberlite Breccia
AS	Academy of Sciences
CCL	Color Cathodoluminescence
CGG	Chemical-Genetic Groups
EPR	Electron Paramagnetic Resonance
FTIR	Fourier Transformations Infrared spectroscopy
IR, IRS	Infrared, Infrared Spectroscopy
LTT	Low-Titanium Type
MNR	Ministry of Natural Resources and the Environment of the Russian Federation
MTT	Medium-Titanium Type
PGA	Production Geological Association
PJSC	Public Joint Stock Company
PL	Photo Luminescence
RS	Raman Spectroscopy
SCR	State Commission on Reserves
SEM	Scanning Electron Microscopy
TR	Craterous tuffaceous Rocks
TsNIGRI	Central Research Institute of Geological Prospecting
UV	Ultraviolet
XLS	X-ray Luminescence Separator
XTB	Xenotuff Breccia
YDP	Yakut Diamondiferous Province

Chapter 1
Geological Summary of Kimberlites and Related Rocks in the Archangelsk Diamondiferous Region (ADR)

1.1 Historical Perspective of ADR Discovery

The founder of the Russian mining science, mineralogy and chemistry, renowned scientist Mikhail Lomonosov, in his treatise "The First Foundations of Metallurgy or Ore Mining"[1763] and in the book "On the Earth's Layers" predicted the possibility of discovering diamonds in the north of Russia ("midnight lands," as dubbed by Lomonosov [1954, republished]), having stated: "In such reasoning and imagining the time when elephants and southern herbs inhabited the north we cannot doubt that diamonds, sapphires, rubies and other gems could have happened there, and, as recently, silver and gold can be found of which our predecessors were unaware."

Reviewing the archives of the Mining Department of Russia, prominent geologist V. O. Ruzhnitsky discovered a variety of documents dated 1823 and related to the diamond potential of northern Europe. However, geologists S. A. Godovan and M. P. Plotnikova found seven diamond crystals only 50 years later (in 1875), while prospecting the placers of the Mezenskaya Pizhma, Pechorskaya Pizhma and Tsilpa rivers.

Following a pause that lasted half a century, the search for diamonds was resumed in Arkhangelsk region in the late 1930s. Acclaimed prospector V. S. Trofimov sampled river sediments and some ultrabasic igneous rocks for diamond, but gained no success (Kharkiv and Zinchuk 1997).

The first diatremes were spotted in Arkhangelsk region (Onega peninsula) as far back as mid-1930s, although they emerged on the geological map only in the 1980s. One of them—Lyvozero—was tapped by a well in 1936 and described by Soviet geologist N. F. Koltsov. The modern views classify the rocks of this pipe as breccias of olivine melilitites (Sinitsin et al. 1992). In 1973, A. F. Stankovsky et al. qualified this rock type as kimberlite-like (Stankovsky et al. 1973). No diamonds were detected in the first diatremes, but their discovery made a paramount incentive to further explore the diamond potential of the Southeastern Belomorye (Kharkiv et al. 1998).

V. Garanin et al., *Diamonds from the Arkhangelsk Province, NW Russia*, Springer Mineralogy, https://doi.org/10.1007/978-3-030-35717-7_1

For a long while all diamond potential prospects in the Southeastern Belomorye relied on the Onega peninsula, while the other part of the region, including the Zimny Bereg (Winter Coast), remained unexplored.

In 1974, the survey unit of Kuloy uncovered pyrope in stream-bed deposits and beach sands of the White Sea, in the Winter Coast. In July 1975, geologist E. M. Verichev of PGA Arkhangelskgeologiya discovered a kimberlite sill containing pyrope and chromite grains while laying surveying routes in the coastal cliffs of the Mela river. In 1976, two small diamond crystals were found in the samples drawn by E. M. Verichev at the Padun river from sediments of the Urzug suite. Pomorskaya, the first kimberlite pipe, was in February 1980 uncovered by the Tova survey unit while verifying magnetic anomaly by drilling in the central part of the Zimny Bereg area; the pipe diamond potential was established in October (Bogatikov et al. 1999). From that moment on, vehement prospecting and exploration started in the region. The Arkhangelsk Diamondiferous Region (ADR) was identified due to discovery of Europe's first diamond non-placer named in honor of M. V. Lomonosov and comprised by six diamondiferous kimberlite pipes (Lomonosov, Arkhangelskaya, Karpinskogo-1, Karpinskogo-2, Pionerskaya, Pomorskaya) within the Zolotitsa kimberlite field, northwards of the East European craton.

In late 1982, geologists of the 17th expedition from PGA Nevskgeologiya discovered the Vesennyaya pipe that crowned the interval of dashing discoveries in the ADR (Bogatikov et al. 1999]. Already by the mid-1980s numerous kimberlite bodies were tapped in the Zimny Bereg area, mainly weakly diamondiferous and non-diamondiferous. The scope of prospecting shrank drastically in the ADR in late 1980s—early 1990s, but the search was underway, despite the poor economic health of the nation.

The search efforts across the region paid off in 1996 by discovery of a new V. Grib highly diamondiferous pipe (nowadays the biggest diamond non-placer in Europe) located in the Chernoozerskoe field of ADR's Zimny Bereg area, named after Vladimir Pavlovich Grib, Chief Geologist of PGA Arkhangelskgeologiya, a discoverer of the Arkhangelsk Diamondiferous Province (Bogatikov et al. 1999). That discovery gave rise to rethinking of the Arkhangelsk province's diamond potential. Serious challenges in diamond prospecting from the date of V. Grib deposit discovery were largely associated to numerous reorganizations of exploration firms. SEVERALMAZ was set up in early 2000s, to be subsequently reorganized into a wholly-owned subsidiary of Russia's well-reputed diamond company ALROSA. Arkhangelsk-based division of ALROSA—ALROSA-Pomorye exploration firm—was also founded at the start of the 2000s, to take charge of diamond prospecting in ADP's Zimny Bereg area. V. V. Vezhak, Ph.D. in Geology and Mineralogy, honored geologist of Russia, the original discoverer of many diamondiferous bodies in the ADR, headed that division from its early days.

The ADR is characterized by no prominent discoveries in 1997–2004. In 2005–2007, geologists of ALROSA-Pomorye, ALROSA's Arkhangelsk-based division, discovered three bodies and one sill: Galina (An-478) (named after G. P. Kudryavtseva, leading researcher in the Faculty of Geology at the Lomonosov Moscow State University), Rozhdestvenskaya (An-162), An-746b, and sill An-495v. In this very period, TsNIGRI at MNR of Russia uncovered new K3 band K8 pipes. All those discoveries were made in the ADR's Kepina field.

As of now, SEVERALMAZ (PJSC) has been mining diamonds from the Arkhangelskaya and Karpinskogo-1 pipes in the Lomonosov deposit of the ADR. A few large diamonds (up to 50 ct) were mined from the Arkhangelskaya pipe. More than 2 million carats of diamonds were produced and 4+ MMT of ore were handled at the site in 2016. Preparations have been underway to develop mining at Karpinskogo-2, and Pionerskaya pipes. The PFS and Development Programs are being prepared for those kimberlite bodies.

The feasibility study for V. Grib pipe deposit (Chernoozerskoe field) was completed and since November 2013 Arkhangelskgeoldobycha, a part of LUKOIL corporation started mining than. Open pit mining has been underway at the deposit, a factory has been built and diamond mining continues. A 188 ct diamond was unearthed at the deposit in spring 2017.

As of now, Chernoozerskoe field (northwards of the V. Grib deposit) has been the focus of prospecting efforts since the chances exist that a new deposit similar to the V. Grib one could be discovered, as confirmed by finding of recently originated picroilmenite grains without intense erosion that migrated from an unidentified source, rather than from the Grib pipe. High chromium content (3.5–9 wt% of Cr_2O_3) is the typomorphic feature of that picroilmenite, typical solely of the Grib pipe. In diamondiferous pipes of the Kepina field picroilmenite with chromium content below 3.5 wt% of Cr_2O_3 is common. The above mineral is a decisive sign for occurrence of a new deposit similar to the V. Grib one in Chernoozerskoe field.

1.2 ADR Geological Location and Structure

The Arkhangelsk diamondiferous kimberlite-picrite region lies in the north of European Russia, occupying a large part of the Arkhangelsk region. The name of the area—Zimny Bereg—is linked to its confinedness to the White Sea coast. Discovery of two commercial diamond non-placers in the early 1980s and mid-1990s (the M. V. Lomonosov and V. Grib deposits) at Zimny Bereg and geographical belonging of the area to Arkhangelsk region in 1980 placed a name to a whole geological region—Arkhangelsk Diamondiferous Province. Although the area of all known magmatic bodies in Arkhangelsk region is less than 20,000 km^2, which is far below the area of the Yakutian province, the region should be viewed as an igneous site, but the petrographic diversity of magmatic—and diamond-bearing—rocks enabled geologists in late twentieth century to regard the Arkhangelsk igneous site as the Arkhangelsk Diamondiferous Province. Implied here is definitely its further future formalization as a province consolidating other alkaline ultrabasic magmatism sites in the north of European Russia. The multiple discoveries of kimberlites, lamproites and related alkaline ultramafic rocks throughout the Russian plate permit to predict expansion of the region covered by the term "province." However, the presence of commercial diamond deposits precisely in Arkhangelsk region and in Finland suggests it would be improper to use the definition "Arkhangelsk Diamondiferous Province" to the north of European part from Finland to Tver region (from north to south), and from Karelia to Timan or even farther(from west to east) (Bogatikov et al. 1999). For this reason, after heaps of new data related to magmatism in the north of European Russia have recently been accumulated, we've seen attempts to develop a new mineragenic zoning of the territory. Achievements of NIGP ALROSA (PJSC) seem to be most interesting in this respect. V. V. Tretyachenko is likely to hold the lead here after having contributed immensely to the understanding of evolution in mantle and crustal rocks formation in the north of the East European craton (Tretyachenko 2008).

Geographically, the area of alkaline ultrabasic and basic magmatism in north-western Russia spans some 150 km northwards and north-east of Arkhangelsk, primarily occupying the Primorsky district of Arkhangelsk region in the scope of the active continental margin (north of the East European craton), which stipulates the peculiarities in formation and evolution of magmatic bodies uncovered therein. Kimberlite magmatism and associated diamond deposits in the Arkhangelsk Diamondiferous Province (nowadays being Arkhangelsk diamondiferous kimberlite-picrite region, ensue V.V. Tretyachenko, ADR in short) should be viewed as a development of regional alkaline magmatism that embraces the entire north of the East European craton.

The ADR location is dictated by its confinedness to the junction zone of the Russian plate and the Baltic shield. Zimny Bereg is part of the Russian plate, an intermediate structure between the Kola geoblock of the Baltic shield and the Mezen syneclise of the Russian plate (Khain 1998; Frolov et al. 2005). Two structural levels are identified: the Archaean–Early Proterozoic crystalline basement and a sedimentary cover comprised by Riphean and Vendian terrigenous formationsand, to a lesser extent, by Paleozoic and Cenozoic carbonaceous and terrigenous sediments. The sedimentary cover thickness in the basement depressions reaches 3–4 km, and dwindles to 0.5–1.2 km in inliers (Bartoshinsky 1992).

The main basement structures that determine location of the alkaline ultramafic magmatism occurrences are shown in the tectonic pattern layout of the Zimny Bereg area (Fig. 1.1). The following are discriminated: Arkhangelsk inlier, Kerets graben, Tova inlier, Padun graben, Ruchyi inlier, Leshukonskoe graben and Polta-Ezhuga uplift zone. Kimberlites are confined to the Tova and Ruchyi uplifted blocks of the basement. The Tova inlier hosts commercial diamond deposits. Olivine melilitite pipes are located within the above inliers along with kimberlites, in their continuation to the Polta-Ezhuga uplift zone, and in the Kerets graben. Tholeiitic basalt pipes were documented within the Kerets and Leshukonskoe grabens and in the Polta-Ezhuga uplift zone of the Mezen syneclise. All known ADR pipes lie in junction zones of the deep faults meridional orientation and dislocations in the direction close to the latitude, which dictates their shape protruding northeastwards.

Four areas showing diamond potential are distinguished in the ADR: Zimny Bereg, Onega, Kanin-Timan, and Dvina. The main diamondiferous area is Zimny Bereg, hosting two deposits (M.V. Lomonosov and V. Grib) (Garanin 2004) (Fig. 1.1).

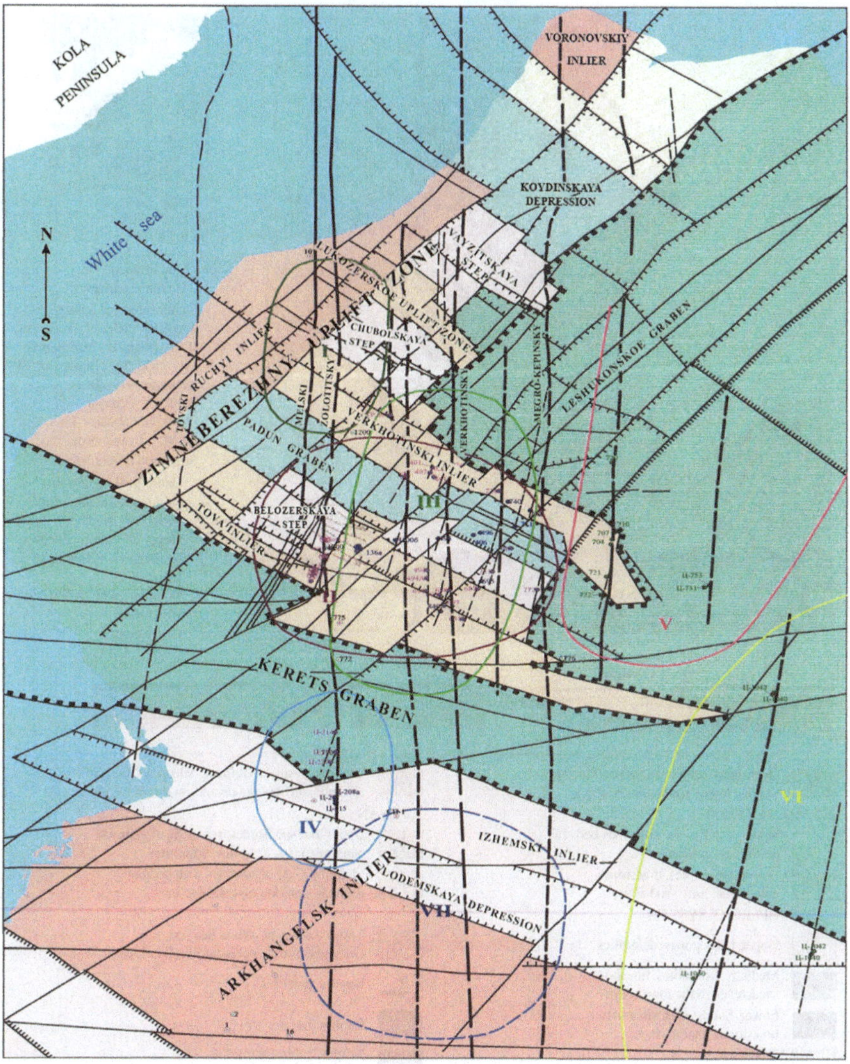

Fig. 1.1 Zimny Bereg tectonic structure schematic, by ALROSA-Pomorye. *Kimberlite fields*: I—*Mela*, II—*Zolotitsa*, III—*Kepina*, IV—*Izhmozero*, VII—*Yuras (predicted)*; *Basalt fields*: V—*Soyana*, VI—*Pinega*. *Scale*: 1 cm—10 km

Nine fields of alkaline ultramafic rocks were identified in the Zimny Bereg diamondiferous area, by spatial distribution of magmatic bodies (Figs. 1.1, 1.2 and 1.3; Table 1.1). The following occurrences are distinguished in the Baltic shield: Zolotitsa kimberlite field; Verkhotina and Kepina kimberlite, picrite and melilitite fields; Izhmozero and Nenoksa melilitite fields; Mela kimberlite-carbonatite magmatism field; Turiya, Polta, Pinega tholeiitic basalt fields (Bogatikov et al. 1999).

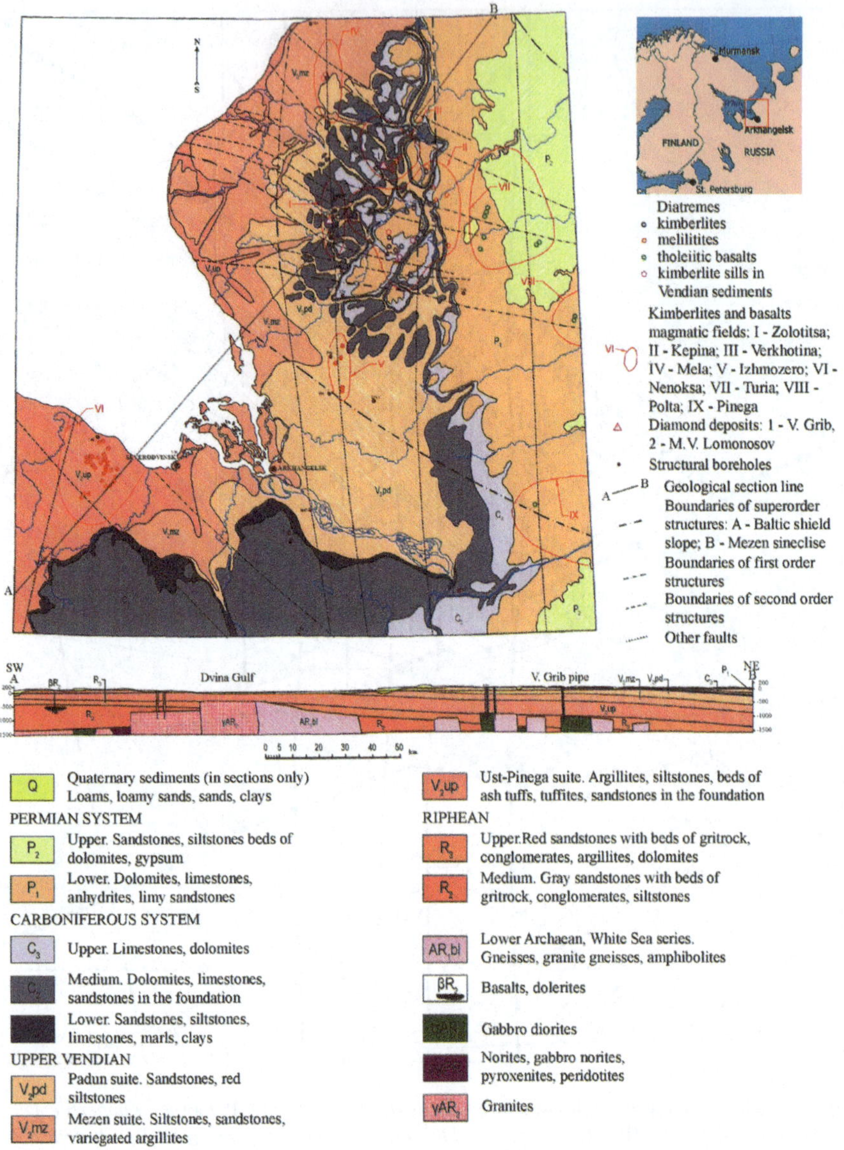

Fig. 1.2 Geology structure map of Zimny Bereg area, with marked magmatism fields. *Fields*: I—*Zolotitsa*; II—*Kepina*; III—*Verkhotina*; IV—*Mela*; V—*Izhmozero*; VI—*Nenoksa (Onega peninsula)*; VII—*Turiya*; VIII—*Polta*; IX—*Pinega* (Verichev 2002)

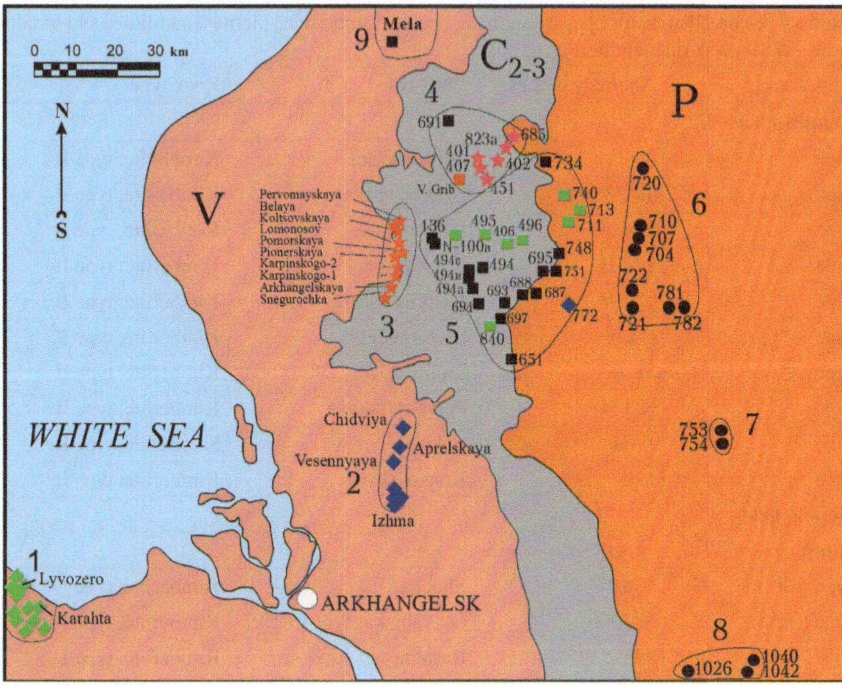

Fig. 1.3 Schematic map with layout of alkaline ultramafic magmatism bodies and basic types of igneous rocks in the Arkhangelsk diamondiferous region. *Fields*: 1—*Nenoksa*; 2—*Izhmozero*; 3—*Zolotitsa*; 4—*Verkhotina and Chernoozerskoe*; 5—*Kepina*; 6—*Turiya*; 7—*Polta*; 8—*Pinega*; 9—*Mela* (Bogatikov et al. 1999)

Medium- and low diamond-bearing kimberlite pipes are concentrated in the Zolotitsa (including the M. V. Lomonosov deposit) and Verkhotina fields (the V. Grib occurrence in Chernoozerskoe field, as per the new zoning) (Figs. 1.2 and 1.3).

The pipes and sills made by kimberlites, picrites, olivine melilitites and basalts of other fields are weakly and non-diamondiferous with diamond content from single fine crystals to 0.1 ct/t.

Pursuant to the paper (Golovin 2003) that accounted for the planar distribution of the magmatic bodies in the fields, mineral composition of rocks, and data of the seismic, gravity, magnetic survey and the entire suite of tectonic structures that control magmatism occurrences in Zimny Bereg, only two potentially diamondiferous fields shall be identified: the Verkhotina (confined to the Ruchyi inlier) and Zolotitsa-Kepina (confined to the Tova inlier). Such zoning has been outdated.

As per Tretyachenko (2008), two areas are distinguished in the kimberlite region of the Southeastern Belomorye: *the Zimny Bereg kimberlite area* that covers commercial diamond-bearing Zolotitsa (the M. V. Lomonosov deposit) and Chernoozerskoe (the Grib pipe) kimberlite fields and barren Verkhotina, Kepina, Megorskoe (pipe 691) and Mela kimberlite fields—pyroxene-free alkaline picrites and *Arkhangelsk*

Table 1.1 Main alkaline ultrabasic magmatic bodies (kimberlites, picrites, melilitites and basalts) of the Arkhangelsk diamondiferous region

Appearance	Anomaly number	Body title	Rock type
Zolotitsa field			
Pipe	323	Snegurochka	Kimberlite, type II
Pipe	Ts-23	Arkhangelskaya	Kimberlite, type II
Pipe	24	Pomorskaya	Kimberlite, type II
Pipe	120-B	Karpinskogo-1	Kimberlite, type II
Pipe	120-V	Karpinskogo-2	Kimberlite, type II
Pipe	N-26	Pionerskaya	Kimberlite, type II
Pipe	Yu-6	Lomonosov	Kimberlite, type II
Pipe	N-154	Pervomayskaya	Kimberlite, type II
Pipe	176a	Belaya	Kimberlite, type II
Pipe	176	Koltsovskaya	Kimberlite, type II
Kepina field			
Shocha group			
Pipe, sill	136a	Shocha	Kimberlite, type I
Pipe	N-100b	Oktyabrskaya	Kimberlite, type I
Pipe	495	Rusalka	Kimberlite, type I
Pipe	406	Pobeda	Kimberlite, type I
Pipe	496	Olginskaya	Kimberlite, type I
Pachuga group			
Pipe	748	–	Kimberlite, type I
Pipe	751	Solokha	Kimberlite, type I
Pipe	695	Solozero	Kimberlite, type I
Sill	687	–	Kimberlite, type I
Pipe	688	Stepnaya	Kimberlite, type I
Pipe	693	–	Kimberlite, type I
Sill	697	–	Kimberlite, type I
Sill	840	–	Kimberlite, type I
Klyuchevskaya group			
Sill	494	Zvezdochka	Kimberlite, type I
Sill	494b	–	Kimberlite, type I
Pipe	494a	Yuras	Kimberlite, type I
Pipe	694	Klyuchevskaya	Kimberlite, type I
Pipe	651	Pachuga	Kimberlite, type I
Pipe	772	Suksoma	Olivine melilitite, type II
Soyana group			

(continued)

Table 1.1 (continued)

Appearance	Anomaly number	Body title	Rock type
Pipe, sill	734	–	Kimberlite, type I
Pipe	740	–	Olivine melilitite, type I
Pipe	713	–	Olivine melilitite, type I
Pipe	711	–	Olivine melilitite, type I
New discoveries of 2005–2007			
Pipe	An-478	Galina	Kimberlite, type I
Pipe	An-162	Rozhdestvenskaya	Kimberlite, type I
Pipe	An-746b	–	Kimberlite, type I
Pipe	K3b	–	Kimberlite, type I
Pipe	K8	–	Kimberlite, type I
Sill	495v	–	Kimberlite, type I
Verkhotina fields			
Pipe	451	Mayskaya	Picrite, olivine melilitite
Pipe	407	Osetinskaya	Picrite, olivine melilitite
Pipe	401	Volchya	Picrite, olivine melilitite
Pipe	402	Verkhotina	Picrite, olivine melilitite
Pipe	823a	–	Picrite, olivine melilitite
Pipe	685	–	Picrite, olivine melilitite
Pipe	691	Verkhnetovskaya	Kimberlite, type I
Chernoozerskoe field			
Pipe	441	V. Grib	Kimberlite, type I
Izhmozero field			
Pipe	Ts-208, Ts-208a	Izhma (Krutikha-south., Krutikha-north.)	Picrite, olivine melilitite
Pipe	Ts-206	Aprelskaya	Picrite, olivine melilitite
Pipe	Ts-214	Chidviya	Picrite, olivine melilitite
Pipe	Ts-237	Vesennyaya	Picrite, olivine melilitite
Turiya field			
Pipe	720	–	Olivine tholeiitic basalt
Pipe	710	–	Olivine tholeiitic basalt
Pipe	707	–	Olivine tholeiitic basalt
Pipe	704	–	Olivine tholeiitic basalt
Pipe	722	–	Olivine tholeiitic basalt
Pipe	721	–	Olivine tholeiitic basalt
Pipe	782	–	Olivine tholeiitic basalt
Pipe	781	–	Olivine tholeiitic basalt

(continued)

Table 1.1 (continued)

Appearance	Anomaly number	Body title	Rock type
Polta field			
Pipe	753	–	Olivine tholeiitic basalt
Pipe	754	–	Olivine tholeiitic basalt
Pinega field			
Pipe	1026	–	Olivine tholeiitic basalt
Pipe	1040	–	Olivine tholeiitic basalt
Pipe	1042	–	Olivine tholeiitic basalt
Mela field			
Sills	–	–	Kimberlite-carbonatite rock

area of feldspathic picrites—olivine melilitites, comprised by Nenoksa, Chidviya–Izhmozero and Suksoma fields. Such mineragenic zoning of the ADR has been generally recognized nowadays. Further on, we'll stick to such mineragenetic zoning of the ADR territory.

Noteworthy is the acceptability of identifying fields (in its simple version) pursuant to Golovin (2003) (Figs. 1.1, 1.2 and 1.3; Table 1.1), in view of planar distribution of bodies in the fields and peculiarities linked to mineral composition of the magmatic rocks, that clearly divides igneous clusters of magmatites and matches the data of seismic, gravity and magnetic surveys that enabled to replicate the suite of tectonic structures that control magmatism occurrences in the Zimny Bereg area. As of now, this is a somewhat simplified and outdated version of zoning the ADR's Zimny Bereg area. From this point on, the ADR field zoning is given as per Bogatikov et al. (1999), with additions from Tretyachenko (2008).

Zolotitsa field lies within the Tova inlier of the crystalline basement and incorporates 10 kimberlite pipes (Pervomayskaya, Belaya, Koltsovskaya, Lomonosov, Pomorskaya, Pionerskaya, Karpinskogo-1, Karpinskogo-2, Arkhangelskaya, Snegurochka) with linear-chain layout, due to being confined to the deep fault zone meridional orientation (Figs. 1.2, 1.3 and 1.4; Table 1.1) (Bogatikov et al. 1999). The distance between individual pipes varies from 100 m to 2.5 km, the total length of pipe chain being 14 km. The pipes have their shapes protruding northeastwards and confined to the junctions of the fault with northeastern and latitude orientation dislocations (Bogatikov et al. 1999; Verichev and Garanin 1991; Garanin 2004; Kharkiv et al. 1998).

Eastwards of the Zolotitsa field lies the Kepina field, comprised by 31 bodies (including new sites) (Table 1.1; Figs. 1.3 and 1.5): 19 kimberlite pipes, 4 olivine melilitite pipes and 8 kimberlite sills unassociated with pipes. By their layout, the pipes and sills create four groups: Shocha, Pachuga, Klyuchevskaya and Soyana, confined within the Tova inlier of the craton basement, although some bodies are in the Chubala and Kerets depressions.

Fig. 1.4 Layout of kimberlite pipes in ADR's Zolotitsa field

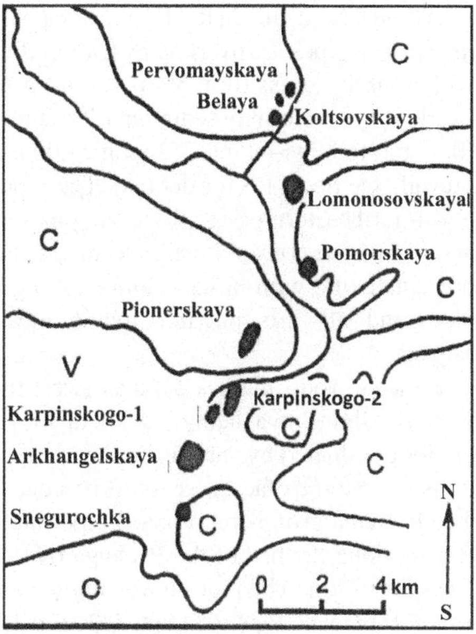

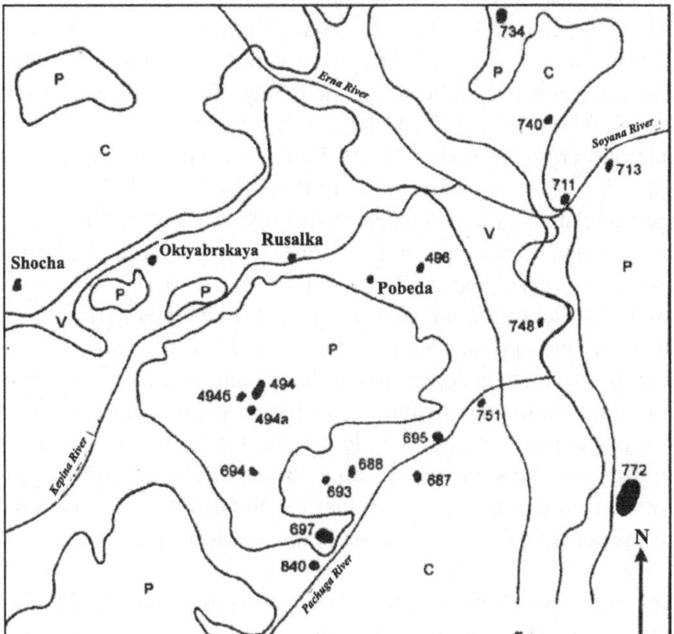

Fig. 1.5 Layout of pipes and sills in ADR's Kepina field

All pipes and sills in the Kepina area burst through weakly lithified Vendian sediments. The pipes are overlaid by Carboniferous, Permian and Quaternary sediments, with total thickness from 50 to 120 m. Anomaly pipe 748 in a paleo valley is only overlaid by Quaternary sediments. The pipes vary in size—from small, like Pipe 693 (0.36 ha) to huge—Pipe 772 (some 200 ha). Altogether, small and medium bodies prevail (see Sect. 1.5 for details). The pipes are isometric by shape, primarily with north-northeastern orientation. The pipes are mainly single-phase. Pipes 136a, 694 and 651 can be conventionally deemed as multi-phase, due to occurrence of xenotuff breccias along with blind columns of autolithic kimberlites. By spatial layout the pipes and sills are split into four groups: Shocha, Pachuga, Klyuchevskaya and Soyana.

Shocha group merges 5 pipes (west-to-east): Shocha, Oktyabrskaya, Rusalka, Pobeda, Olginskaya, being 3–8 km apart from one another (Fig. 1.4; Table 1.1). The bodies are shaped by various kimberlite rock types—tuffaceous sedimentary formations of craterous facies, xenotuff breccias and autolithic kimberlites of vent facies. The Pachuga group comprises 8 pipes and sills with linear-chain layout northeastwards, along the bed of the Pachuga river. The distance between separate bodies is from 2 to 7 km. The pipes have simple internal structure, are formed by kimberlite xenotuff breccia. Pipes 651 and 772 (Fig. 1.5) of the Pachuga group are located separately, in the southeastward part of the Kepina field. By the nature of rocks these could be assigned to the Klyuchevskaya pipe group. Klyuchevskaya group is comprised by two pipes and two sills, lies in the Kepina-Pachuga interfluve and is characterized by cluster-wise distribution of the magmatic bodies (Bogatikov et al. 1999; Garanin 2004; Golovin 2003) (Fig. 1.5; Table 1.1).

Nine diamond crystals, grade $-2 + 0.5$ mm, with the total weight of 41.5 mg (the biggest being 18.2 mg) were found in Pipe 694 of the Klyuchevskaya group. The average calculated diamond content in the pipe is 0.03 ct/t. The Soyana group is made up by four standalone pipes in the northeastern area of the Kepina field. They mark the Soyana river bed and are 2–5 km apart (Fig. 1.5).

The Verkhotina field with the area of some 300 km^2 lies northeastwards of the Zolotitsa field and incorporates 8 pipes (Table 1.1; Figs. 1.3 and 1.6a). Structurally, it is confined by the Chubala depression of the crystalline basement. The pipes form two short (1.5 and 6 km) chains running northwest and northeast, at the distance of 1.5 km from one another. As per the old zoning, the Verkhotina field contains the V. Grib deposit with the same-named pipe (Bogatikov et al. 1999; Garanin 2004; Golovin 2003). Please note once again that the V. Grib pipe deposit has now been assigned to the separate Chernoozerskoe field, standing apart from the Verkhotina field.

The Izhmozero field contains 5 olivine melilitite pipes that form two independent groups (Figs. 1.2, 1.3, 6b; Table 1.1).

The northern area of the Izhmozero field hosts the Chidviya pipe group: Chidviya, Aprelskaya, and Vesennyaya; the southern part of the field (12 km eastwards of Izhmozero lake) is occupied by the Izhmozero group: Izhma pipe of narrow shape elongated northwards and two-vent internal structure; it is often broken down into two pipes: Krutikha-Northern and Krutikha-Southern (Garanin et al. 2001; Golovin

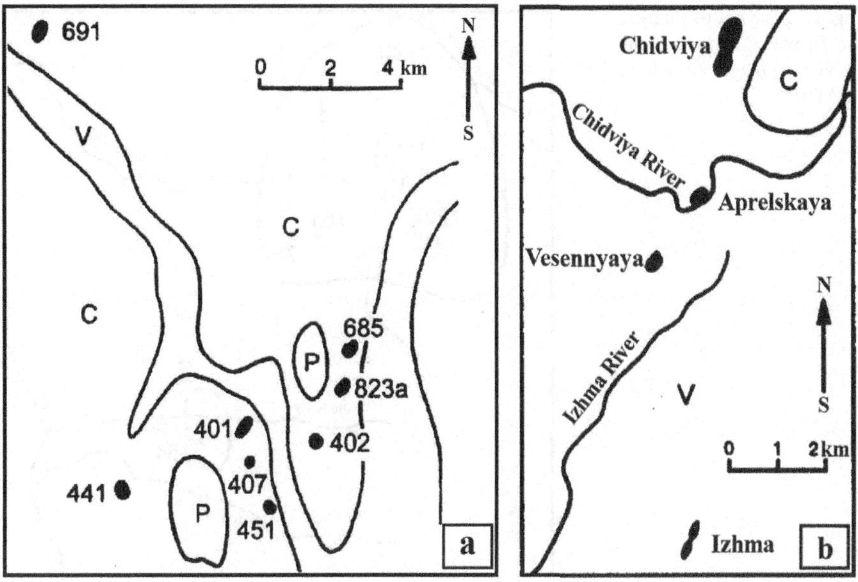

Fig. 1.6 Layout of pipes in the Verkhotina (**a**) and Izhmozero (**b**) fields

2003) (Fig. 1.6b; Table 1.1). The field spans some 20 km. The Northern body group (Chidviya) is confined to the Kerets graben, whereas the southern (Izhmozero)—to the Arkhangelsk basement inlier (Bogatikov et al. 1999; Garanin 2004) (Figs. 1.2, 1.3 and 1.6b).

The Turiya field merges 8 basalt diatremes, six of them forming an arch-shaped chain of some 30 km long (Figs. 1.3 and 1.7; Table 1.1). Two pipes (781, 782) lie separately, approximately 14 km eastwards of the main body group (Bogatikov et al. 1999; Garanin 2004) (Fig. 1.7).

Southwards of the Turiya field there are Pinega and Polta fields that comprise five basaltoid diatremes (Fig. 1.7; Table 1.1). The Polta field consists two uncovered pipes of tholeiitic basalts in the area of the Polta-Ezhuga uplift, confined southeastwards of the Turiya field. The Pinega field embraces three basalt pipes and is located to the south of the Polta field (Fig. 1.7; Table 1.1). The field is confined to the Kerets depression (Bogatikov et al. 1999; Golovin 2003; Garanin 2004).

The Mela field of kimberlite-carbonatite sills lies in the northwest of the Zimny Bereg area, 50 km northwards of the Zolotitsa field, at the Ruchyi inlier of the Kuloy uplift of the basement (Figs. 1.1 and 1.2; Table 1.1). The sills are lavish in the left bank of the Mela river, in the area of some 1.5 km^2. They show subhorizontal occurrence and form five isolated vents to the erosion truncation surface (Bogatikov et al. 1999; Garanin 2004; Golovin 2003).

The age of ADR magmatism was ascertained based on studies of fossil flora remnants in vent and craterous sections of diatremes (Larchenko 1993; Sablukov 1995; Verichev et al. 2000; Verichev 2002; Tretyachenko 2008). Remnants of the

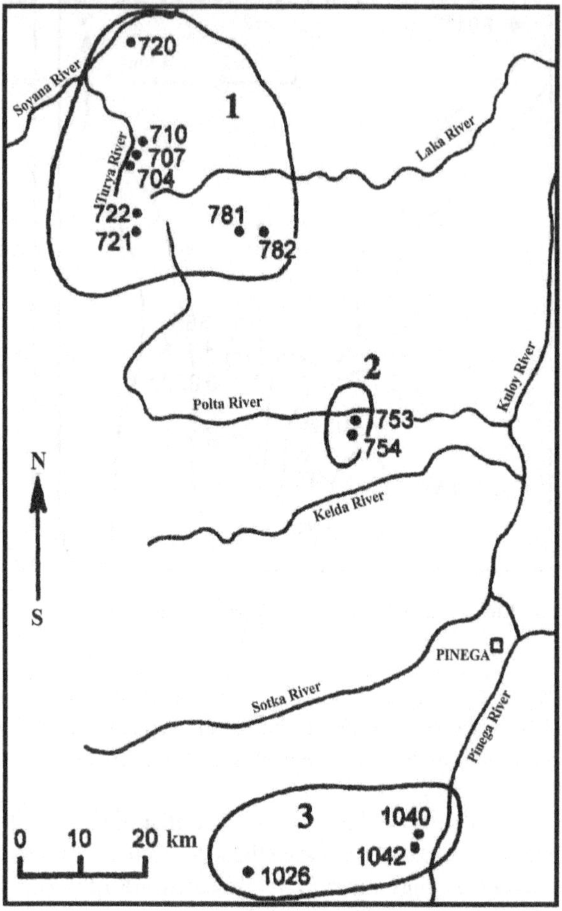

Fig. 1.7 Layout of pipes in the Turiya (1), Polta (2) and Pinega (3) magmatism fields of ADR

charred wood of higher plants from Callixylon sp specieswere detected in the Pomorskaya, Lomonosov, Karpinskogo-1 and 2, Snegurochka pipes of the Zolotitsa field. Fragments of Lycopodiophyta division plant stems were discovered in the Grib pipe. Said plant species are typical of the Upper Devonian. In Pipe 691 (Verkhnetovskaya) the age of charred plant residues is determined as Middle Devonian-Lower Carboniferous, and pteridosperm fragments found in Pipe 840 are allegedly dated Lower Carboniferous (Larchenko 1993). Plant residues resembling wood tissue and known by Lower Devonian were discovered in pipes 691 (Verkhotina field), 651 and 688 (Kepina field), 214 Chidviya (Izhmozero field). The kimberlite breccias of the Shocha pipe offers rare inclusions of charred wood which proves it had formed no earlier than Late Devonian.

Upper Devonian plant residues typical of the Fransian stage were also found in the tholeiitic basalt pipes 735 (Polta field), 1040 (Pinega field).

Absolute age was determined for the olivine melilitite in Pipe 208—Krutikha Northern (Izhmozero field) by the Institute of Minerals Geochemistry and Physics at the AS of the Ukrainian SSR by K–Ar dating method to stand at 350 ± 10 Ma (Larchenko 1993).

The absolute age determination of carbonatite kimberlite sills at the Mela river by K–Ar dating method enabled to date igneous activity occurrences for those type of rocks at the Mela field—393 ± 8 Ma (Larchenko 1993).

Thus, the studies of fossil residues of flora and fauna, and absolute age data enable to identify the total age of igneous bodies formation in the Zimny Bereg area as Late Devonian-Early Carboniferous: 400–340 Ma (Tretyachenko 2008). For igneous rocks of the Kepina, Izhmozero, Verkhotina and Zolotitsa fields the age of magmatism may be determined by fossil residues and preliminary absolute age calculations in the range of 370–340 ma. The age of rocks comprising alkaline, alkaline ultramafic blocks, kimberlites and olivine melilitites of the Kola peninsula northwestward of the ADP shows similar data (Kramm et al. 1993; Arzamastsev 1998).

The chronological sequence of pipe formation in the Zimny Bereg area within the united Early Hercynian stage of tectonomagmatic activation (Bogatikov et al. 1999; Verzhak 2001; Verichev 2002) is as follows: kimberlite and melilitite pipes of the Kepina field, kimberlite pipe 691-Verkhnetovskaya, pipes Ts-214 (Chidviya) and Ts-206 (Aprelskaya) of Izhmozero field were the earliest to originate; formation of the Zolotitsa field pipes occurred afterwards; melilitites of the Verkhotina field, southern pipes of the Izhmozero field and V. Grib kimberlite pipes of Chernoozerskoe field are the youngest (Tretyachenko 2008).

Lateral zoning of the ADR was established (Verichev 2002; Verzhak 2001), expressed in shifting the diamond-bearing grade of rocks from center to periphery of the province. All known diamondiferous bodies are confined to central parts of the province (Zolotitsa and Verkhotina fields) and genetically linked to kimberlites. Non-diamondiferous fields (Fig. 1.2) were revealed in the province periphery, made up by olivine melilitites (partially in the Kepina and Verkhotina fields, Izhmozero field) and olivine tholeiitic basalts (Turiya, Pola, Pinega fields).

Central type ADR was also established (Sablukov 1995; Sablukov et al. 2004) based on mantle substrate features. The province central part shows well-developed eclogite-peridotite substrate, and homogeneous dunite substrate in the periphery (Fig. 1.8).

Such a pattern is associated with the irruptive mantle diapir geometry (Sablukov 1995; Sablukov et al. 2004) and based on results of studying mantle xenoliths from ADP bodies (Garanin et al. 2005; Sablukov et al. 2000). This is a dunite-based homogeneous substrate with clino- and orthopyroxene, pyrope and chrome-spinelide as accessory minerals, with rare occurrence of lherzolites, harzburgites and typical eclogites, absence of ilmenite ultrabasites. These offer intense occurrences of mantle metasomatism (amphibolitization, phlogopitization). It stands out sharply among other kimberlite regions of the world by depleted basalt components (Sablukov et al. 2000, 2004). The heterogeneous eclogite-peridotite substrate offers a wide array of

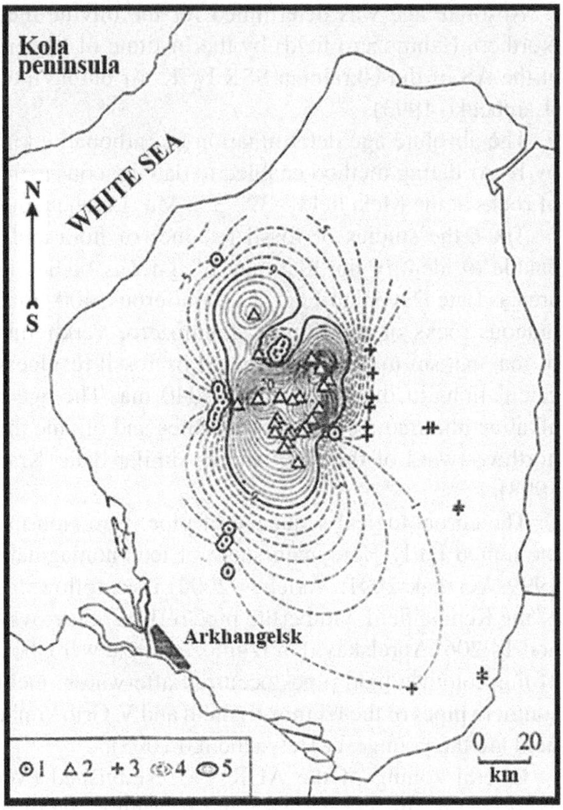

Fig. 1.8 Distribution of mantle substrate types across Zimny Bereg kimberlite area. *Legend* 1—series Al kimberlite rocks, 2—series Fe-Ti kimberlite rocks; 3—basalts; 4, 5—tantalum content isolines in volcanic rock autoliths (interval—1 ppm); 4—homogeneous 'dunite' mantle substrate development area; 5—heterogeneous 'eclogite-peridotite' mantle substrate development area

rocks, with substantially developed ilmenite and pyrope-ilmenite olivinites, peridotites, pyroxenites, ferriferous-magnesian and magnesian eclogites (Sablukov et al. 2000, 2004).

Two substrate types comply with igneous rocks of two kimberlite series of Zimny Bereg area: the homogeneous type with aluminous rocks (Lomonosov deposit), the heterogeneous type with ferriferous-titaniferous series (Grib deposit) (Fig. 1.8; Table 1.2) (Sablukov et al. 2000, 2004).

Two kimberlite types are identified in the ADR based on mineral composition features (Bogatikov et al. 1999; Garanin et al. 2001).

Type I rocks are primarily represented by kimberlites in some bodies of the Kepina field and Grib pipe (Chernoozerskoe field). By quantitative and mineral composition, they resemble group I kimberlites I of South Africa (Table 1.2). Picroilmenite prevalence, presence of pyrope and chrome-diopside is common for accessories of diamond (Bogatikov et al. 1999; Garanin et al. 2001; Golovin 2003; Sablukov et al. 2004), given low chrome-spinelide content. The rocks are characterized by abundant titaniferous chrome-spinelides, ilmenite, perovskite and rutile in the rock basis, and in some aphanitic kimberlites–by plenty of apatite (sill 697).

Table 1.2 Comparison of diamond-bearing type I and II kimberlites of the Arkhangelsk diamondiferous region (Russia) and South Africa, based on published data (Smith et al. 1985; Mitchell 1986; Bogatikov et al. 1999, 2007; Golubeva et al. 2006; Kononova et al. 2007)

No.	*Characteristics*	ADR kimberlites		South African kimberlites	
		Type I (V. Grib pipe)	Type II (M. V. Lomonosov deposit)	Type I	Type II
1	**Isotopic ratios** $^{87}Sr/^{86}Sr$ $^{143}Nd/^{144}Nd$ $^{206}Pb/^{204}Pb$	0.707834 ± 17 0.512357 ± 7 18.03–18.08	0.705709 ± 53; 0.707488 ± 39 0.512277 ± 11; 0.512144 ± 10 18.13–18.27	0.703 – 0.705 0.51268–0.51276 18.3 – 20.0	0.7075 – 0.710 0.51206–0.51227 17.2 – 17.7
2	**Geochemical peculiarities**	REE and RE poor	REE and RE poor	–	–
3	**Kimberlite age, Ma**	340–360	340–360	80–100; 114; 150; 240; 340–360; 1200	114–127; 145; 200
4	**Mineralogy Prevalent Heavy fraction minerals Minerals of the binding mass of kimberlites**	Ilmenite, garnet, clinopyroxene with low chromite content Ti-chromite, magnesian Cr-ulvospinel, ilmenite, perovskite, titaniferous magnetite calcite, minor quantities of phlogopite and rutile	Chromite, with ancillary concentrations of garnet and clinopyroxene Ilmenite is absent Ti-chromite, ulvospinel, sphene. Ilmenite, perovskite, monticellite are absent Phlogopite is widespread	Ilmenite, garnet, clinopyroxene with low chromite content Perovskite, magnesian ulvospinel, ilmenite, calcite, monticellite, low phlogopite contents Ti-chromite	Chromite, with ancillary concentrations of garnet and clinopyroxene Ilmenite is absent Magnesian ulvospinel, ilmenite and monticellite are absent. Perovskite presents in low concentrations. Phlogopite is widespread
5	**Dominating diamond morphology type**	Dodecahedral, with higher percentage of octahedrons	Dodecahedral	Dodecahedral, with higher percentage of octahedrons	Dodecahedral
6	**Xenoliths** Integrity Cataclased lherzolites Megacrystals	Present Good Present Substantial presence	Present Advanced alterations Absent Absent	Present Present: picroilmenite, phlogopite, low-chromium clinopyroxene	Present Absent

(continued)

Table 1.2 (continued)

No.	*Characteristics*	ADR kimberlites		South African kimberlites	
		Type I (V. Grib pipe)	Type II (M. V. Lomonosov deposit)	Type I	Type II
7	**Presumed source** one reservoir-deep plume	Ancient enriched subcontinental lithosphere REE and RE poor	Ancient weakly enriched subcontinental lithosphere REE and RE poor	Asthenospheric. Depleted and enriched mantle. REE, RE and Rb poor, characterized by higher conc. of Nb and Ti	Ancient enriched subcontinent. lithosphere. Enriched with light REEs and Rb, higher conc. of SiO_2, K_2O, Pb, Ba

Type II is represented by highly diamondiferous kimberlites of the Zolotitsa field. These are characterized by prevalence of titaniferous chrome-spinelides over other oxide phases in the binding rock mass, very low concentrations of typical diamond accessories—garnet and chrome-diopside, and widespread occurrence of chrome-spinelides, absence of picroilmenite, and minor quantities of eclogites and peridotites xenoliths with porphyroblastic structures (Verichev and Garanin 1991; Sablukov 1995; Bogatikov et al. 1999; Garanin et al. 2001; Golovin 2003; Sablukov et al. 2004). Combined with geochemical survey results, such features approach them with group II kimberlites II of South Africa (Smith et al. 1985; Mitchell 1986) (Table 1.2).

ADR kimberlites vary by geochemical features and isotopic composition (Bogatikov et al. 2007; Sablukov et al. 2000). Two major types of kimberlites have been identified (Sablukov et al. 2000): Zolotitsa and Kepina, plus an intermediate type—kimberlites of the Grib pipe and the Mela river sills (offer intermediate mineralogical, geochemical properties and isotopic ratios).

Kimberlites from igneous bodies of the Kepina field are LREE-rich—the $(La/Yb)_n$ ratio therein is 70–130; they show steadily positive ε_{Nd} values (+2.8… + 1.2), high following ratios: $^{206}Pb/^{204}Pb = 18.46–19.03$, $^{207}Pb/^{204}Pb = 15.53–15.65$, $^{208}Pb/^{204}Pb = 38.43–38.77$.

The V. Grib pipe kimberlites have intermediate isotopic composition. They resemble BSE by isotope abundance of Nd, ε_{Nd} values in them vary from −1.0 to + 1.5. At the same time, by primary isotopic composition $^{206}Pb/^{204}Pb = 18.03–18.08$, $^{207}Pb/^{204}Pb = 15.49–15.52$, $^{208}Pb/^{204}Pb = 37.89–38.02$ these are shifted towards Zolotitsa kimberlites (Bogatikov et al. 2007) (Table 1.2). They show transition $(La/Yb)_n$ value of 38–87.

Kimberlites of the Zolotitsa field are less LREE-rich—the $(La/Yb)_n$ ratio is 18–44. Unlike the Kepina field kimberlites they have negative ε_{Nd} values (−2.2…−5.3) at low ε_{Sr} values (−4.5 to +29) (Bogatikov et al. 2007). They have less radiogenic lead composition $^{206}Pb/^{204}Pb = 18.13–18.27$, $^{207}Pb/^{204}Pb = 15.50–15.60$, $^{208}Pb/^{204}Pb = 37.69–38.14$. Such geochemistry emphasizes the significant role of the EMI type ancient enriched mantle in the rocks (Bogatikov et al. 2007) (Table 1.2).

Kimberlites of the Mela river sills offer low ε_{Nd} values (−6.1…−5.0) and ε_{Sr} values (−4.1… + 2.9), but have high isotopic ratios of lead $^{206}Pb/^{204}Pb = 18.0$, $^{207}Pb/^{204}Pb = 15.53$, $^{208}Pb/^{204}Pb = 38.05$, comparable to those of the Kepina kimberlites (Bogatikov et al. 2007) (Table 1.2). By the $(La/Yb)_n$ ratio at 56–81 these are shifted towards Zolotitsa kimberlites (Table 1.2).

So, by isotopic composition of Nd, Pb and Sr they match the South-African group I kimberlites (Bogatikov et al. 2007) (Table 1.2).

The petrography and mineralogy of rocks, structural and formation peculiarities of diamond-bearing diatremes and other igneous bodies in the ADR have been detailed in publications (Sablukov 1995; Kharkiv et al. 1998; Bogatikov et al. 1999; Verzhak 2001; Verichevm and Garanin 2001; Garanin et al. 2001; Verichev 2002; Garanin 2004; Sablukov et al. 2004; Tretyachenko 2008).

This paper summarizes the main petrographic, mineralogical data on diamondiferous kimberlite pipes at the V. Grib and M.V. Lomonosov deposits, and on representatives of alkaline ultramafic rocks from the Zimny Bereg area of the ADR, not covered by the deposits.

1.3 V. Grib and M. V. Lomonosov Deposits

V. Grib pipe deposit. The Chernoozerskoe field (Figs. 1.2, 1.3 and 1.6a) hosts the average diamondiferous (0.5–1.5 ct/t) V. Grib pipe (most magnesian of all ADR bodies)—the only commercial-scale site which can be deemed as a separate deposit by its parameters (Bogatikov et al. 1999; Verichev 2002; Garanin 2004) (Fig. 1.9). The non-uniform diamond-bearing grade of bodies in the area, inter alia the pipes of the Verkhotina field, relies on the varying depth and dynamics of kimberlite melt hoisting, slow movement of which caused intense dissolution of diamond crystals (Bogatikov et al. 1999; Verichev 2002). The diatremes are characterized by small dimensions (0.5–16.0 ha). The overlying Carboniferous and Quaternary sediments are 45–80 m thick.

E. M. Verichev discovered the V. Grib kimberlite pipe in 1996; it was named after geologist V. P. Grib of Arkhangelsk, who had been the first person to predict its existence (Verichev et al. 2000). The reserves were duly approbated by the SCR of the MNR of Russia in 2002. The pipe is situated within the Verkhotina uplift of the Ruchyi basement inlier, close to the deep fault protruding northwestwards (Fig. 1.1). The geotectonic position of the Grib pipe and its age (≈355 Ma) point to formation of the alkaline ultramafic magmatism in the ADR (390–340 Ma) at the Kola Archaean craton, in line with the common time–space link with all other occurrences.

The pipe penetrates the weakly lithified Upper Riphean and Upper Vendian sedimentary rocks, overlaid by the Middle Carboniferous terrigenous carbonate rocks and Quaternary loose sediments with the average thickness of 67 m (Fig. 1.9) (Verichev 2002). The data on geological structure of the pipe, mineralogical, petrochemical and geochemical features of rocks in the pipe are detailed in the publications (Verichev 2002; Garanin 2004).

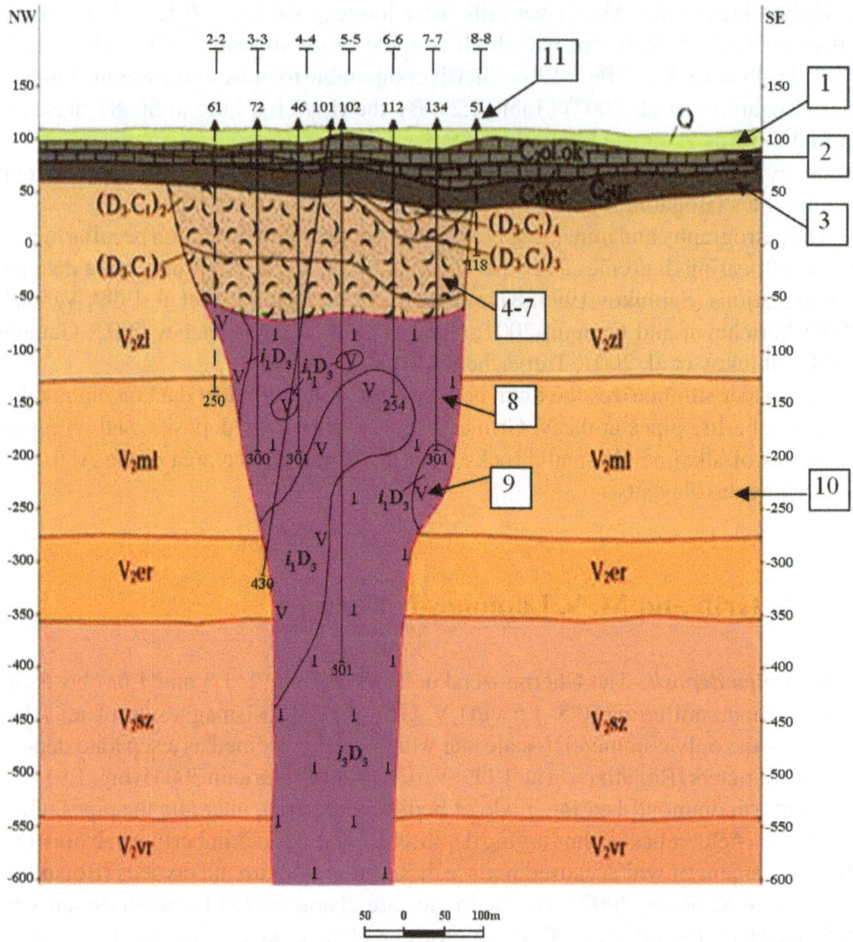

Fig. 1.9 Geological plan and section of the V. Grib kimberlite pipe (Bogatikov et al. 1999; Verichev 2002). *Legend* 1—Quaternary sediments (loams, sands); 2—Carboniferous limestones, dolomites; 3—Urzug suite sandstones; 4–7—craterous facie rocks: 4—clay sandstones; 5—tuffaceous sandstones; 6—tuffs and tuffites; 7—surrounding rock breccias, conglomerates with tuff and tuffites inclusions; 8–9—vent facie rocks: 8—tuffisitic kimberlites, tuff and xenotuff breccias; 9—kimberlite; 10–Vendian age surrounding rocks, 11—drill wells. Pipe dimensions: 570 × 480 m

In the layout the pipe shows rhombic-round shape, elongated northeastwards, and acquires pear-shaped form with narrowed southern end as the depth increases. In the upright section the pipe is an inverted cone with a cup-shaped mouth in its upper part (Fig. 1.9). The craterous part is made by the volcaniclastic-sedimentary and sedimentary rocks: quartz-argillaceous sandstones, sedimentary breccias, kimberlite tuffs and tuffites. The thickness of craterous facie sediments varies from 67 to 145 m (Bogatikov et al. 1999; Verichev 2002).

The pipe vent reveals two-phase structure, made up by basic kimberlite varieties: xenotuff breccia and massive autolithic kimberlite (Fig. 1.9). For the first time for ADR pipes, the vent of the Grib pipe revealed the weathering crust ("blue earth") of some 0.7–2.4 m thick, primarily comprised by large chrysotile crystals (Bogatikov et al. 1999; Verichev 2002; Garanin 2004].

Kimberlites of the Grib pipe belong to the highly magnesian rocks (25–37 wt% of MgO, 3–12 wt% of Fe_2O_3) of ferrous-titaniferous series (Verichev 2002). They are moderately titaniferous (1.5–3.0 wt% of TiO_2) versus the low-titaniferous rocks of the Zolotitsa field pipes (0.0–1.3 wt% of TiO_2) and high-titaniferous kimberlites of the Kepina field (>3.0–4.0 wt% of TiO_2) (Sablukov et al. 2000). Kimberlites of the Grib pipe show relatively low alkali, CaO and Al_2O_3 concentrations as compared to the Lomonosov deposit kimberlites, feature elevated MgO, TiO_2, Fe_2O_3 content (Verichev 2002). They drastically differ from the rocks of the Lomonosov deposit by high concentration of heavy minerals (up to 50 kg/t). Most widespread are the pyrope + picroilmenite association, ilmenite (up to 19.5 kg/t) (including picroilmenite with high Mg and Cr content), garnet (up to 5.7 kg/t) and clinopyroxene (up to 7 kg/t). Chromite contents are low (up to 40 g/t) (Bartoshinsky 2005; Bogatikov et al. 1999; Kudryavtseva et al. 2005).

The binding mass oxides are dominated by titaniferous chromite, magnesian chromiferous ulvospinel, titaniferous magnetite, chromiferous picroilmenite. Mineralogical features of the binding mass for kimberlites of the V. Grib pipe, wide occurrence of cataclased lherzolite xenoliths, prevalence of picroilmenite, garnet and chrome-diopside in the heavy rock fraction with very low concentrations of chromite, presence of picroilmenite, garnet, clinopyroxene, phlogopite megacrystals enable to classify the V. Grib pipe rocks as type I kimberlites (Bogatikov et al. 1999). By the petrochemical properties and mineral composition these are similar to some classic diamond-bearing kimberlites of Yakutia (Internatsionalnaya, Mir, Udachnaya pipes) and close to group I kimberlites of South Africa by those parameters (Bogatikov et al. 1999; Verichev 2002; Garanin 2004) (Table 1.2). Please note that the isotopic and geochemical characteristics (the ratio of rock-forming components—magnesium, iron, titanium and alkali) of the Grib pipe kimberlites are most close to those of the Lomonosov deposit pipes (group II kimberlites).

The utmost similarity is revealed with the deep horizon kimberlite of the Pionerskaya pipe (Verichev 2002).

The Grib pipe rocks commonly have low content of rare and rare-earth elements, which distinguishes them from the majority of kimberlites in the world (Bogatikov et al. 1999; Verichev 2002). The ancient enriched subcontinental lithosphere is presumed to be the mantle source of the V. Grib pipe kimberlites.

Ultramafic rocks (primarily lherzolitic rocks), including ilmenite rocks, prevail in the mantle rocks section below the pipe. Wide occurrence of various xenoliths of metamorphic and mantle rocks (peridotites, pyroxenites, eclogites and crustal rocks: diamondiferous dunites and harzburgites, lherzolites, cataclased ilmenite lherzolites, websterites, pyroxenites) was ascertained (Verichev 2002; Garanin et al. 2005), as well as absence of xenoliths of aluminous and calcareous eclogites, which may evidence faint manifestation of subduction processes in craton formation below the

ADR (Verichev 2002; Garanin et al. 2005; Sablukov et al. 2000). The mantle rocks show signs of mantle metasomatism, exposed by presence of thin celyphitic rims in garnets and altered marginal facies in clinopyroxene grains, and by a wide array of metasomatites of garnet-pyroxene-ilmenite-phlogopite association. It was ascertained (Verichev 2002) that the deep-seated rocks of the pipe originated in conditions of their saturation with volatile components (potassium and titanium), which could be an occurrence of mantle metasomatism and happened in history of the Earth some 990 Ma prior to formation of the Grib kimberlite pipe (some 355 Ma, as per (Verichev 2002)).

PT parameters for xenoliths of pyrope peridotites (T = 1000–1250 °C, P = 38–46 kbar) and eclogites (T = 950–1050 °C, P = 35–39 kbar) were calculated (Verichev 2002; Garanin et al. 2005), which prove the rocks had been formed close to the graphite-diamond transition boundary. The mineral associations of crustal (eclogite-like) rocks show they had been formed in the garnet-pyroxene-plagioclase facie of metamorphism (T = 800–900 °C and P = 6–9 kbar) (Verichev 2002; Garanin et al. 2005).

The pipe formation took three stages, under intrusion of the three major rocks with varying diamond-bearing potential: vent facie—(1) kimberlite tuff and xenotuff breccias, (2) kimberlite; craterous facie—tuffaceous sedimentary rocks (Verichev 2002; Garanin et al. 2005). Diamond presence in various rocks is as follows: total, in the pipe—100%, in vent kimberlites—16.9%, in tuff- and xenotuff breccias—57.5%, in craterous facie rocks—42.5%, in sandstones—11.4%, in sedimentary rock breccias—25.3%. Diamond content in the V. Grib pipe is more than 2.5 times higher than the diamond grade of any pipe in the Lomonosov deposit.

M. V. Lomonosov deposit. The deposit lies in the west of the Zolotitsa field (ADR) and embraces six commercial diamondiferous kimberlite pipes: Lomonosov, Pomorskaya, Pionerskaya, Karpinskogo-1, Karpinskogo-2, Arkhangelskaya (Figs. 1.2, 1.3 and 1.9). The pipes fall under two morphology types: oval—the Lomonosov, Karpinskogo-1, Arkhangelskaya, Pomorskaya, and lenticular—Pionerskaya, Karpinskogo-2 (Bogatikov et al. 1999; Verzhak 2001; Kharkiv et al. 1998) (Fig. 1.10). By the surface size, the Pionerskaya pipe (36.9 ha) belongs to very large bodies; the Lomonosov, Karpinskogo-1 and 2, Arkhangelskaya pipes (10.2–20.2 ha)—to large bodies; Pomorskaya pipe (5.6 ha)—to average bodies. All deposit pipes have been medium diamondiferous to some extent (0.5–1.0 ct/t), although in some pipe vents the content could reach 1.0–3.0 ct/t.

The majority of diamondiferous bodies in the Zolotitsa field formed in several stages, so many pipes are multiphase. Unlike the pipes of other fields, these are made up solely by kimberlite rocks, mainly tuffs, tuffites, tuff breccias, autolithic breccias (Verzhak 2001). The kimberlite rocks of the Zolotitsa field are magnesian and aluminous by structure (MgO content up to 36.1 wt%, Al_2O_3 up to 6.2 wt%, SiO_2 up to 48 wt%) with reduced concentration of titanium (TiO_2 up to 1.3 wt%), which drastically differs their material composition from conventional kimberlites of Yakutian and South African provinces (Bogatikov et al. 1999; Verzhak 2001; Sablukov 1995; Sablukov et al. 2004). Low yield of the heavy minerals is typical of the Zolotitsa field pipes: chrome-spinelide prevails, low concentrations of accessories of

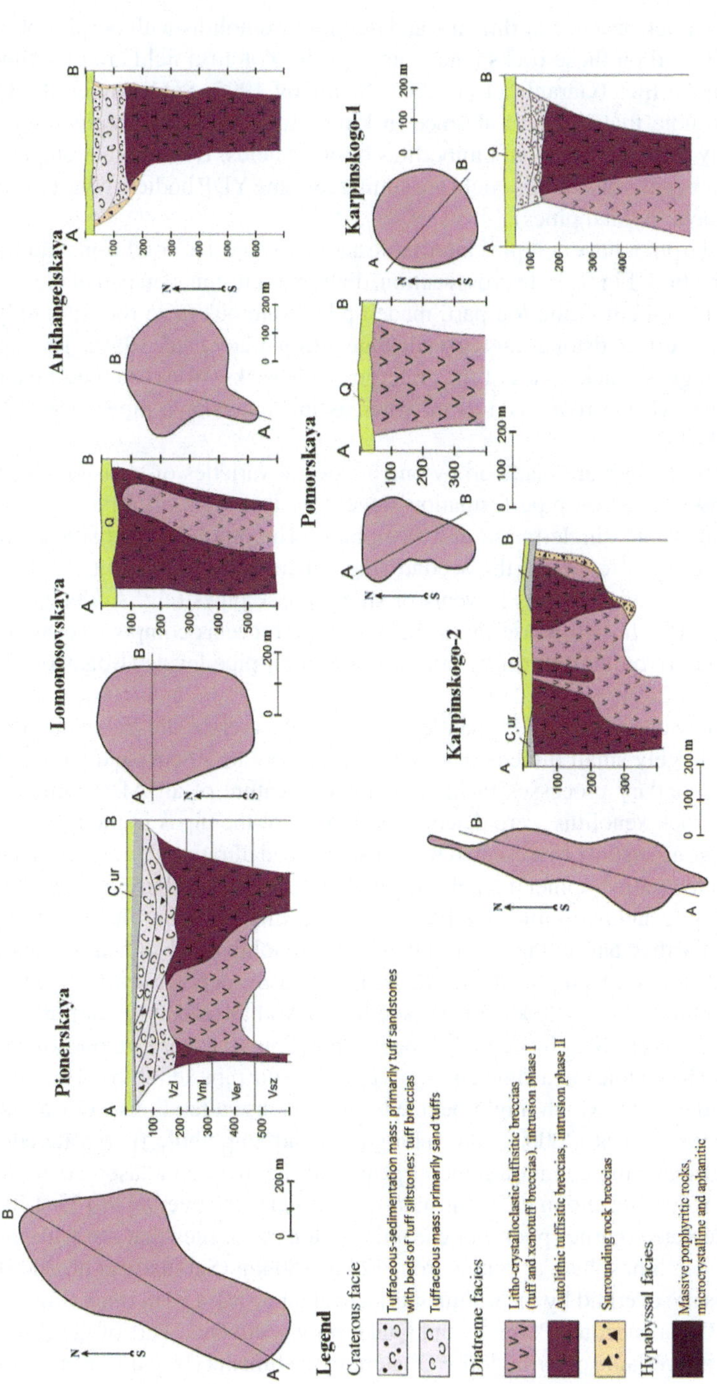

Fig. 1.10 Internal structure, shape and dimensions of kimberlite pipes at M. V. Lomonosov depositin Arkhangelsk diamondiferous region, by ALROSA-Pomorye

diamond—garnet, absence of ilmenite and peridotite xenoliths with porphyroblastic structures. Based on those rocks signs, pipes of the Zolotitsa field are classified as group II kimberlites (Garanin et al. 2001; Sablukov 1995; Sablukov et al. 2004). It stands to note that ilmenite absence in kimberlites is a typomorphic sign and substantially differs those from kimberlites of other fields. By mineral composition the kimberlite rocks of Zolotitsa field are similar to some YDP bodies (Nyurbinskaya, Botuobinskaya, Aykhal pipes).

Most of the pipes have complex internal structures, subject to conditions and stages of their formation (Fig. 1.9). In cross section, diatremes are cone-shaped bodies, often having a wide mouth in the top part, made up by craterous facie rocks, mainly by tuffs, tuffites, tuff sandstones and tuff siltstones (Bogatikov et al. 1999). The largest pipes have biggest thicknesses of the craterous facie rocks. In Arkhangelskaya pipe the thickness of such rocks is 140 m, whereas in Pionerskaya pipe—160–170 m (Verzhak 2001).

The vents of pipes are made up by one or several varieties of tuffisitic breccias; in many cases based on pipe formation stages. By that sign, all pipes in the area are subdivided into single-phase and two-phase. The autolithic tuffisitic breccias in two-phase pipes represent the second stage of melt intrusion (Fig. 1.10). This breccia type less often occurs in vents of single-phase pipes (e.g. Arkhangelskaya pipe) (Fig. 1.10). In two-phase pipes the autolithic breccias compose separate ore columns and are quite uniform by composition in all pipe levels (Bogatikov et al. 1999; Verzhak 2001).

Overall low concentration of mantle rock xenoliths (versus the Yakutian kimberlites), their largely small dimensions (0.5–5 cm, rarely to 10 cm) and strong alterations by secondary processes make a distinctive feature of all ADR kimberlites. The mantle rock xenoliths were discovered in most of the pipes in the Lomonosov deposit; these are mainly represented by coarse-grained ultramafic rocks: spinel lherzolites, dunites, garnet-spinel lherzolites, garnet lherzolites, harzburgites and dunites. Typical eclogite and peridotite xenoliths occur in negligible quantities.

By peculiarities and composition features of xenoliths and indicator minerals, commercial kimberlite pipes at the M. V. Lomonosov deposit are split into two groups (Sablukov et al. 2000, 2004): southern (Arkhangelskaya, Karpinskogo-1 pipes) and northern (Karpinskogo-2, Pionerskaya, Pomorskaya, Lomonosov pipes) (Fig. 1.10). The mantle metasomatism signs in kimberlite pipe xenoliths of the deposit are hardly identified, which might be due to their intense alterations. As was ascertained (Sablukov et al. 2000, 2004), phlogopite and amphibole (pargasite-edenite hornblende) make the common mineral phases of the mantle metasomatism. Most intense phlogopite development in the mantle occurred in the interval of 125–150 km, in highly depleted, garnet-poor, but chromite-rich harzburgites that are widespread in the province lithosphere at depths from 120 to 180 km (Sablukov et al. 2000).

All pipes are overlaid by the sedimentary cover some 50–150 m thick (Fig. 1.10), mainly by Carbonaceous, Permian and Quaternary sediments. Part of the Zolotitsa field pipes is overlaid solely by loose Quaternary sediments (Bogatikov et al. 1999; Verzhak 2001).

The paper (Bogatikov et al. 1999) notes that the efforts performed at Arkhangelskaya, Pionerskaya, Lomonosov, Karpinskogo-1 and 2 pipes revealed these are nearly identical by composition and process behavior of ores, grain size and visual features of diamond (morphology, color, integrity).

1.4 Low- and Non-diamondiferous ADR Bodies

Deemed non-diamondiferous and weakly diamondiferous have been kimberlites, olivine melilitites and tholeiitic basalts in most of ADR fields, such as the Izhmozero, Nenoksa, Turiya, Polta and Pinega fields (Table 1.1; Figs. 1.2, 1.6 and 1.7) (Bogatikov et al. 1999). Bodies with total diamond-bearing grade below 0.1 ct/t were found in prospecting blocks of the fields. No diamonds were discovered in the Nenoksa field.

The Kepina and Verkhotina fields reveal the potential for discovery of weakly diamondiferous bodies, with Chernoozerskoe field being most promising.

The pipes and sills of the Kepina area, comprised by kimberlites and olivine melilitites, are characterized by poor diamond-bearing potential. High-chromium pyrope is widespread in some pipes (Verichev and Garanin 1991; Garanin 2004; Golovin 2003). Picroilmenite was discovered in most of the pipes. The content of garnet belonging to the pyrope-almandine variety is higher than in pipes of the Zolotitsa field. Development of the secondary minerals is typical (serpentine, saponite, hydromica, carbonates) that form pseudomorphs by large mega phenocrysts of olivine and minerals of the binding mass. Nearly no fresh olivine is present. The weakly diamondiferous Fe-Ti kimberlites in the Kepina field are characterized by presence of titaniferous chrome-spinelides, titaniferous magnetites, ilmenite, rutile and perovskite in the binding mass of kimberlites, and picroilmenite, pyrope and chrome-diopside among heavy indicator minerals. Likewise, present are xenoliths of coarse-grained peridotites, eclogites and peridotites with porphyroblastic structures (Garanin 2004; Golovin 2003).

The Verkhotina field contains weakly diamondiferous bodies of olivine melilitites. The major diatremes of those reside in the Mayskaya, Osetinskaya, Volchya, Verkhotina, Verkhnetovskaya, Poludennaya pipes (Fig. 1.6) (Bogatikov et al. 1999; Golovin 2003). They are characterized by elevated content of titaniferous magnetite and rutile, very low content of Al-, Ti-containing chrome-spinelides. As for chrome, picroilmenite from olivine melilitite pipes of the Verkhotina field is less chrome-saturated versus the same mineral from the highly diamondiferous V. Grib pipe at the Chernoozerskoe field (Bogatikov et al. 1999; Garanin 2004; Golovin 2003).

Therefore, the Arkhangelsk diamondiferous region is a potential northern territory for development of the national economy by means of the diamond mining industry. Quite close occurrence of diamond-bearing and other bodies of the alkaline ultramafic magmatism enables to focus the diamond prospecting and geological exploration efforts in a compact site and utilize the uniform prospecting and exploration process in the province, thus boosting efficiency of such efforts. This requires an integrated approach in searching for the diamondiferous bodies, comprised by a wide range of geophysical methods, heavy mineral analysis method and geochemical methods, including the ways to determine HC systems.

Many issues, inter alia those related to zoning of ADR's Zimny Bereg area, are still unsolved. The current most advanced mineragenic ADR territory zoning pattern is derived by V. V. Tretyachenko (2008), accounts for the geological, tectonic and age-related placement of magmatic bodies, their mineral composition, requires to be amended and clarified.

One should bear in mind the existing data on investigation of petrology, geochemistry of kimberlitesand lamproites around the north of the East European Province (Karelian-Kola diamondiferous subprovince, as reported by ALROSA professionals, of the Baltic-Belomorye subprovince of carbonatites and kimberlites, according to V.V. Tretyachenko) in order the materials to be used in evaluating the diamond potential of the entire territory. First and foremost, we should take note of the age of kimberlite and lamproite bodies in the region, i.e. the Karelian-Kola subprovince. The main stages of the kimberlite and lamproite magmatism for the Baltic-Belomorye subprovince of carbonatites and kimberlites could be sequenced as follows: late Paleoproterozoic being the earliest (1750–1800 Ma, Kimozero pipe, Russia), followed by Mesoproterozoic (Medium Riphean, some 1200 million years, Kostomuksha, Russia), then by Neoproterozoic (some 600 Ma, Kuopio and Kaavi fields, Finland), Later—Medium Ordovician (approximately 460 Ma, Kuusamo kimberlites, Finland) and, finally, the last and most productive stage—Devonian (410–360 Ma). Commercial diamond-bearing grade of the bodies (M. V. Lomonosov and V. Grib diamond deposits, Russia) have only been defined for the last stage (Devonian). It appears the deep magmatism was actively evolving at depths down to 200 km, i.e. in the diamondiferous zone of the mantle, only at that time.

The typomorphic peculiarities of diamond as such have been studied and described in multiple papers, including monographs, which helped identify the typomorphism of ADR diamond to use its properties in zoning of the Zimny Bereg area and search for new diamond deposits in the area and beyond—in the Kola peninsula, Karelia, on Timan and other adjacent localities.

The book offers new data on morphology and spectroscopic properties of diamond from pipes of the Zimny Bereg area. Each new part of material facilitates a new step in the path of understanding the genesis of diamond, diamondiferous mantle, formation and evolution of protokimberlite and kimberlite melts. Combined with geological data this enables to expand the diamond prospecting efforts, which will be finally result in discovery of new diamondiferous bodies (Tables 1.3 and 1.4).

Table 1.3 General geological summary of some alkali-ultrabasic rocks at Zimny Bereg (Winter Coast), based on Bogatikov et al. (1999), Verichev et al. (2000), Garanin (2004)

Studied body	Body morphology						Crater facie thickness	Number of interstitial phases	Main varieties of diatreme facies lithology	Series of volcanic rocks
	Body shape	Body dimensions								
		Area, ha	Length, m	Width, m	Isometric coefficient	Elongation azimuth, degrees				
M. V. Lomonosov deposit										
Lomonosov	Pipe	20.2	600	450	0.75	0	–	2	XTB, AK	Mg-A-
Karpinskogo-1	Pipe	18.8	880	250	0.28	15	–	2	XTB, AK	
Karpinskogo-2	Pipe	10.2	575	430	0.75	15	22	2	XTB, AK	
Arkhangelskaya	Pipe	20.6	620	550	0.89	20	43	2	AK	
Pionerskaya	Pipe	38.0	1200	500	0.42	0	55	2	XTB, AK	
Pomorskaya	Pipe	5.6	400	240	0.60	15	–	1	XTB	
V. Grib deposit	Pipe	16	570	480	0.84	25	110	2	XTB, AK, PK	Mg
Shocha										
South-Western Body	Pipe	16.5	485	430	0.89	45	30	2	XTB, AK, PK	Fe-Ti
North-Eastern Body		7.3	350	270	0.77	45	–	2		
Anomaly-734	Pipe	3.3	275	150	0.55	290	30	2	XTB, PK	Fe-Ti
Verkhotina	Pipe	1.66	173	138	0.80	12	40	1	XTB	Fe-Ti (Al)

Legend TB—tuff breccia, XTB—xenotuff breccia, AK—autolithic (tuffisitic) kimberlite, PK—porphyritic kimberlite

Table 1.4 Mineral composition of pipes and sills of alkaline ultrabasic rocks in the ADP (Garanin 2004)

Magmatic field	Body	Appearance form	Rock type	Typical textural types of rock	Typomorphic minerals
Zolotitsa	Lomonosov Karpinskogo-1 Karpinskogo-1 Arkhangelskaya Pionerskaya Pomorskaya	pipe	Kimberlite, type II	Massive porphyritic macrocrystal kimberlite (pipe root), autolithic kimberlite and litho-crystal clastic breccia diatreme)	Ol-I (>20 wt%), Ol-II, Phl, Spl, Prv, ± Cal, ± Mel
Verkhotina	V. Grib	pipe	Kimberlite, type I	Massive porphyritic kimberlite (pipe root), litho-crystal clastic brecciaand autolithic kimberlite (diatreme)	Ol-I (30–50%), Ol-II, Phl, Grt, Ilm, Spl, Crd, Cal
	Verkhotina	pipe	Ol-Phl-melilitite	Porphyraceous phlogopitic olivine melilitite	Ol-I (<10%), Ol-II, Mel, Phl, Spl
Kepina field	Shocha	pipe	Kimberlite, type II	Massive porphyritic kimberlite, autolithic kimberlite and litho-crystal clastic breccia (diatreme)	Ol-I (<20 wt%), Ol-II, Phl, Spl, Crd, Grt
	Anomaly-734	pipe	Kimberlite, type I	Massive porphyritic kimberlite	Ol-I (<10 wt%), Ol-II (30–40%), Cpx, Phl, Spl, Prv, Cal, Ilm, Rt

Note Mineral species abbreviations: Ol—olivine, Phl—phlogopite, Spl—spinelides, Prv—perovskite, Ilm—ilmenite, picroilmenite, Rt—rutile, Px—pyroxene, Cpx—clinopyroxene, Mel—melilite, Plg—plagioclase, Cal—calcite, Grt—garnet, Crd—cordierite

References

Arzamastsev, A.A.: Evolution of Paleozoic alkaline magmatism of the north-western Baltic Shield: abstract of DSc in geology and mineralogy thesis, p. 58. Apatity: Kola Science Center of RAS (1998)

Bartoshinsky, Z.V.: Crystal morphology of diamond from kimberlites of Arkhangelsk diamondiferous province. Miner. Sb. **46**(2), 64–73 (1992)

Bogatikov O.A., Garanin V.K., Kononova V.A. et al. Arkhangelsk diamondiferous province. M.: MSU Publishing (1999), 522p

Bogatikov, O.A., Kononova, V.A., Nosova, A.A., Kondrashov, I.A.: Kimberlites and lamproites of the East European Craton—petrology, geochemistry. Petrology **15**(4), 339–360 (2007)

Frolov A.A., Lapin A.V., Tolstov A.V. et al.: Kimberlites and Carbonatites (Relationships, Minerageny, and Forecasting). M.: NIA-Priroda (2005), 540p

Garanin K.V.: Alkaline ultramafic magmatites of Zimny Bereg: their potential diamond-bearing grade and commercial development prospects. Ph.D. in Geology and Mineralogy Thesis. M.: MSU (2004), 50p

Garanin, V.K., Kudryavtseva, G.P., Posukhova, T.V., Verzhak, V.V., Verichev, E.M., Garanin, K.V.: Two types of the diamondiferous kimberlites from the Arkhangelsk province. Geol. Explor. **4**, 36–49 (2001)

Garanin V.K., Garanin K.V., Vasilieva E.P. et al.: Mineralogy of mantle xenoliths from the V. Grib pipe diamondiferous kimberlite pipe (Arkhangelsk Diamondiferous Province). In: Proceedings of IHLs. Series Geology and Exploration. No 1. 2005, pp. 23–28.

Golovin N.N.: Geological structure, mineral composition and formation conditions of alkaline ultramafic rocks in the Kepina area. Ph.D. in Geology and Mineralogy Thesis. M.: MSU (2003), 45p

Golubeva, Y.Y., Pervov, V.A., Kononova, V.A.: Petrogenesis of autoliths from kimberlitic breccias in the V. Grib pipe (Arkhangelsk district). Dokl. Earth Sci. **411**, 1257–1262 (2006)

Khain, V.E.: The forces which created the unique face of our planet. Soros Educ. J. **11**, 103–110 (1998)

Kharkiv, A.D., Zinchukm, N.N.: The history of diamond. M.: Nedra (1997), 600p

Kharkiv, A.D., Zinchuk, N.N., Kryuchkov, A.I.: Diamond Non-placers of the World. M.: Nedra (1998), 554p

Kononova, V.A., Golubeva, Y.Y., Bogatikov, O.A., Kargin, A.V.: Diamond content of Zimneberezhnoe Field kimberlites (Arkhangelsk region). Geol. Ore Deposits **6**, 483–505 (2007)

Kramm, U., Kogarko, L.N., Kononova, V.A.: Middle and Late Devonian—a short period of magmatic activity in Paleozoic Kola alkaline province. In: Bogatikov, O.A. Kononova, V.A. Zek, Kh.A. Lippolt Kh.I. (eds.) Magmatism of Rifts and Foldbelts, pp. 148–168. M.: Nauka (1993)

Kudryavtseva G.P., Posukhova T.V., Verzhak V.V., Verichev E.M., Garanin V.K., Golovin N.N., Zuev V.M.: Atlas: morphogenesis of diamonds and their mineral-satellites from the kimberlites and other relative rocks from the Arkhangelsk diamondiferous province. 1st ed. M.: Polyarny Krug Publishing (2005). 624p

Larchenko V.A.: Drafting a set of advanced geological maps for Zimny Bereg area … 1993. Archives Arkhangelskgeorazvedka CJSC.

Lomonosov M.V.: The First Foundations of Metallurgy or Ore Mining. / Lomonosov M. V. Works, vol. VII. L.: Publishing House of AS USSR (1934), 612p

Lomonosov M.V.: Works on Mineralogy, Metallurgy and Mining. Selecta. L.: AS of USSR Publishing (1954), 747p

Mitchell, R.H.: Kimberlites: Mineralogy, p. 436. Plenum Press, Geochemistry and Petrology, New York (1986)

Sablukov S.M.: Volcanism of Zimny Bereg and petrological criteria for the diamond potential of kimberlites. Ph.D. in geology and mineralogy thesis. M.: TsNIGRI (1995), 24p

Sablukov, S.A., Sablukova, L.I., Shavyrina, M.V.: Mantle xenoliths from the Zimny Bereg kimberlite deposits of rounded diamonds (Arkhangelsk diamondiferous province). Petrology **8**(5), 518–548 (2000)

Sablukov, S.M., Sablukova, L.I., Verichev, E.M.: Types of mantle substrate in the Zimny Bereg region in connection with formation of kimberlites hosting rounded and flat-faced diamonds (Arkhangelsk diamondiferous province). Wors of TsNIGRI **218**, 134–149 (2004)

Sinitsin, A.V., Dauev, Yu.M., Grib, V.P.: Structural position and productivity of kimberlites in the Arkhangelsk province. Geol. Geophy. **10**, 74–83 (1992)

Smith C.B., Gurney J.J., Skinner E.M.W.: Geochemical character of Southern African Kimberlites: a new approach based on isotopic constraints. Trans. Geol. Soc. South Africa **88**, 267–280 (1985)

Stankovsky, A.F., Danilov, M.A., Grib, V.P., Sinitsin, A.V.: Diatremes of the Onega peninsula. Soviet Geol. **8**, 69–79 (1973)

Tretyachenko V.V.: Mineragenic zoning of kimberlite area of South Eastern White Sea region. Ph.D. in Geology and Mineralogy Thesis. M.: (2008), 28p

Verichev E.M.: Geological conditions of origination and exploration of the V. Grib diamond deposit. Ph.D. in geology and mineralogy thesis. M.: MSU (2002) 36p

Verichev, E.M., Garanin, V.K.: Geological structure, mineralogical and petrological features of kimberlites from Arkhangelsk province. Geol. Explor. **4**, 23–28 (1991)

Verichev E.M., Golovin N.N., Zaostrovtsev A.A.: Geological structure and matter composition of the V. Grib pipe. Essays on geology and minerals of Arkhangelsk region. Arkhangelsk (2000), 127p

Verzhak V.V.: Geological structure, matter composition, formation conditions and methods for exploration of the M.V. Lomonosov diamond deposit. Ph.D. in Geology and Mineralogy Thesis. M.: MSU (2001), 36p

Chapter 2
Typomorphism of Diamond from Deposits of the Arkhangelsk Diamondiferous Region

Industrial, genetic and gemological mineralogy aspects of diamond from major types of ores in the studied pipes embrace qualitative parameters: average concentration, grain sizing, morphological peculiarities, and a row of other typomorphic properties, largely summarized below.

2.1 Mineralogical Characteristics of Diamonds from the M.V. Lomonosov Deposit Pipes

The M. V. Lomonosov diamond deposit is comprised by the Northern and Southern groups of bodies identified based on their distinctions in the mineralogical and petrographic composition of kimberlites and degree of their diamond-bearing potential. Optical characteristics and structural features of diamonds for crystals from Arkhangelskaya and Karpinskogo-1 pipes (Southern pipe group), and Pionerskaya, Pomorskaya pipes (Northern pipe group) were studied.

The diamond study results substantiate the classification and division of deposit pipes by their mineralogical criteria, and by typomorphism of the external and internal morphology of the diamond as such, and appear crucial for studying the genetic aspects of diamond formation.

The average diamond contents in the pipes of the Northern group are substantially lower than in similar rocks of the Southern body group.

The diamond content in the principal most diamondiferous rock—autolithic kimberlite breccias of the Northern group—has been 0.8 ct/t in Pionerskaya, 0.7 ct/t in Lomonosovskaya, and 0.4 ct/t for Karpinskogo-2 pipes; for the Southern group the figures are 1.8 ct/t for Karpinskogo-1 and 1.4 ct/t for Arkhangelskaya pipes.

A drastic difference is observed in the ratio of tuff breccias: 0.4–0.6 ct/t in Pionerskaya, 0.2–0.3 ct/t in Karpinskogo-2, and only about 0.1 ct/t in Karpinskogo-1.

Xenotuff breccias of the Northern and Southern pipe groups are similar by their diamond content: elevated concentrations of 0.15–0.2 ct/t are typical of the

V. Garanin et al., *Diamonds from the Arkhangelsk Province, NW Russia*, Springer Mineralogy, https://doi.org/10.1007/978-3-030-35717-7_2

Karpinskogo-2, Pionerskaya and Lomonosovskaya, whereas the minimum of 0.1 ct/t is proper for the Karpinskogo-1 and Pomorskaya pipes.

Meanwhile, tuffaceous rocks of craterous origin are characterized by essentially close diamond concentrations—0.7 ct/t for Arkhangelskaya and 0.2 ct/t for Karpinskogo-1 and Pionerskaya pipes.

On the quantity side, grade −1+0.5 mm prevails in the deposit, which is also common for other known kimberlite occurrences. As reported by Makhin et al. (1990), their content is up to 78% in the Lomonosovskaya pipe and minimum (57%) in the Karpinskogo-1 pipes, being 20 and 8% by weight, respectively.

The crystal distribution by weight is alike in all pipes, and dominated by individuals weighing more than 0.05 ct. In terms of weight, grade −4+2 mm individuals prevail in all pipes, with maximum concentration in the Pomorskaya pipe (48%), close values in the Pionerskaya pipe (42%), and minimum in the Karpinskogo-2 pipe. In the scope of the deposit, elevated average weight of one crystal (0.04 ct) is observed in the Karpinskogo-1 pipe. In a large-volume prospecting sample dated 1997–2000, crystals weighing 1–3 ct total 0.1 wt% of all diamonds. The biggest single crystals weigh up to 20–40 ct; these were extracted from the Karpinskogo-1 and Arkhangelskaya pipes (21.00 and 32.83 ct, respectively). A big 46-carat diamond has been the latest extraction of 2017 from the Arkhangelskaya pipe.

2.1.1 Crystal Morphology

During diamond evaluation, the notion of "shape" shall mean the morphological shape of crystals, crystal integrity, potential large cleavages, stone distortion and ovalization degree, i.e. the extent of plane-faced (flat-faced) octahedron transformation into a curve-faced individual (dodecahedroid). The most precious flawless crystals are shaped as isometric smooth-faced sharp-edged octahedron. Ceteris paribus, prominent ovalization (for instance the examined diamond is a dodecahedroid with no inclusions) reduces the value by 15–20%.

As evidenced by the study data, each pipe of the M. V. Lomonosov deposit is distinguished by individual distribution of crystals by habit (Fig. 2.1). For results of these studies and previous data (Makhin et al. 1990) see Table 2.1.

As compared to other pipes of the Lomonosov deposit, the Pionerskaya pipe has higher quantity of combined (both plane-faced and curve-faced) and flat-faced crystals.

In the Karpinskogo-1 and Arkhangelskaya pipes, the crystal distribution by habit types is very similar. Curve-shaped crystals of dodecahedral habit prevail (>60%), and the share of combinatory diamonds with octahedral faces and dodecahedral surfaces is much lower (Figs. 2.2, 2.3 and 2.4). Elevated cubic and combined tetrahexahedral habit diamond content is observed.

The majority of cubic and tetrahexahedral crystals are transparent or translucent (Fig. 2.3). About a third of cubic habit diamonds are colorless with yellow tint (33%) and yellow (29%). Amber-yellow (3%), gray (2%) and greenish-yellow (2%)

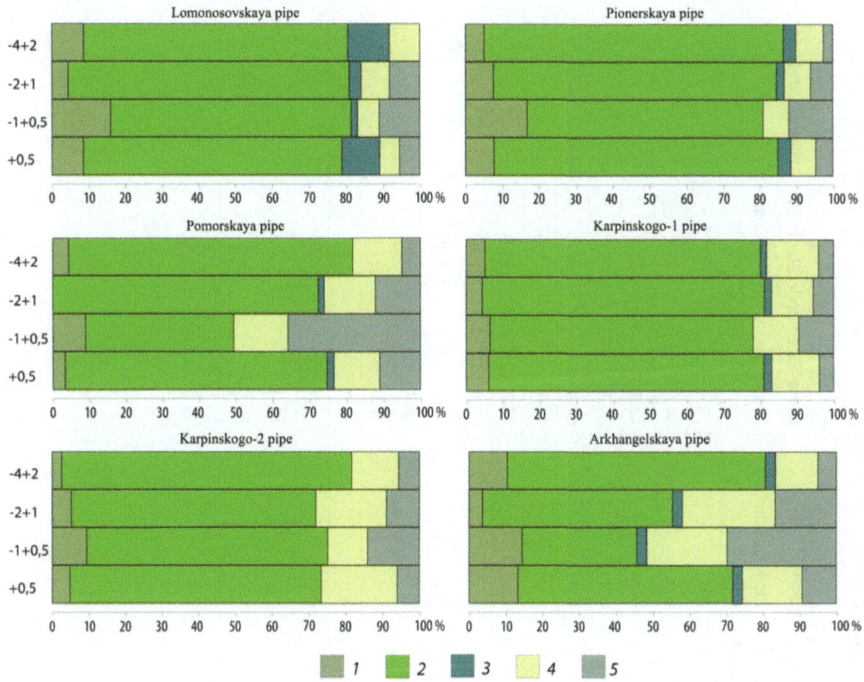

Fig. 2.1 Diamond distribution by habit in various size grades. Habit: 1—octahedral, 2—rhombic-dodecahedral, 3—combined, 4—cubic, 5—undefined (Kudryavtseva et al. 2005)

diamonds occur less frequent. Small grades are dominated by cubes with "saddle-shaped" faces, i.e. growth of large step-like depressions instead of faces (100). Drop-like hackly surface relief is common for tetrahexahedrons.

Cubic crystals often have zoning. The core commonly tends to change from colorless to yellow or brown, and the coat—from light-gray to dark-gray or yellow-green. Complex zoning crystals also occur, with alternating light-gray transparent and dark opaque zones that originated in multiply changed crystal growth mechanisms.

The Karpinskogo-1 and Arkhangelskaya pipes (Southern group), and Karpinskogo-2 and Pomorskaya pipes (Northern group of bodies) are most similar by distribution of diamonds with varied crystal morphology.

Dodecahedral habit diamonds with tangential and zonal-sectorial face growth mechanisms prevail in all the studied pipes. The quantity of such diamonds is much higher in pipes of the Northern group (up to 82% in the Pomorskaya and Karpinskogo-1 pipes), since the share of tetrahexahedroids and cubes is reduced to 5% in all size grades. An elevated content of octahedral crystals (up to 4–6% in −4 +2 mm grade) is noted in the Pomorskaya, Karpinskogo-2, Pionerskaya pipes, versus the Karpinskogo-1 and Arkhangelskaya pipes (1–3% in −4 +2 mm grade).

The quantity of octahedral habit diamonds is reduced and varies from 10 to 15%, most of them falling under −1.0 +0.5 mm grade by size and weight. Octahedron

Table 2.1 Distribution of diamonds by crystal morphology types in Pionerskaya, Arkhangelskaya, Pomorskaya, Karpinskogo-1, and Karpinskogo-2 pipes

Pipe	Sample	Occurrence frequency of diamonds by habit (%)						
		Flat-faced octahedron		Combinatory crystal		Curve-faced dodecahedroid		Undefined shape
		Sharp-edged	Round-edged	Octahedron + dodecahedroid	Dodecahedroid + cube	Urals type	Laminar	
Pomorskaya	TB	2.7	5.4	2.7	0.0	78.4	2.7	2.7
	XTB	2.6	0.0	0.0	0.0	84.6	0.0	7.7
	PS	11.2	0.6	9.1	–	23.1	28.9	–
Pionerskaya	TR	0.7	8.6	16.4	0.0	69.4	0.0	0.0
	AKB	0.7	10.7	16.4	0.0	67.9	0.0	0.0
	PS	10.6	4.9	25.4	–	24.7	19.6	–
Karpinskogo-2	XTB	8.0	3.8	28.4	–	22.8	19.0	–
	PS	8.0	3.8	28.4	–	22.8	19.0	–
Karpinskogo-1	TR + AKB	1.1	1.1	11.7	0.3	22.1	54.9	1.6
	AKB	0.0	0.7	5.8	0.7	10.4	44.1	18.8
	PS	1.3	1.0	14.6	–	37.0	21.9	0.8
Arkhangelskaya	TR	3.0	0.0	15.3	3.0	52.2	–	3.0
	AKB	1.4	0.0	7.1	0.0	70.0	–	4.3
	PS	8.8	0.8	12.2	–	15.6	22.3	–

(continued)

Table 2.1 (continued)

Pipe	Sample	Occurrence frequency of diamonds by habit (%)								
		Entire octahedron—dodecahedroid row	Cube	Tetrahexahedroid	Cube + tetrahexahedroid	Dodecahedroid (dissolved cube)	Total	Coated diamonds	Microblock dodecahedroids	Polycrystalline aggregates
Pomorskaya	TB	94.6	0.0	2.7	0.0	2.7	5.4	0.0	0.0	0.0
	XTB	94.9	0.0	0.0	2.6	2.6	5.1	0.0	0.0	0.0
	PS	72.9	0.9	19.3	–	–	20.2	–	7.2	–
Pionerskaya	TR	95.1	2.1	2.1	0.7	0.0	4.9	0.0	0.0	0.0
	AKB	95.8	2.1	0.0	2.1	0.0	4.2	0.0	0.0	0.0
	PS	85.2	0.9	5.9	–	–	6.8	–	6.6	1.3
Karpinskogo-2	XTB	82.0	0.8	10.9	–	–	11.7	0.0	5.6	0.7
	PS	82.0	0.8	10.9	–	–	11.7	0.0	5.6	0.7
Karpinskogo-1	TR + AKB	92.8	0.5	1.4	0.6	4.1	6.6	Sing	0.6	–
	AKB	80.5	0.7	8.0	3.6	6.5	18.8	0.0	0.7	–
	PS	86.1	0.7	12.9	–	–	13.6	Sing	10.3	0.3
Arkhangelskaya	TR	76.5	2.1	6.0	10.3	2.1	20.6	0.0	0.0	3.0
	AKB	82.8	2.9	8.6	5.7	2.9	20.0	0.0	0.0	0.0
	PS	59.7	1.0	27.4	–	–	28.4	–	11.0	0.9

Note Diamond samples from rocks: XTB—xenotuff breccia, TR—craterous tuffaceous rocks, AKB—autolithic kimberlite breccia, PS—prospecting samples of rough diamonds, as reported by Makhin et al. (1990). Sing—singular grains

Fig. 2.2 Sample morphological types of octahedral and rhombic-dodecahedral habits from the M. V. Lomonosov deposit: **a**—flat-faced octahedron; **b**—octahedron with polycentric ragged face growth; **c**—dodecahedron with large olivine-graphite inclusion below the surface

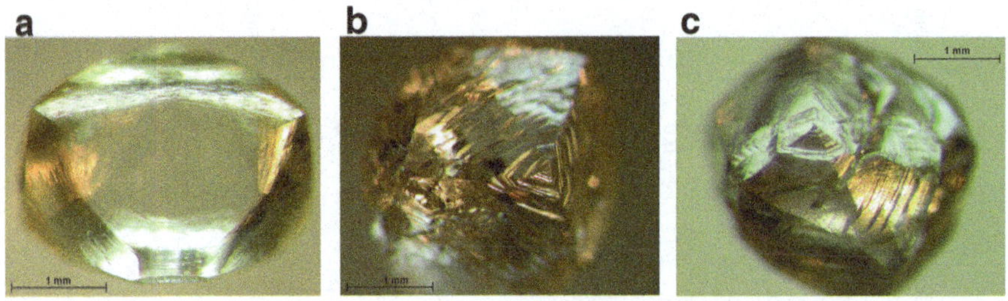

Fig. 2.3 Combined habit diamond crystals with octahedral faces and dodecahedral surfaces, 15–0.30 ct: **a**—flat-faced—curve-faced crystal with hackly relief on dodecahedron surfaces; **b**—crystal with octahedral face relics and dodecahedral hillocky and blocky surfaces; **c**—blocky crystal with an octahedral face relic, dislocation of blocks and plastic deformation bands

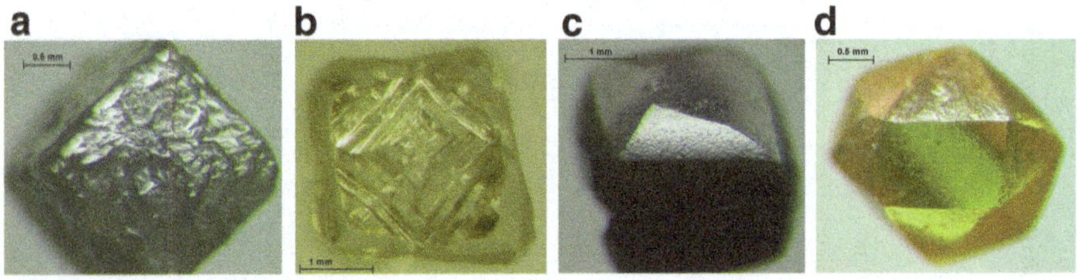

Fig. 2.4 Cubic habit diamond crystals, 15–0.30 ct: **a**—gray cube with dissolved surface; **b**—yellow-tinted cube with saddle-shaped faces; **c**—dark-gray cubic crystal with dodecahedral surfaces; **d**—canary yellow tetrahexahedroid with drop-like pattern

content is nearly the same in the Pomorskaya, Karpinskogo-2 and Arkhangelskaya pipes (12.6–13.2%), somewhat higher in the Lomonosov and Pionerskaya (15.1–18.1%), and minimal in the Karpinskogo-1 pipe (4.8%). Transition octahedral-dodecahedral crystals are uncommon (maximum 5%). Most of combinatory crystals are below >2 mm in size.

The diamond crystal morphology features enable to identify two groups of bodies differing for quantitative ratios of mineral that varies by habit (Zakharchenko et al. 2002). Lomonosov and Pionerskaya pipes fall into group I, with all the rest being in group II (Pomorskaya pipe as the typical representative). Group I is dominated by dodecahedral and combined habit diamonds. The tetrahexahedron content is drastically reduced (4–6%). Group II, with essentially the same habit spectrum of crystals (equal content of octahedrons, slightly reduced concentration of dodecahedrons), has higher tetrahexahedron content (7–27%), which may indicate a longer diamond crystallization and evolution process in the pipes. The pipes show low cube content (1.2% on average). Pseudo-hemimorphic crystals are rare (to 0.9%) and represented by a combination of plane-faced octahedron and curve-faced dodecahedron.

A sharp rise in the octahedron concentration in small grade −1 +0.5 mm is common for the Northern and Southern groups, due to presence of the "kimberlite" association crystals: up to 20% in the northern pipes, including Pionerskaya, and 15–17% among microcrystals of the southern group, as reported (Makhin et al. 1990). In most of the deposit pipes microdiamonds (<0.5 mm) are represented by smooth-faced crystals and their aggregates, mainly of octahedral and cubic habits with itching traces, skeleton crystals. Tetrahexahedron content in all size grades remains nearly constant for each pipe: 5–13% for the Northern group and 14–27% for the Southern group of bodies.

Crucially, a minor increase in the amount of curve-shaped dodecahedral crystals and tetrahexahedrons is noted when comparing diamonds of various rocks types from the same pipe in the tuffaceous crater cluster with the autolithic kimberlite breccia.

As established in studying diamonds from one and the same pipe, despite varying diamond content the tuffaceous rocks and AKB of the Pionerskaya pipe (tuff breccias and xenotuff breccias of the Pomorskaya pipe) show affinity of crystals by prevalence of habit forms, minor differences by color range, luminescence, defect-impurity composition. All those features prove the genetic affinity of diamond-bearing rocks, while varying conditions of diamond delivery and post-growth evolution define a drop in crystal quality by increasing the share of corroded crystals, etching degree, fracturing, integrity deterioration, presence of epigenetically colored crystals.

Meanwhile, tuffaceous rocks from various pipes in the same body group show similarity by parameters of the diamond post-crystallization history. So, one can predict that in deeper horizons of various pipes diamond of autolithic kimberlite breccias will have close characteristics.

Integrity degree for diamonds of tuffaceous craterous rocks and xenotuff breccias is much lower than for AKB. For data on various size grades of diamonds from the studied pipes see Table 2.2. For instance, Pionerskaya pipes is distinguished by average (deposit-wide) crystal integrity degree, typical of kimberlites. The tuffaceous craterous rocks the integrity degree is very low, whole undamaged crystals in small sizes make a third of the sample (28%), the main portion (41 and 15%) falling on chips and splinters. The autolithic kimberlite breccia rocks contain 48% of whole crystals and slightly more than a third (altogether 25.5 and 19.4%) of chips and splinters.

Table 2.2 Integrity-based distribution of diamonds from Pionerskaya, Arkhangelskaya, Pomorskaya, Karpinskogo-1, Karpinskogo-2 pipes

Pipe	Size-class (mm)	Quantity (pcs)	Whole	Compromised crystals (%)					
				Integrity degree			Cleavage nature		
				Minor cleavage	Chips	Splinters	Natural	Man-made	Combined
Lomonosovskaya	−4 + 2	342	41.8	9.1	47.4	1.8	50.0	4.1	4.2
	−2+1	564	32.2	4.8	57.0	6.0	49.8	7.1	11.0
	−1+0.5	782	30.4	4.7	60.1	4.7	49.0	9.6	11.0
Pomorskaya	−4+2	233	41.6	8.6	49.8	0.0	49.8	6.0	2.6
	−2+1	481	37.0	8.6	49.0	6.4	41.0	12.7	9.3
	−1+0.5	639	25.8	5.8	29.4	39.0	61.3	11.0	1.9
Pionerskaya	−4+2	369	46.1	8.7	44.7	0.5	42.0	4.3	7.6
	−2+1	501	30.9	6.4	59.1	3.6	41.1	10.2	17.8
	−1+0.5	867	27.9	4.6	63.7	3.8	50.6	13.6	7.8
Karpinskogo-1	−4+2	374	60.7	7.2	31.6	0.5	33.3	4.0	1.9
	−2+1	541	48.4	6.3	42.5	2.8	44.0	6.1	1.5
	−1+0.5	749	42.7	4.4	49.5	3.3	50.6	4.7	2.0
Karpinskogo-2	−4+2	344	43.9	5.5	49.1	1.5	41.6	8.7	5.8
	−2+1	651	37.9	5.8	52.4	3.9	44.3	9.8	8.0
	−1+0.5	668	30.7	5.7	59.6	4.0	46.9	13.0	9.4
Arkhangelskaya	−4+2	399	55.6	6.5	36.6	1.2	33.1	7.0	4.3
	−2+1	500	37.0	6.2	35.4	21.4	49.4	8.4	5.2
	−1+0.5 mm	700	22.8	8.9	28.6	39.7	60.3	9.3	7.6

Note Table based on data by Makhin et al. (1990)

Diamonds from southern bodies of the M. V. Lomonosov deposit have high natural material integrity degree. Whole crystals and crystals with minor apex cleavage make 55–60% in the large grades and 22–42% in −1+0.5 grade (Makhin et al. 1990). Chips (more than 1/3 crystal is absent), largely of natural protomagmatic origin, total 31–36% and 28–49%, respectively (values for the Karpinskogo-1 and Arkhangelskaya pipes).

Integrity is somewhat worse for diamonds of the Northern group. The share of whole individuals is some 42% in −4+2 mm grade and up to 28% in small grade, where amount of chips with undefined shape and splinters (individuals confined by cleavage surfaces) increases drastically.

Aggregates. All pipes of the deposit show elevated content of irregular aggregates from 2 to 5 individuals, and of diamonds twinned under the spinel law. Twins are commonly made up by 2–3 crystallites, severely flattened by the triad axis, which shapes them as triangular plates or lentils. The quantity of twins grows from the Pomorskaya pipe (9%) to Arkhangelskaya pipe (26%). In small grade −1+0.5 mm the concentration of twins and aggregates rises nearly twofold, as reported by Makhin et al. (1990).

The polycrystalline aggregates, both medium-grained and micro-grained, are very rare (some 0.5%) and only in the Lomonosov pipe their content reaches 3%. Widespread have been crystals with small intergrowths that partially surface with a clear-cut composition plane, and diamonds with composition planes, i.e. intergrowths, fit into the relief, but can complicate the diamond cutting process.

Presence of composition planes substantially reduces rough diamond quality and quite often visually pure diamonds (with no inclusions) are graded as near-gem items under industrial classification of rough diamonds (maccles, cleavage, makeable).

2.1.2 Internal Structure

Diamonds from the Southern group of kimberlites show prevalence of plane-faced and curve-faced crystals with polycentric face growth pattern, which evidences the crystallization environment is oversaturated with hydrocarbons (Punin 1981; Krasnova and Petrov 1997), and based on sources (Zinchuk and Koptil 2003) proves that such individuals were formed in thermodynamic conditions of the eclogitic paragenesis.

Diamonds from kimberlites of the Northern group have higher share of crystals with anti-skeleton face growth mechanism, which is often revealed in presence of grooved edges on combinatory crystals and apices elongated in several directions and matching the direction of the triad axis.

For the most part, crystals with layered and combined growth mechanism, with their frequent zoning and interleaving, are widespread in the deposits among the dodecahedral habit diamonds, which is caused by growth in unstable, oscillatory changing conditions. Presence of diamonds with fibrous coating and cubic habit

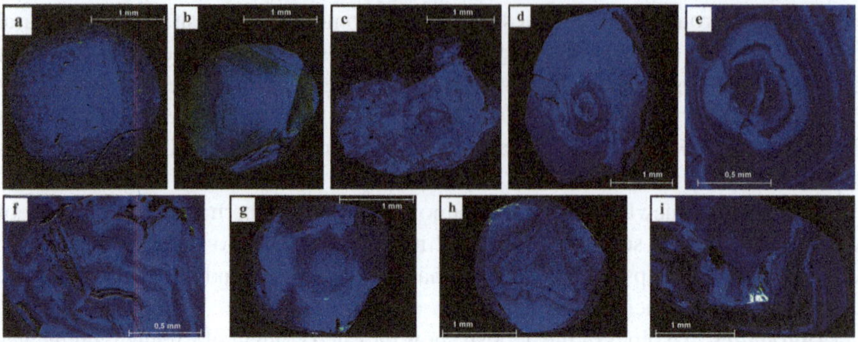

Fig. 2.5 Crystal internal structure (CCL images): **a**—homogeneous, {100}; **b**—straight-layered octahedral zoning, {100}; **c**—spotted, {100}; **d**–**f**—agate-like zoning, {111}; **g**—sectorial, {111}; **h**—fibrous growth, {111}; **i**—layered undulating zoning, {111}. T—tangential, N—normal, T + N—combined growth mechanism; crystal section is given in the parentheses

crystals with normal and combined growth mechanisms evidences high carbon oversaturation and drastic thermal gradient in the growth environment of such crystals. All of those evolved in generation of the defective structure and capture of a spate of inclusions by cubic growth sectors, and to quick transformation of the cubic forms into dodecahedral under post-growth oxidative dissolution of diamonds.

Color cathodoluminescence (CCL) and local IR spectroscopy (FTIR) were used to study internal structure of diamond crystals and allocation of nitrogen, hydrogen impurity centers and platelets by zones in crystals of diamond from the Karpinskogo-1 pipe.

Diamond offers blue to pale-blue cathodoluminescence. Half of the studied crystals (50%) have homogenous or octahedron-shaped zonal layered internal structure, originating in the tangential (T) growth mechanism (Fig. 2.5). The growth layers often fail to close in, which proves the altering diamond crystallization conditions. The rest of crystals have more diverse internal structure: zonal-sectorial, sectorial and complex oscillatory structure under normal (N) and combined (T+N) growth mechanisms. Parallel zoning was found in cubic crystals {111}. More complicated curved zoning was revealed in tetrahexahedroids.

Combinatory internal structure originated by, most often multiple, alterations of the tangential growth to normal and resulting in layered octahedral zoning that changes into sectorial structure or agate-like zoning, or else fibrous growth, is peculiar for diamond crystals from the Karpinskogo-1 pipe. The alternation sequence of normal and tangential growth zones varies. The central region of most crystals offers layered octahedral structure. So, in a rounded dodecahedral crystal (Fig. 2.5b) the layered octahedral internal region gets overgrown with fibrous diamond, which reflects the alteration of tangential to normal growth mechanism under the T→N scheme. Complicated T→N→T→N sequence was revealed in the cubic habit crystal (Fig. 2.5c). In two rhombic-dodecahedral habit crystals (Fig. 2.5d, e) the central region has agate-like structure and subsequently gets overgrown with octahedral

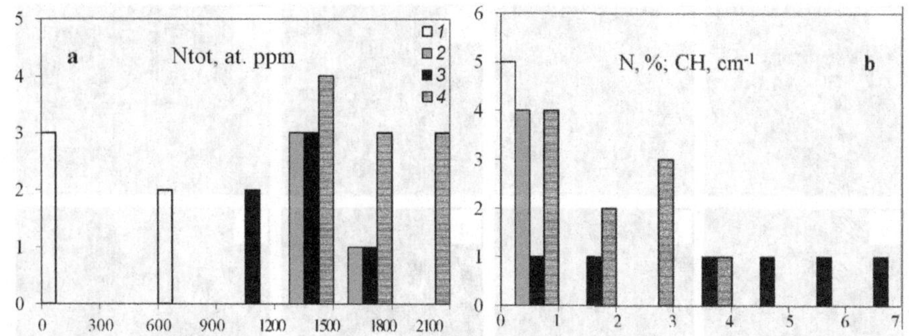

Fig. 2.6 Distribution of structural nitrogen (**a**), hydrogen (**b**) in diamond crystals from the Karpinskogo-1 pipe, subject to their color and internal structure features: 1—colorless individuals with layered octahedral structure; 2—gray crystals with layered octahedral structure; 3—fibrous structure crystals; 4—crystals with combined layered and fibrous structure; N—quantity of crystals

faces. In the initial stage the crystals were apparently growing under the normal (N) and combined (N+T) mechanisms, that afterwards altered to tangential (T) (Fig. 2.5f–i).

We see the correlation between distribution of nitrogen, hydrogen impurity centers and platelets in the studied crystals and peculiarities of their internal structure.

The colorless individuals formed under the tangential growth mechanism with homogenous and low-contrast layer-by-layer closed octahedral structure are characterized by reduced total concentration of structural nitrogen at 270 at. ppm maximum, actual absence (in the scope of method quantification level) of hydrogen centers and presence of platelets 1.9–3.2 cm^{-1} (Fig. 2.6a).

Specimens with layered octahedral and microblock structure, with abnormally high nitrogen concentration (1204–2218 at. ppm) make a special group of gray-colored individuals. They differ from colorless crystals of similar structure by elevated hydrogen centers content (0.6–2.9 cm^{-1}) (Fig. 2.6b). By nitrogen and hydrogen centers distribution the studied gray-colored specimens are comparable with crystals from the placers of northern Yakutia that were detailed in the work (Afanasiev et al. 2010).

Fibrous structure crystals exhibit elevated nitrogen (1137–1643 at. ppm) and hydrogen (0.5–6.4 cm^{-1}) concentrations, and nearly no platelets (Fig. 2.6).

As noted, more nonuniform defect distribution by zones is typical of diamonds from the Karpinskogo-1 pipe. The crystal core may exhibit both higher and lower nitrogen concentration versus the intermediate or marginal zone: N_{tot} core/margin = 971–1446; 2000–2200; 2500–1746; 1564–1438 at. ppm. This pattern is linked to polyphase crystal formation, which is reflected in their internal structure (Fig. 2.7).

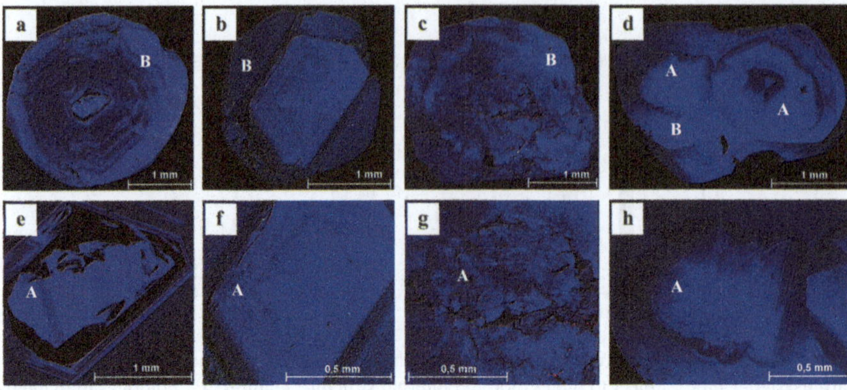

Fig. 2.7 Internal structure of crystals with seeds: **a**, **d**—main view, {111}; **c**, **d**—main view, {100}; **e**–**h**—seed enlarged view. A—seed, earlier generation; B—host diamond, a later diamond generation

2.1.3 Diamond Coloration

Color and transparency are among major criteria determining the value of rough diamonds. Mineral coloring, its visual perception and spectral identification are typomorphically crucial for many deposits. Diamond color is dictated by impurities that are isomorphically captured by the carbon diamond lattice or influence the structure and aggregation of nitrogen centers under exposure of temperature, pressure and radiation.

Most precious crystals (with highest estimated value) are totally colorless and "blend into the standard white paper background." Diamonds of fancy colors, such as pink and pale-blue, are currently most high-valued.

Hence, we need to pin down the diamond's *color origination nature*. A perfect diamond is totally colorless; the mineral color is linked to presence of defect-impurity centers in its structure that originate during crystal growth and are potentially affected by the post-growth processes. The syngenetic coloring is caused by nitrogen atoms penetrating the diamond structure and replacing carbon atoms, and origination of nitrogen-vacancy defects (Vins et al. 2008; Vins 2011). High concentration of isolated nitrogen atoms (C defect) causes yellow, yellow-brown or yellow-orange colors. During growth at high temperatures nitrogen atoms are transformed and generate N3 (three nitrogen atoms and a vacancy), whereby crystals acquire lemon-yellow color of varying intensity. Origination of A and B (two nitrogen atoms, four nitrogen atoms and a vacancy, respectively) defects does not cause coloration.

Syngenetic coloring (from colorless to fancy yellow) may be overlaid by epigenetic one. Epigenetic color is related to crystal plastic deformation processes occurring after (or in the final stages of) diamond crystallization, whereby the crystals acquire smoky-brown or violet color.

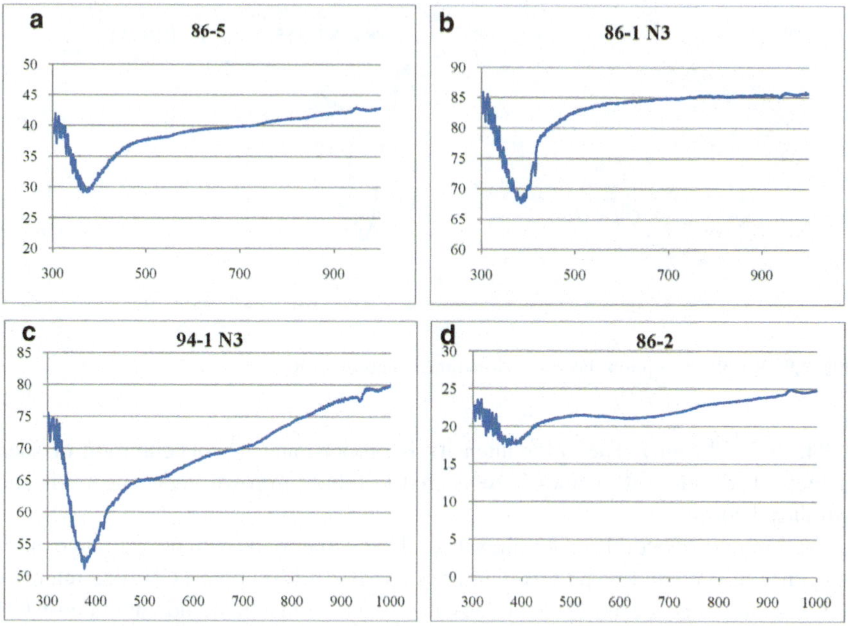

Fig. 2.8 Visible absorption spectra, colorless diamonds (1C)

Optical and spectrometric features and its correlation with coloration nature for crystals from the Karpinskogo-1 and Arkhangelskaya pipes was studied to enable diamond classification by nature of color.

Colorless and slightly yellow or gray tinted (1C, 2C)—diamonds of mostly dodecahedral shape (Fig. 2.8) give pale-blue luminescence.

The photoluminescence spectrum and visible range spectrum exhibit absorption lines related to presence of N3 center (three nitrogen atoms and a vacancy, major peak 415 nm).

An extra peak of 925 nm is registered in many colorless diamonds, which could be a diagnostic sign of Arkhangelsk diamonds. This feature was mentioned in the work by Makhin et al. (1990).

Yellow-tinted diamonds offer high concentrations of N3 centers. In some of N3-containing diamonds IR spectroscopy identified also a C-center, which is also contributing a tint into perception of yellow color. Such crystals were revealed solely in the M. V. Lomonosov deposit (Fig. 2.9). The overlay of pale-blue and yellow luminescence visually results in yellow-green perception of such glow. It is believed that N3 and C-centers cannot co-exist in the structure, so zonal structure can be proposed for such yellow cubes and dodecahedroids.

Straw-colored diamonds (5C, cape, color) of dodecahedral and tetrahexahedral habit have the following peaks in the absorption spectrum and visible wavelength range: 470 (BROAD), 685, 705 (?), 843 and N3, 455, 475. The UV range absorption spectrum reveals the peaks that we presumably classified to C, S2, S3 defects

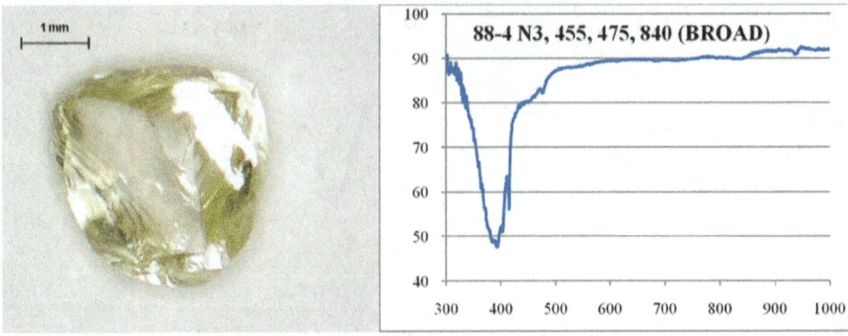

Fig. 2.9 Visible absorption spectra, yellow-tinted diamonds (3C)

(384, 497, 523 nm). The color intensity is proportional to concentration of such defects. The same set of lines is registered in tobacco-brown dodecahedroids and tetrahexahedroids.

Yellow dodecahedral and tetrahexahedral habit diamonds with weak orange luminescence in UV band. Type IaA diamonds (A=1020–1460 ppm), CH (line intensity 1.1–8.2 cm^{-1}). B2 center (1.3 cm^{-1}) is also observed in a diamond with intense CH absorption band.

Orange diamonds—this coloration type is observed in cubic amber-yellow crystals (Fig. 2.10). Cubes are represented by low-nitrogen (A<50 ppm) Ib type diamonds. CH center fluctuation lines revealed (intensity 0.12–0.23 cm^{-1}). C defect concentration is 325–350 ppm, C+ 15–20 ppm. These are inert in UV.

Brown diamonds are broken into two color rows: smoky-brown (pink-brown with fancy colors), tobacco-brown (yellow–brown) diamonds with varying color intensity (2 Brn-6 Brn).

Diamonds with ***slight smoky-colored tint*** (1Brn) in the absorption spectrum in visible range have N3 center absorption peaks and a low broad absorption maximum in the range of 500–700 nm, linked with broken C-C links due to plastic deformation traces. Visually, no plastic deformation traces exhibited by glide lines or shagreen surface are observed.

Smoky-brown diamonds are represented both by rounded dodecahedroids, and by octahedron-dodecahedron combined polygons; these are often crystal chips and crystals with composition planes. The visible absorption spectrum has the main broad peak with the maximum of 550 nm (500–700 nm) linked to existing plastic deformation traces (Fig. 2.11).

Pink-brown diamonds of dodecahedral habit are represented by high-nitrogen (Ntot = 1040 ppm) IaAB type crystals, with nitrogen aggregation in B1 form to 47%, intense B2 center absorption band (22.7 cm^{-1}) and CH band (1.7 cm^{-1}). In the UV it reveals pale-blue luminescence caused by N3 center.

Brown-tinted combined shape diamonds reveal intense pale-blue UV luminescence caused by a N3 center. In the visible absorption spectrum has 451, 477, 550 (BROAD) peaks. It is represented by a medium-nitrogen (N_{tot}=470 ppm) IaAB type

Fig. 2.10 Visible absorption spectra, colored diamonds: straw-yellow (5C, cape) and amber-yellow, orange (colored), cubic and tetrahexahedral habits

diamond, with intense absorption band of the B2 center (12.38 cm^{-1}) and weak CH band (0.09 cm^{-1}).

Tobacco-brown dodecahedroidsand tetrahexahedroids are similar to the straw-yellow stones by spectrum properties. They also reveal C, S2, S3 peaks (384, 497, 523 nm), and a broad peak with the maximum of 550 nm (500–700 nm). Color intensity is proportional to defect concentration.

Green, yellow-green diamonds exhibit intense absorption on the verge of the visible short-wave range, revealing H3 centers with zero-phonon line of 503.2 nm, H4 with zero-phonon line of 495.8 nm. H3, H4 are complex aggregated nitrogen-vacancy centers manifesting in diamonds in natural radioactive exposure.

An N3 center usually adds to H3, H4 flaws, and the crystal reveals more greenish-yellow color (Fig. 2.12). Yellow-green diamonds of varied shapes, octahedron-dodecahedroid series, tetrahexahedroids reveal yellow and yellow-green luminescence in UV. Such diamonds are represented by two types of visible range spectra:

1. medium-to-high nitrogen IaA (A = 300–1500 ppm), nitrogen in B2 form (line intensity 0.0–0.67 cm^{-1}), with CH link lines (2.0–11.6 cm^{-1}). A set of extra, N3, 451, 477, 700 (BROAD), 830 (BROAD) bands is registered in the visible range spectrum.

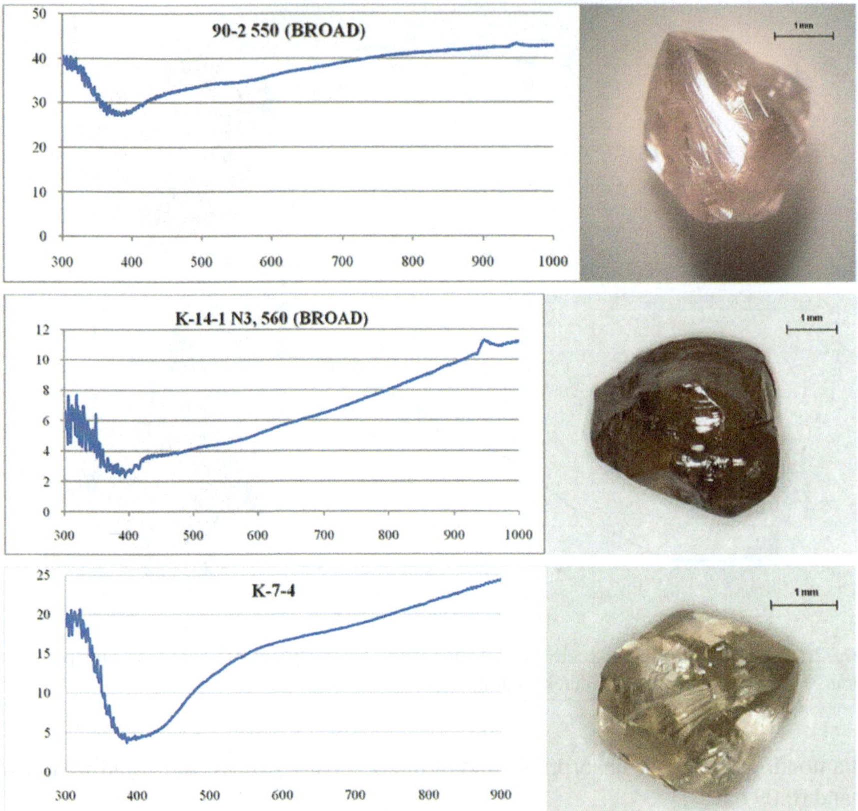

Fig. 2.11 Visible absorption spectrum, brown diamonds

2. high-nitrogen IaAB (Ntot = 1100–1700 ppm) with nitrogen aggregation in B1 form to 50%, nitrogen in B2 form (line intensity 0.3–20.0 cm^{-1}) and CH centers (intensity 0.04–13.20 cm^{-1}).

Gray-green diamond of dodecahedral habit (Fig. 2.12). In UV reveals intense yellow-green luminescence. It is represented by high-nitrogen (N_{tot} = 1040 ppm) type IaA diamond, with intense CH band (10.4 cm^{-1}).

Gray color of diamond crystals is traditionally due to multiple scattered microscopic dark-colored inclusions (graphite). Recent studies, including covered by this work, indicate the presence of the second type of gray crystals: uniform gray diamonds have high hydrogen impurity concentration, registered in the IR absorption spectra, in visible range—a typical spectrum of colorless crystals. These are represented by IaA type diamonds. Gray diamonds are inert under UV exposure.

Greenish-yellow combined-shape crystal, with tetrahexahedroidal and cubic faces. Sp. K-1–3, Karpinskogo-1 pipe.

Fig. 2.12 Visible absorption spectrum, yellow-green diamonds

It is possible now to discriminate the diamond features that are typomorphic for the entire M.V. Lomonosov deposit:

Diamonds of yellow and gray tints prevail in all size classes, and the share of totally colorless crystals is about 20% in each pipe.

Gray individuals prevail among colored crystals and their chips. Fancy-colored crystals (yellow, tawny-brown, pink) occur quite rarely.

Unfortunately, fancy-colored often have evident fractures, so their investment appeal to increase the deposit valuation is yet unsolved.

Diamonds from the Southern group of bodies in the M.V. Lomonosov deposit are characterized by the maximum spectrum of colors and occurrence frequency of colored stones among the octahedron-dodecahedroid diamonds. Yellow, gray, smoky-brown crystals prevail; it is crucial to stress that fancy-colored individuals occur: pink-violet, pale-blue, orange (red-brown), yellow. The share of colorless stones is minimum (below 10%), totals 20–25% together with slight tinted diamonds, and dominated by the crystals with yellow (30–54%) tint of varying intensity.

As shown in the collection mineralogical description, the majority of the Lomonosov deposit diamonds are fractured and contain epigenetic dark-colored graphite inclusions. Therefore, the rough diamond samples look gray, which is particularly true for the Arkhangelskaya pipe.

To identify the genetic features, we need to "deduct" the gray component and define the syngenetic color of mineral.

The mineralogical study of diamond with reference to rock types enabled to reveal the following trends.

Diamonds from AKB and TR of the studied pipes have the same genetic reservoir, which is confirmed by affinity of their properties determined by diamond matter crystallization conditions. Both AKB and TR exhibit similar quantity of diamonds with syngenetic yellow color. AKB and TR show affinity by distribution of octahedron-dodecahedroid habit crystal types, although they differ by the number of tetrahexahedroids generated in the final diamond formation stages.

Many post-crystallization properties of diamonds from AKB and TB differ notably, due to varying features of intrusion and formation of different kimberlite rock types. Low integrity, widespread plastic deformation traces, higher quantity of epigenetically colored smoky-brown crystals is more typical of the TB-derived diamonds. In the Arkhangelskaya pipe TR-based crystals are more fragmented, dominated by chips and splinters. AKB-based diamonds from that pipe reveal higher integrity degree (multiple whole crystals) and higher quality, as less fractured and containing less local etching traces.

2.1.4 Diamond Gemological Characteristics

The rough diamond quality is based on the popular 4C system (carat, cut, color, clarity); for uncut crystals it can be conventionally interpreted like size (weight), shape, color and clarity. In this paper we'll study those parameters from the mineralogical and gemological viewpoints, aiming at economically viable deposit development, relying on the diamond concentration and quality data.

As is known, identity of mineral individuals is caused by specifics of their crystallization and post-growth evolution. The study of diamond, its defect-impurity composition, inclusions is essential for learning the diamond formation environment. Internal structural features and crystal impurities dictate the color, transparency, and all toughness, sizing characteristics, which subsequently drastically affects the quality and valuation of rough diamonds.

The pattern of diamonds from the studied pipes by crystal morphology, integrity degree, coloration and other signs, properties, is approximately the same, with slight variations. On the whole, by a set of typomorphic features of diamonds, the Lomonosovskaya, Pionerskaya, Pomorskaya pipes are merged into one group, while the others are placed into the second, given the intermediate nature of the Karpinskogo-2 pipe (Makhin et al. 1990).

The study of diamond mineralogical characteristics, corroborated by review of data from references, enables to conclude that despite low-level diamond content in pipes of the Northern group, many quality parameters of rough stones are much better than those for Southern pipes.

Pionerskaya pipe offers the highest rough diamond quality of all studied bodies. It shows prevalence of near-gem quality crystals (up to 65–70%), with a slight share of industrial grade diamonds.

The lowest numbers of highly graphitized fractured crystals are observed in the Pionerskaya pipe diamonds, which offer an elevated share of highly transparent colorless isometric dodecahedroids, classified as enlarged Sawables Light items. Clastic gem quality and near-gem quality diamonds prevail (Chips, Maccles, Cliv

Mb), whereas the dominating share of diamonds from the Arkhangelskaya and Karpinskogo-1 pipes is classified as industrial (Rejection stones, Rejections).

In ***Pomorskaya pipe,*** colorless highly transparent dodecahedroids of high-temperature ultramafic origin comprise most of crystals. However, those diamonds are prone to strong oxidative dissolution and local etching. These reveal much higher concentration of crystals with multiple graphite inclusions, versus the Pionerskaya pipe diamonds. Tetrahexahedroids, which are typomorphic for the pipe, largely contain a multitude of scattered and finely dispersed inclusions and are not gem-quality.

Low integrity, elevated concentration of crystals with graphite inclusions, intergrowths and aggregates, and individuals with fractures, feathers and deep etch channels drastically reduce the cost parameters of the Pomorskaya pipe crystals.

Karpinskogo-2 pipe diamonds. We studied rough diamonds from the craterous xenotuff breccia and tuff breccia samples. The crystals exhibited substantial corrosion, but our studied showed a similar alteration was observed in small crystals from craterous formations of the Pionerskaya pipe. We do not have an adequate set of data to evaluate diamonds of the Karpinskogo-2 pipe, but under the references (Makhin et al. 1990) and mineralogical features of coloration and crystal morphology diamonds of the Karpinskogo-2 pipe are close to those of other Northern group bodies.

We may suggest, relying on such guidance, that the quality of rough diamonds in the Karpinskogo-2 pipe shall be comparable with the Pionerskaya pipe. That conclusion is interim, whereas its validation requires a study of mineralogy and defect-impurity composition on a statistically representative material, for inclusion into the morphogenetic groups typical of diamonds from other pipes of the deposit.

In ***Arkhangelskaya*** and ***Karpinskogo-1 pipes*** gem-quality diamonds make about 25% of the sample, near-gem quality—14%, industrial grade—60%.

The mineralogical and gemological characteristics of diamonds highlights deterioration of diamond quality in Arkhangelskaya and Karpinskogo-1 pipesowing to defective structure of crystals, due to prevalence of dodecahedral and tetrahexahedral diamonds with zonal-sectorial structure, high stresses with multiple fractures, graphite and fluid inclusions. Composition plane diamonds and aggregates (Maccles, Cliv Mb), are widespread in the pipes, whereas monocrystals are often irregular.

2.1.5 Diamond Typomorphic Characteristics

The mineralogical characteristics of diamonds shows the general deterioration of diamond quality in the *pipes of the Southern group* owing to initially more defective structure of crystals, due to prevalence of dodecahedral and tetrahexahedral diamonds with zonal-sectorial structure, high stresses with multiple fractures, graphite and fluid inclusions.

The Arkhangelskaya and Karpinskogo-1 pipes are characterized by a substantial share of eclogitic paragenesis diamonds, with subordinate role of ultramafic paragenesis diamonds. The dominating diamonds of high-nitrogen groups, both dodecahedrons, and tetrahexahedroids, have zonal-sectorial structure. The low defect aggregation degree proves reduced formation temperatures for most of crystals and a short time of post-crystallization annealing.

High nitrogen concentrations evident prevalence of A centers over B centers make the diamond structure hard, but brittle. The highly defective structure of diamonds from the Southern group of bodies favored capture of impurities (nitrogen, hydrogen and carbonates), microinclusions, melt and fluid inclusions.

By contrast, domination of colorless high transparency diamonds with layer-by-layer growth mechanism is noted in the *Northern group pipes,* which is typical of the high-yielding classical Yakutian kimberlites. Unlike the latter, pipes of the Northern groups are dominated by dodecahedral habit diamonds with high etching and corrosion degree.

Largely ultramafic genesis of diamonds can be derived from prevalence of low- and medium-nitrogen diamonds in the Pionerskaya pipe. The diamonds exhibit low hydrogen impurity concentration, with distribution maximum registered at 0–5 cm^{-1} in the large range, and at 3–10 cm^{-1} for diamonds of the Southern group.

Diamonds of the Pionerskaya pipe are characterized by high concentrations of structural nitrogen impurities and centers caused by plastic deformation. Unlike in diamonds of the Arkhangelskaya and Karpinskogo-1 pipes, structural nitrogen is higher aggregated (%N_B—as reported by IRS, P2, N2 as reported by EPR), which proves longer stay in high-temperature conditions and contribute to oxidative dissolution.

The absolute majority of crystals in the Pomorskaya pipe are colorless highly transparent, severely corroded dodecahedroids of high-temperature ultramafic generation. But these diamonds are exposed to strong oxidative dissolution and local etching. As compared to the Pionerskaya pipe diamonds, these offer more crystals with multiple graphite inclusions. Tetrahexahedroids, typomorphic for the pipe, mainly contain multiple scattered finely dispersed inclusions and are not of gem quality.

Despite low diamond content in the Northern group pipes, quality characteristics of the rough diamonds drastically exceed such for the Southern group in many parameters.

A much higher yielding capacity of the major ore type—autolithic breccias in the Karpinskogo-1 and Arkhangelskaya pipes versus the kimberlite rocks in the Karpinskogo-2 and Pionerskaya pipes—is caused by a substantially lower exposure to the melt mantle metasomatism, whose maximum occurrence versus the Northern pipe group kimberlites caused destruction of a substantial portion of diamonds during dissolution, and a drastic rise in the share of dodecahedroids, especially in large grades, accompanied by a loss of their weight. Besides, the share of eclogitic paragenesis diamonds increases right in the highest yielding pipes of the Southern group.

2.2 Mineralogical Characteristics of Diamonds from Pipe of the V. Grib Deposit

The V. Grib pipe is represented by magnesian-ferriferous kimberlite rocks, which draws it closer to pipes of Yakutia's Mirniy and Daldyn-Alakit fields. By their typomorphic characteristics, diamonds of the deposit resemble crystals from the M.V Lomonosov deposit; there are also populations whose properties are similar to diamonds of classical kimberlite, like Mir and Udachnaya (Yakutia).

2.2.1 Crystal Morphology

Diamonds of the V. Grib pipe are represented by various mineralogical types (Fig. 2.13), with substantial domination of gem quality and near-gem quality octahedral and dodecahedral individuals (87%), with subordinate quantities of dark-gray and black dodecahedroids (2.12%), cubic habit diamonds (4.4%); coated diamonds (1.37%) and polycrystalline aggregates (5.16%).

Octahedrons. Prevalence of octahedral and combined habit with octahedral faces and dodecahedral surfaces was revealed. The quantity of regular smooth-shaped octahedrons and those with slight parallel striations is low—5%. Most of the octahedrons are fine- and medium-laminar with sheaflike, sheaflike-splintery striations, drop-shaped relief, with negative trigons in the faces (111) (Fig. 2.13). Most frequent are the plane-faced and curve-faced octahedral crystals with slightly rounded edges

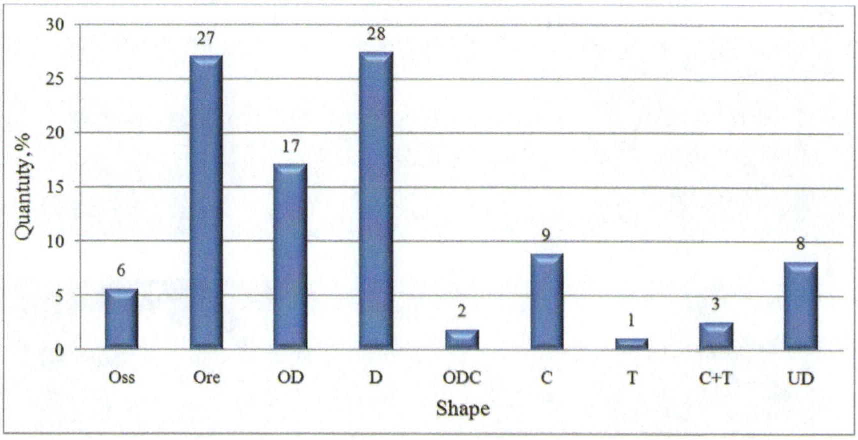

Fig. 2.13 Distribution of studied diamond crystals from the V. Grib pipe by shape, quantity in %. Oss—smooth-shaped octahedron; Ore—octahedron with rounded edges; OD—combined crystal with octahedral faces and dodecahedral surfaces; D—dodecahedroid; ODC—combined crystal with octahedral, dodecahedral and cubic faces; C, T—tetrahexahedroid; C + T—combined crystal, cube and tetrahexahedroid; UD—crystal of undefined shape

(27%) and combined crystals with octahedral faces and dodecahedral surfaces (17%), distinct for presence of faces (111) of ditrigonal shape, hummocky and blocky (roll-shaped) relief, right to formation of pyramid-shaped apices in the center of faces (Fig. 2.14).

Octahedrons offers a bunch of specific features:

- relatively high number of spinel and cyclic twins;
- distinctive high and medium flattening by L_3;
- high share of colorless individuals;
- pale-blue, green, lemon-yellow are most widespread among colored crystals.

Fig. 2.14 Standard shapes of diamond crystals from the V. Grib pipe

The small grades show lower concentration of octahedrons and twins, distorted diamonds and crystals with blue—pale-blue, violet photoluminescence. The quantity of smooth-faced flawless crystals increases.

Dodecahedroids. The content of dodecahedral diamonds is nearly the same (28%) as octahedron concentration (Fig. 2.13). Of these, 17% falls to regular dodecahedroids with smooth rhombic surfaces, sharp edges and thin parallel striations, 10% to dodecahedroids with minor octahedral face relics and 2% to combined crystals with prevailing dodecahedral surfaces and sharply subordinate cubic and octahedral elements.

Most of the dodecahedrons are medium- and fine-laminar with splintery and shagreen sculptures (Fig. 2.14). As a rule, dodecahedral crystals are rounded with fine shagreen relief or sheaf like striations. It differs for high integrity degree, seldom has blemishes and caverns on the surface. Also encountered are coarse-laminar dodecahedroids with tessellated and blocky sculpture and cryptolaminar individuals with drop-like and blocky sculpture (Fig. 2.14).

Dodecahedroid features:

- higher number of individuals with complex distortion;
- higher relative share of brown crystals;
- increased amount of plastically deformed individuals;
- high concentration of dodecahedrons with hummocky and blocky structure.

Combined-shape diamond with octahedral habit faces and dodecahedral habit surfaces are similar to octahedral by nature of distortion and twinning, higher concentration of individuals with pigmentation spots.

The main distinctions: very few crystals comprised by ditrigonal layers, much more individuals with signs of coarse-laminar policentric growth, increased relative number of fine-laminar crystals with sheaflike splintery striations and combination of splintery and shagreen sculptures.

Combined-shape crystals have smooth surfaces without visible traces of dissolution.

Cubic habit diamonds are not abundant in the V. Grib pipe and comprise a total of 4.4%. These individuals vary by color characteristics: gray, black, yellow and colorless (Fig. 2.14). Common features for cubic habit diamonds:

Cubic crystals are represented by rounded individuals, often with pits and tetragonal pyramids of dissolution;

- poor integrity (high content of individuals with natural cleavages and fractures);
- most crystals are isometric;
- high proportion of cubic penetration twins;
- relatively elevated number of colored crystals, especially yellow ones, with a gray and yellow tint.

The content of tetrahexahedral crystals remains low compared to the M.V. Lomonosov pipes (around 1%).

Crystals with zoning are numerous among cubic habit individuals. In UV rays, zoning is expressed by alternation of bands with canary yellow, bright-green and weak featureless glow.

Crystals are found that combine external peculiarities of cubes with fibrous and combined growth mechanism with octahedron sectors that make cube apices. Some individuals are represented by pseudo-cubes with layered growth mechanism: they have thin parallel striations on edge parts, blocky sculpture on the cube edges with microsculpture in the form of thin shagreen and small tetragonal etching piramyds and relics of triangular octahedral faces at apices.

Coated diamonds are very rare for this deposit, they account for only 1.4%. As a rule, such diamonds have saturated yellow-green or yellow color. In the studied collection they are represented by individual crystals of octahedral habit. The tinted translucent coat surrounding the transparent core can be seen on cleavage surfaces. Photoluminescence is zonal, yellow or green.

Vertex-type diamonds are represented by gray, less frequently, black combined-shape diamonds with octahedron faces and dodecahedroid surfaces with transparent apices overhanging the faces surface. The color is dark due to graphite microinclusions that fill the whole body of crystals. The proportion of such specimens does not exceed 2.2%, which is much lower than in the M. V. Lomonosov deposit pipes (5–11%).

Polycrystalline aggregates are characterized by diversity of shapes of individuals they include. Apart from smooth-faced sharp-edged octahedrons and combined-shape individuals with octahedron faces and dodecahedroid surfaces, peculiar spherulites and poorly crystallized individuals can be found in polycrystalline aggregates. All polycrystalline aggregates are characterized by high content of black inclusions, low transparency, gray or black color.

Micromorphology of crystals. The diamond specimens from the V. Grib pipe covered by the study were divided into two groups: without traces of local dissolution and with the surface bearing traces of local etching. The first group includes diamonds with smooth faces and sharp edges of octahedral, dodecahedral and combined shapes with octahedron faces and dodecahedroid surfaces. The second group is formed by crystals with traces of deformation and dissolution processes of various intensity degree.

Two dissolution types were singled out: (1) soft systemic that leads to formation of round-shaped diamonds, smoothing of apices and edges and emergence of fine-laminar relief in the form of sheaflike and concentric striations, drop-shaped relief or weak shagreen; (2) aggressive local dissolution of surface which is developed in weak crystal zones, places of dislocation outcrop or near-surface locations of inclusions. Local dissolution leads to formation of deep etch channels, caverns, separate pits and strongly corroded surface; formation of strongly dissected relief, coarse layered dissolution in the form of “fir-shaped” figures.

Growth processes in the form of layered faces growth, directed tangential growth in the form of separate “pencil-shaped” were noted.

There are blocky sections on many crystals, the crystal relief in such sections is hillocky or lenticular. Presence of parallel plastic deformation bands with traces of crystal displacement (shrinking) between bands indicates stress pressure in the diamond environment.

Crystals integrity and nature of formation. The diamonds from the V. Grib pipe are characterized by high degree of the crystal integrity. Whole diamonds account for 71%, chips make 15%, cleaved (cleavage does not exceed 1/3 of the whole diamond) and damaged crystals comprise 14%. About 70% of all crystals are fractured. There are several varieties of fractures: small ones from the surface in the peripheral zone; major fracture planes reaching the the crystal core; accumulation of fractures splitting the crystal into blocks. In many fractures there are black epigenetic inclusions or fragments of surrounding rocks.

Diamonds from the V. Grib pipe are mostly represented by monocrystals (84%), less frequently irregular aggregates and twins. Twins account for 5% of all crystals: spinel (maccles), less frequently cyclic crystals or penetrations.

Specimens of cubic and octahedral habits with signs of skeletal growth were found. The characteristic feature of skeletal growth is that the crystal edges and apices develop more intensively than its faces (Fig. 2.15a–c).

In the pipe there are crystals with sharp polycentrism of faces which grew uniformly in all directions, as a result of which the diamond acquired a rounded or close to that shape (Fig. 2.16). Some crystals acquired a rounded shape as a result of further systemic surface dissolution (Fig. 2.16b, c). The initial octahedral shape

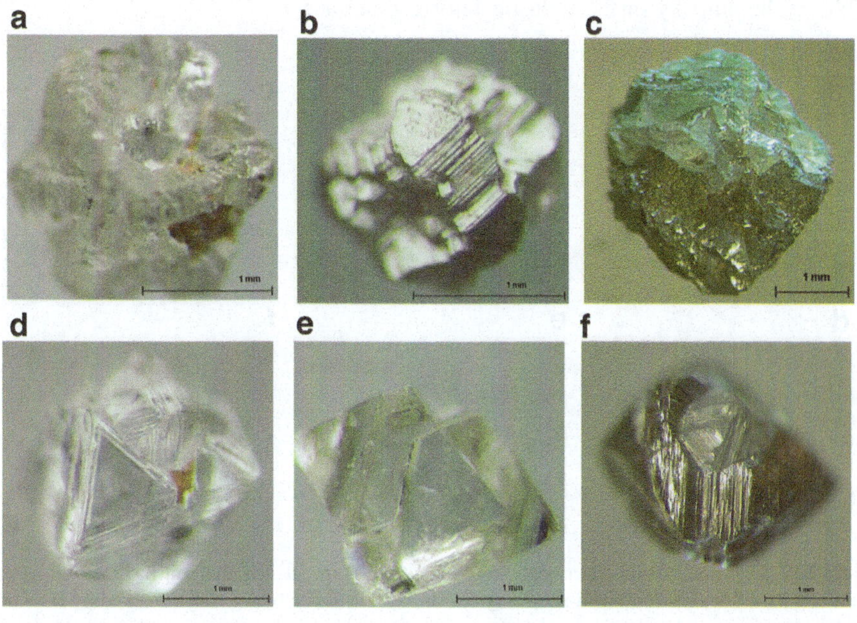

Fig. 2.15 Crystals formed as a result of penetration and splitting with distinct skeletal and anti-skeletal growth: **a**—skeletal growth, cubic crystals penetration; **b**—skeletal growth, octahedral crystal, **c**—cubes penetration; **d**–**f**—octahedrons penetration

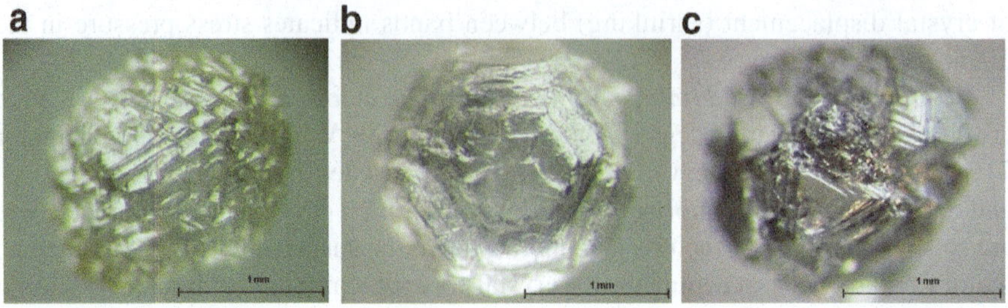

Fig. 2.16 Diamond crystals of a rounded or close to spherical shape: distinct polycentric and anti-skeletal nature of face formation

of such crystals is recognized by relics of surfaces at places of cubic and octahedral faces development.

Crystal deformation degree. Over 30% of crystals are isometric, 18% and 11%, respectively, are weakly and moderately deformed. Strongly deformed crystals account for 22%. Deformation, as a rule, shows in flattening of diamond crystal by L_3 axis (Fig. 2.17), less frequently in simultaneous flattening by L_3 axis and elongation by L_2 axis. As a rule, in case of strong deformation, the crystal evolves into a triangular plate of spindle like shape, "wolf tooth" type (Fig. 2.17f). Strong deformation results in changes of homogeneity of the internal and external structure (blocky, hummocky surface, plastic deformation traces).

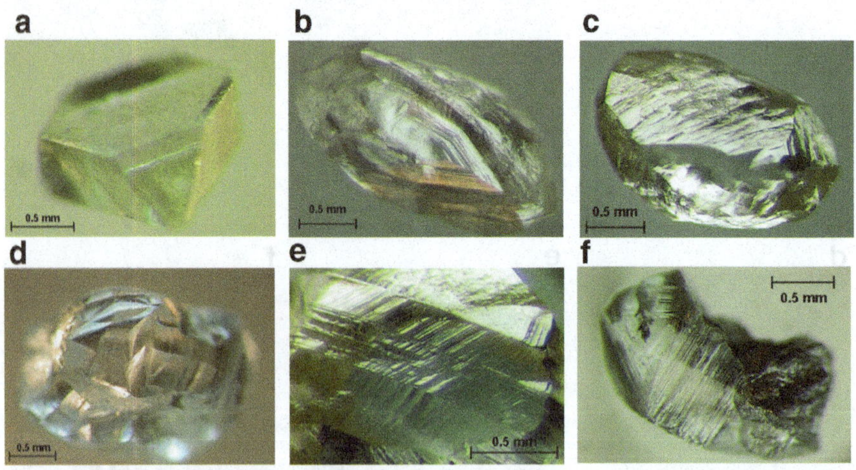

Fig. 2.17 Deformed diamond crystals from the V. Grib pipe: **a**—octahedron flattened by L_3 to a triangular plate; **b**—combined-shape crystal strongly flattened by L_3 and elongated by L_2 with octahedron faces and dodecahedroid surfaces with drop-like-splintery relief on edges; **c**—dodecahedroid, a crystal elongated by L_2 and L_3; **d**—crystal of blocky structure, aggregate; **e**—dodecahedroid with distinct plastic deformation bands; **f**—rhombic dodecahedron elongated by L_3, as a consequence of intensive anti-skeletal growth

2.2.2 Internal Structure

The diamonds from the V. Grib pipe, as well as from other ADR bodies offer pale-blue and blue cathodoluminescence. As noted, most studied crystals (75%) have homogeneous (non-zonal) or layered octahedral internal structure formed as a result of tangential (T) growth mechanism in the most favorable conditions (Fig. 2.18a–c).

In some crystals, alternation of zones with straight-layered structure and layered undulating structure occurs. A few dodecahedral crystals with sectorial zoning were found: in CCL rays, bright blue sectors with higher nitrogen content and dark sections filled with numerous small inclusions (Fig. 2.18d–f). There are also crystals with complicated internal structure formed due to oscillatory growth as crystallization conditions changed (oversaturation, impurities concentration, etc.). Crystals with oscillatory internal structure, as a result of which layered growth under the tangential (T) mechanism is replaced by fibrous (N) under normal growth mechanism or by combined one (T + N) indicate that the process of diamond formation in the pipe included multiple stages.

As noted, in all studied diamond crystals, the alternation sequence of zones formed under T, N or T + N growth mechanisms varies. As a rule, the core zone offers layered octahedral structure, T → N growth mechanism (layered octahedral zoning alternate with fibrous growth) is registered more frequently. The following most frequently registered schemes of crystals oscillatory growth were determined based on the internal structure: T → N, T → N → T → N, T → T + N, less frequently N → N → T, N → T + N → T, N + T → T + N, etc.

Many crystals (~5%) in the core zone have one or several seeds: crystallazation center, cut or dissolved micro-diamond, a chip (Fig. 2.19). The seed differs from the

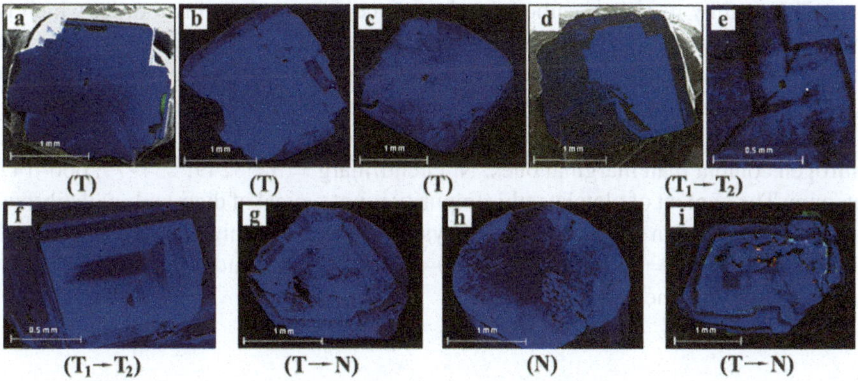

Fig. 2.18 Internal structure of diamond crystals from the V. Grib pipe (CCL images): **a**, **b**—homogeneous, {100}; **c**—spotted, {100}; **d**–**f**—closed straight-layered octahedral zoning, {100}; **g**—layered-open-ended zoning replaced by fibrous growth, {111}; **h**—sectorial structure, {111}; **i**—layered-open-ended zoning replaced by fibrous growth, {110}. Filming in CCL + secondary electrons mode. T—tangential, N—normal, T + N—combined growth mechanisms

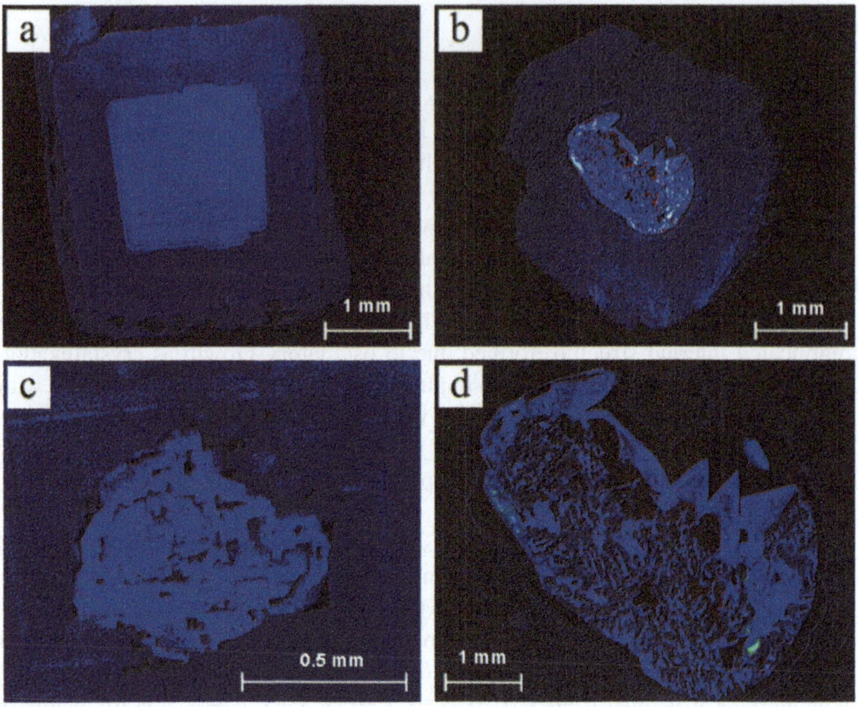

Fig. 2.19 Internal structure of diamond crystals with a seed: **a**, **b**—host diamond of a later generation (B) the seed overgrows, {100}; **c**, **d**—diamond-seed of an early generation (A) with blocky structure

host diamond by its shape, cathodoluminescence color and distinct border which is often marked by inclusions. It should be pointed out that, as a rule, the seed has fibrous or blocky internal structure. The crystal-seed is a diamond of an earlier generation (A), which the diamond of a later generation autoepitaxically overgrows (B).

As noted, the core parts of zonal crystals from the V. Grib pipe have higher nitrogen content than marginal ones: N_{tot} centr/marg = 325–249; 254–79; 606–147 at. ppm. The content of platelets (this impurity is not typical of diamonds with fibrous internal structure) in the core parts of crystals is two times higher than in marginal ones: P centr/marg = 20–6.6; 8.6–5.2; 10–6.6 cm^{-1}, which indicates predominantly tangential growth mechanism.

2.2.3 *Diamond Coloration*

The V. Grib pipe gives high quality indicators of rough diamonds: 80% of crystals are transparent, 18% are translucent diamonds, and 2% are opaque.

Fig. 2.20 Diamond crystals, octahedron—dodecahedroid with varying color: a—pale–blue smooth-faced octahedron; **b**—green octahedron, round-stepped; **c**—yellow-colored dodecahedroid; **d**—brown smooth-faced octahedron, elongated by L_2 axis; **e**—dark-brown diamond of combined habit, sharp-stepped with deep trigonal depressions (pyramids); **f**—near-colorless cubic crystal

The following crystal groups were discriminated: colorless, yellow, gray, colored (visually pale-blue, black, magenta, etc.). Each group was split into subgroups by primary color tint—i.e. colorless crystal with slight yellow tint. Colorless and near-colorless crystals prevail in the V. Grib pipe (65%) Table 2.1), with low, versus diamond crystals from other ADR bodies, content of yellow individuals (12%) and approximate content of gray diamonds (16%).

The breakdown of colored crystals (7%): pale-blue—2%, black—1.4%, and colors (Figs. 2.20 and 2.21). Some 2% of diamonds have intrinsic coloring, 5% are colored due to surface color films, etc. As a rule, the milk-white color of crystals (1.7%) is caused by surface peculiarities – rough relief scatters the light, which defines the apparent color (Table 2.3).

As a rule, diamond crystals have uniform color throughout the body. Crystals (3%) with non-uniform (spotted, layered) color distribution and zonal crystals occur. In some diamonds (1%) the black and gray colors are dictated by tiniest black spotted inclusions that fill in the entire body or separate crystal zones.

2.2.4 Diamond Microcrystals

Electric pulse disaggregation method that does not destroy minerals was used to "uncover" the kimberlite of the V. Grib pipe (Garanin 2004), where 17 diamond

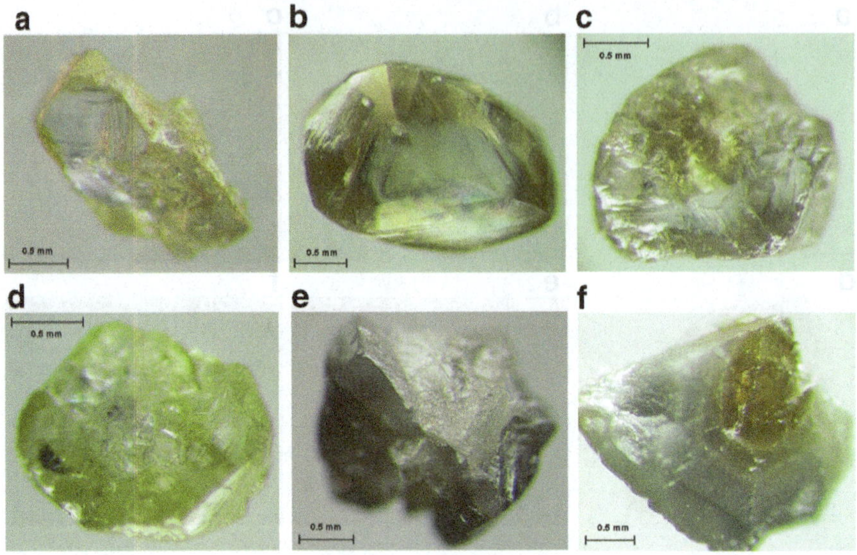

Fig. 2.21 Diamond crystals with heterogeneous and zonal distribution of color: **a**—thin-coated octahedron chip; **b, c**—yellow–brown (3brn) dodecahedroid with fancy-colored core zone; **d**—yellow chip of cubic crystal with near-colorless core zone; **e**—coated diamond: colorless core, gray intermediate and dark-gray peripheral zones; **f**—cube chip: yellow core zone, dark-gray coat

microcrystals below 500 μm were found and scrutinized using a JSM–820 scanning electron microscope by Jeol (Fig. 2.22).

Microcrystals are represented by (1) man-caused splinters and chips of larger diamond crystals; (2) whole crystals (mainly of octahedral habit), their logical aggregates and twins under the spinel law. All crystals are transparent, colorless, some splinters with slight yellow tint.

The V. Grib pipe showed the trend for prevalence of smooth-faced octahedrons among microcrystals, contrary to less diamondiferous Pionerskaya pipe (deep horizons) where microcrystals revealed dissolved octahedrons and box skeleton diamonds. The presence of well-preserved microcrystals of some 200 μm (largely smooth-faced octahedrons) in the V. Grib pipes may prove their formation as a separate generation of kimberlite genesis.

The data obtained corroborate a conclusion about interlinking morphological features of diamond macro- and microcrystals, enabling a preliminary estimate of rough diamonds in rocks of diamond-bearing bodies by typomorphism of diamond microcrystals at initial geological prospecting stages, when evaluating potential of field horizons at relatively minor testing scope. We note the potential of studying diamond microcrystals from kimberlite bodies with varying diamond-bearing capacity and their halos for wider adoption of the mineral microcrystals typomorphism in prospecting for new deposits.

Table 2.3 V. Grib pipe, distribution of diamond crystals by color and integrity

Diamond color		Quantity of crystals (%)
Fancy	Color	
Colorless	Colorless	34.0
Tint	Yellow tint	16.8
	Visible yellow	2.2
	Yellow-green	1.9
	Gray-yellow	4.1
	Brown	6.5
Total, tinted crystals		31.5
Color	Light-yellow	3.5
	Lemon-yellow	3.0
	Yellow	2.5
	Yellow–brown	2.0
	Smoky-brown	5.7
	Milk-white	1.7
	Pale-blue	2.0
Total, colored crystals		20.4
Industrial grade, gray colors	Light-gray	5.5
	Gray	4.5
	Dark-gray	2.7
	Black	1.4
Total, gray crystals		14.1
Integrity degree		
Whole		71.2
Cleaved crystals		11.1
Chips		14.7

2.2.5 Diamond Gemological Characteristics

Mineralogical characteristics of the V. Grib deposit diamond evidences prevalence of plane-faced and curve-faced crystals with layer-by-layer growth mechanism, dominated by near-colorless and slightly yellow tinted crystals, with a minor share of smoky-brown crystals.

Internal structure and defect-impurity composition features of crystals are as follows: prevalence of moderate-nitrogen diamonds with average aggregation of nitrogen defects (%B 25–40) and high platelets indicators, which is similar for the main crystal population from the Udachnaya pipe.

It is noteworthy that the pipes are similar by petrochemical characteristics of kimberlite rocks and belong to the moderate-titaniferous magnesian-ferriferous

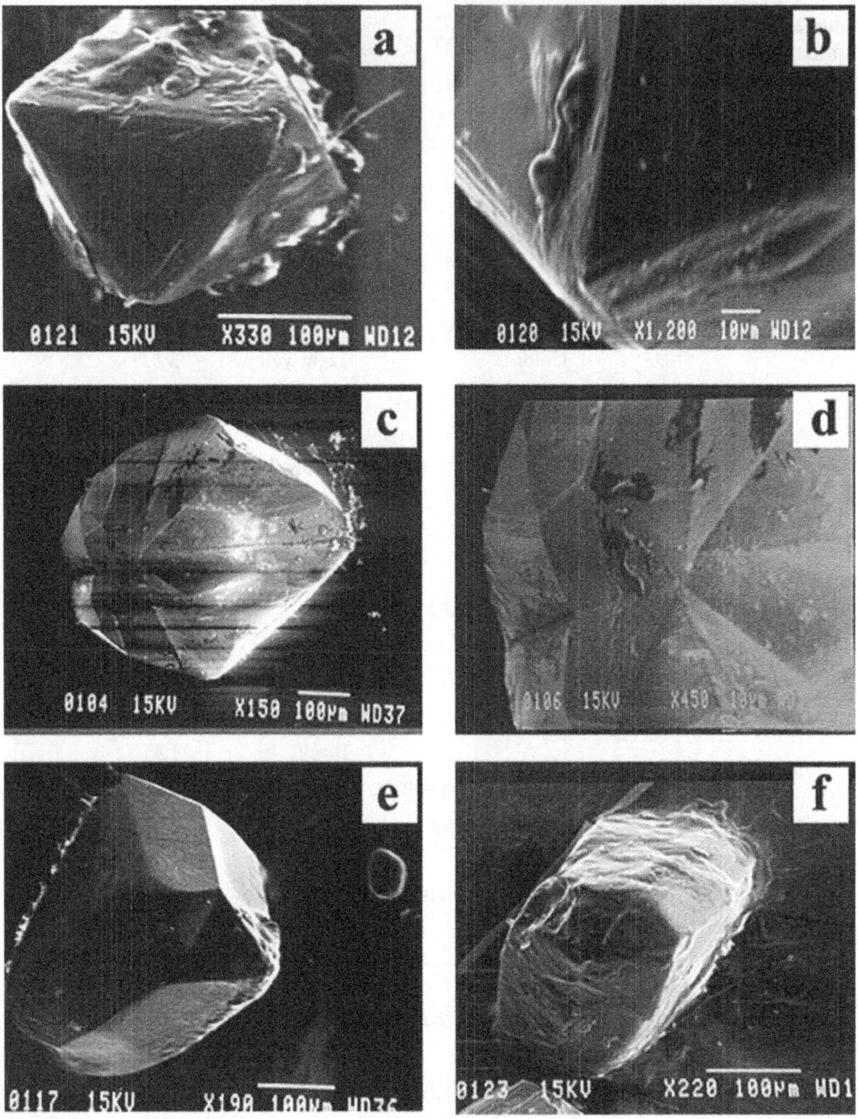

Fig. 2.22 Microdiamonds from the V. Grib pipe: **a**, **b**—octahedron (~250 µm), 330x; **c**, **d**—combined crystal with octahedral faces and dodecahedral surfaces with grooved edges, 150x; **e**—spinel twin of smooth-faced micro octahedrons, 190x; **f**—plane-faced and curve-faced spinel twin of micro octahedrons, 220x. Photo images in secondary electrons

type of kimberlites and, as shown in Chap. 5, have very similar diamond crystallization conditions favorable for formation of diamonds with high gemological characteristics. Hence, high gem quality crystals can be predicted in the V. Grib pipe.

All of this proves higher quality of rough diamonds as compared to that from the Arkhangelskaya and Karpinskogo-1 pipes.

Altogether, the V. Grib pipe shows average grain size of diamond crystals, with occurrence of separate big near-gem quality crystals.

A very big diamond, unique for the ADR, was extracted in commercial open pit mining in early 2017. The diamond weight is 181.68 ct (Fig. 2.23).

Just to compare: the biggest gem quality diamond recently extracted from M. V. Lomonosov deposit's Arkhangelskaya pipe weight 46 ct.

The V. Grib pipe has been developed since 2014, and summary production stands at 9 million carats. By 2018 the company is expected to reach the output of 4.5 million carats annually.

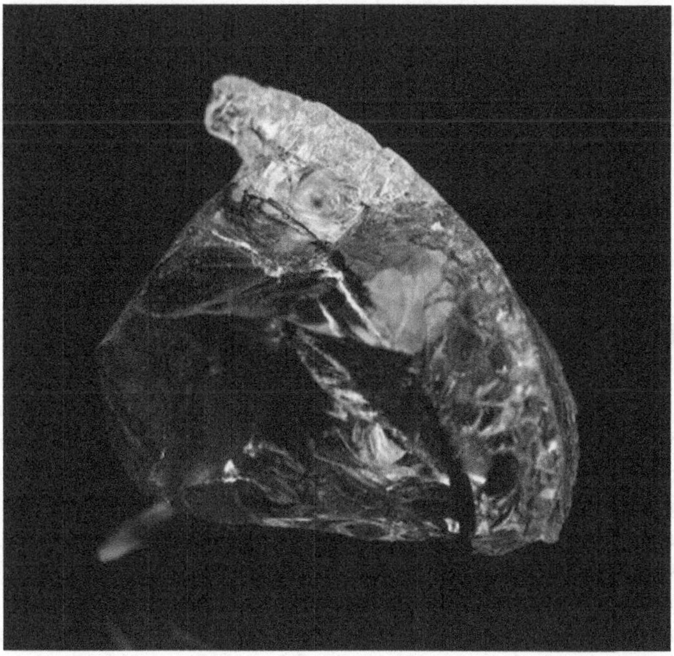

Fig. 2.23 A 181.68 ct diamond, extracted in late February 2017 at the open pit of the V Grib pipe

2.2.6 Typomorphic Characteristics of the V. Grib Pipe Diamonds

The following could be outlines for the V. Grib pipe.

Octahedron—dodecahedroid diamonds comprise 85%. Cubic habit diamonds make some 4%, with isolated coated diamonds and polycrystalline aggregates.

The share of octahedrons in all morphological types in the Grib pipe is the biggest of all ADR pipes, with a substantial share of dodecahedroids.

Crystal deformations were largely revealed by the triad axis.

Quality indicators of rough diamonds are high: substantial share of whole crystals (70%), with up to 15% of chips and relatively many diamonds without inclusions.

The studied specimens show high transparency degree (80%), with only 2 totally opaque crystals.

Colorless and slightly tinted diamonds comprise 65% in the V. Grib pipe, whereas gray specimens prevail among the colored samples.

Judging by intrinsic inclusions, most of the diamonds relate to the ultra-basic paragenesis minerals with the subordinate quantity of eclogitic paragenesis diamonds.

Diamond-formation intermittency was discovered for the V. Grib pipe diamonds, exhibiting in presence of "diamond-in-diamond" inclusions, and availability of microcrystals without dissolution traces in the pipe, along the resorbed macrocrystals.

2.3 Morphology and Other Major Characteristics of ADR Diamond from Low and Poor Diamond-Bearing Kimberlites Zolotitsa and Kepina Fields

Snegurochka, Pervomayskaya, Koltsovskaya pipes lie in the Zolotitsa field, but are not part of the M. V. Lomonosov deposit since have non-commercial diamond content in the bodies, below 0.3 ct/t. Snegurochka pipe is the southernmost of all pipes in the Zolotitsa field, most proximate to Arkhangelskaya pipe. The former is small by dimensions and unprofitable for development at the current stage by total reserves. Koltsovskaya and Pervomayskaya pipes are confined to the same submeridional fault and lie far to the north of the M. V. Lomonosov deposit bodies.

The mineralogical description of diamonds from low diamond-bearing potential bodies was done using crystals obtained from prospecting samples (3–5 tons of ore) during mining exploration. The data on those crystals enabled to estimate the diamond-bearing capacity and commercial potential of the sites (Snegurochka, Pervomayskaya, Koltsovskaya pipes).

Good quality of the Snegurochka pipe diamonds was noted in initial prospecting, in the 1980s. It is highly likely that as further investigation of diamonds and large-volume trial run take place, Snegurochka pipe may be reclassified as profitable for development, of the relevant infrastructure is put into place. For this reason, now we need to have a complete set of data about diamonds from low diamondiferous bodies that are close to the deposit main pipes by location.

Kepina field is the largest, very uniform by structure, so is of great interest in terms of prospecting. The currently known pipes and sills of the Kepina field are comprised by kimberlites and olivine melilitites with poor diamond-bearing potential. In our work we studied a collection of diamond crystals from pipes in the main clusters of the Kepina field: Kepina, Shocha, and Pachuga clusters.

2.3.1 Zolotitsa Field Diamonds: Snegurochka, Pervomayskaya, Koltsovskaya Pipes

In studying non-commercial deposits of ADR's Zolotitsa field, we examined Snegurochka, Pervomayskaya, Koltsovskaya pipes. The total quantity was 1336 diamonds from 3 size and sieve grades: −4+2 mm, −2+1 mm, −1+0.5 mm (Table 2.4). The diamonds of those bodies have much in common by mineralogy.

Crystals of the Zolotitsa field pipes are characterized by best integrity versus the diamonds from bodies of the Verkhotina and Kepina fields – the share of whole individuals is 60%, with $\leq$ 10% of chips.

Formation nature and integrity degree. In the studied collection, the ratio of isometric and slightly distorted diamonds is roughly the same, whereby the number of severely deformed diamonds in the Snegurochka pipe is 20%, which is lower compared to other bodies of Verkhotina and Kepina fields.

Crystal morphology. Most of the studied diamonds belong to the octahedron—dodecahedroid series. Noteworthy is rather essential content of highly graphitized microblock crystals, typical of all pipes in the Zolotitsa field.

Also of note is the quantity of cubic habit individuals—up to 8.6%, which is typical of the M. V. Lomonosov deposit pipes and appears rather high versus crystals from weakly diamondiferous bodies of the Kepina and Verkhotina fields. Half of the cubic crystals from Snegurochka, Pervomayskaya, and Koltsovskaya pipes exhibit tetrahexahedral surfaces. The studied diamonds have tetrahexahedroids with drop-shaped relief, primarily yellow-colored with varying intensity and translucent. We observe crystals that are isometric, elongated along one side of the octahedron faces

Table 2.4 Quantity of studied diamonds from non-commercial deposits of the Zolotitsa field

Pipe name	Total number of specimens, pcs	Content of diamonds from various size and weight groups, %		
		−4+2	−2+1	−1+0.5
Snegurochka	1065	3.2	25.7	71.1
Pervomayskaya	167	4.8	15.6	79.6
Koltsovskaya	97	5.2	28.9	66.0
Total	1329	3.3	24.7	72.0

(111), and complex distortion crystals. Emergences of axes (100) frequently exhibit extra face seams, with occasional minor cube faces with slight tetragonal depressions.

Diamonds from the studied pipes of the Zolotitsa field exhibit average quantity of rhombic and dodecahedral habit crystals (some 62%) and high concentration of undefined habit individuals (maximum 11.6%) (Kopchikov 2009) due to heavily corroded surface.

By the ratio of morphological groups Snegurochka, Pervomayskaya and Koltsovskaya pipes are similar (Table 2.5; Fig. 2.24). The quantity of big crystals (−4+2 mm) in the examined bodies is below 5%. Comparing the crystal morphology of diamonds from various size grades we observe the trend that is typical for all diamondiferous bodies of the Zolotitsa field: the role of octahedral and cubic habit crystals increases as the crystal size decreases. Together with the combined habit crystals, octahedral diamonds make up to 45% of all grade −1+0.5 mm crystals. The content of dodecahedral and undefined habit diamonds declines in the small grade (see Table 2.5).

Medium- and fine-laminar dodecahedroids with splintery striations or shagreen sculptures are typical of the Zolotitsa field. Several dodecahedroid types were distinguished based on crystal surface features: smooth-faced (or with very fine striations), with sheaf like and splintery striations, with c drop-shaped and shagreen relief, with blocky structure.

The diamonds collection of the Snegurochka and Pervomayskaya pipes exhibit the biggest number of dodecahedroids with fine splintery striations, which is formed in dissolution of fine-laminar crystals grown in stable high-temperature conditions.

Smooth-faced dodecahedroids are represented by colorless individuals or those with slight gray or yellow tint; they commonly contain no internal defects (inclusions, fractures). Dodecahedroids with sheaf like (concentric) striations are quite widespread in the studied collection and make up to 15% of all crystals in that habit. Coarse-laminar dodecahedroids with tessellated and blocky structure and cryptolaminar individuals with drop-like and blocky sculptures occur less frequent.

By formation nature crystals of the studied collection are dominated by monocrystals (68.4–75.5%), diamond crystal aggregates total 9.0–21.5% of all individuals. Irregular aggregations are most widespread among aggregates in bodies of the Zolotitsa field. In the Snegurochka, Pervomayskaya and Koltsovskaya pipes regularly twinned crystals comprise a fifth part of all aggregations.

Coloration and transparency. Most of the diamonds in the studied collection are colorless (55% of the total, on average), and with slight gray or smoky-brown tint. The Snegurochka, Pervomayskaya and Koltsovskaya pipes reveal a considerable content of high and medium transparency diamonds (40–50%).

The diamond color and visual transparency degree are linked to crystal habit. Absence of brown crystals is typical of the octahedral habit diamond; the share of colored individuals among them is low.

Combinatory diamonds with octahedral faces and dodecahedral surfaces show the average share of colorless and slightly tinted individuals of 60%. Tetrahexahedroids of the Zolotitsa field pipes offer the largest variety of colors, especially in small size grades. In small size grades the quantity of colored crystals is drastically higher, with

Table 2.5 Diamond distribution by habit in pipes of the Zolotitsa field

Pipe	Quantity of crystals (pcs)	Diamond distribution by habit (%)						
		Dodecahedral		Octahedral	Combined	Cubic		Undefined
		Rounded	Laminar			Cube	Tetra-hexahedroid	
Snegurochka	1065	34.4	28.2	13.7	8.2	0.5	7.8	7.6
Pervomayskaya	167	31.9	26.1	17.8	4.2	0.3	8.3	11.6
Koltsovskaya	97	35.8	29.3	15.1	2.5	0.7	7.0	10.2
Average, for Zolotitsa field	−2+1 mm	61	7	14	7	4		6
	−1+0.5 mm	51	5	21	8	9		6

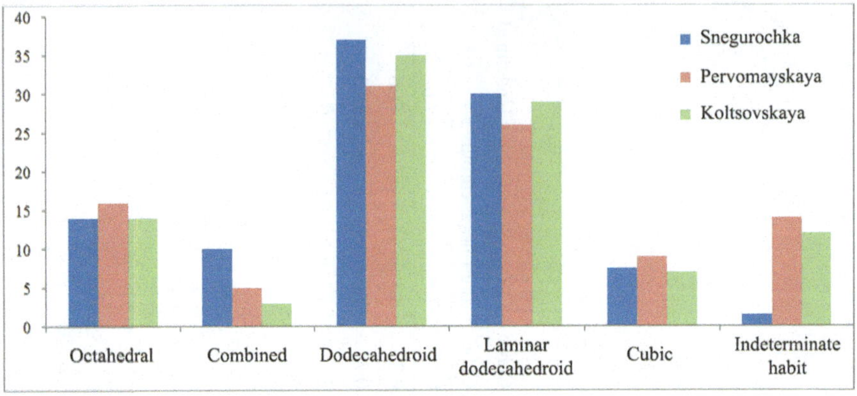

Fig. 2.24 Distribution of diamonds by habit, weakly diamondiferous pipes of the Zolotitsa field (Kopchikov 2009)

relatively more frequent red-brown, canary-yellow diamonds and pale-blue tinted crystals. The share of transparent polygons with slight yellow-gray tints and gray colors increases as the size grows.

2.3.2 *Kepina Field Diamonds*

The Kepina field lies eastwards and southeastwards of the Zolotitsa field, and these two belong to the same deep fault series.

The studied collection of the Kepina field bodies includes 117 crystals with the total weight of 1.38 carats, of which 38 (0.41 ct) falls within the Kepina group of bodies and 64 (0.67 ct) to the Pachuga cluster. The Shocha cluster is represented by 12 specimens (0.18 ct).

Diamonds of the studied collection are represented by crystals and their chips with varying integrity. Dodecahedroids (50–71%) prevail among crystals, followed by combinatory individuals with octahedral faces and dodecahedral surfaces (10–35%) plus some 11% falling to octahedral diamonds (Fig. 2.25). Octahedral shape is largely typical of small size grade crystals—up to 0.01 ct, whereas only 25% of combined crystals weigh more than 0.01 ct. No individuals of hexahedral and tetrahedral habits occur. Similar ratio of habit forms is preserved for diamonds of all groups in the Verkhotina and Kepina fields.

Integrity degree. Diamonds of the Kepina group bodies are characterized by low integrity. More than a half of specimens from the studied collection are crystal chips and splinters (Table 2.6). The absolute majority of cleavages is of protomagmatic nature and hillocky surface relief, due to dissolution.

Crystal morphology. Out of 41 crystals from the Kepina cluster, whose bodies are comprised by poorly diamondiferous and non-diamondiferous kimberlites, only

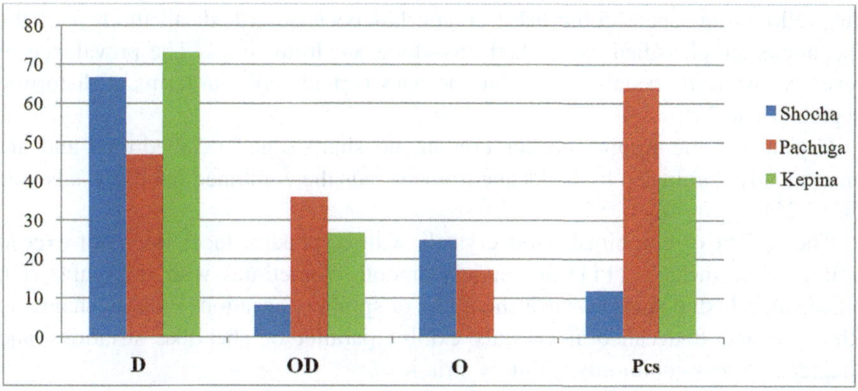

Fig. 2.25 Diamond distribution by crystal morphology forms, quantity is given in pieces. Diamond from the Shocha, Kepina, and Pachuga clusters. O—octahedron, OD—combined habit octahedron-dodecahedroid, D—dodecahedroid, Pcs—pieces

Table 2.6 Main mineralogical characteristics of a diamond from Kepina field pipes

Diamond characteristic	Quantity of crystals within a cluster (%)		
	Cluster		
	Shocha	Kepina	Pachuga
Integrity degree			
Crystal	25.0	28.9	34.4
Damaged crystal	25.0	7.9	6.3
Chip	41.7	55.3	57.8
Aggregate	8.3	7.9	1.6
Habit			
Octahedron	16.7	0.0	14.1
Octahedron—rhombic dodecahedron	8.3	26.3	29.7
Rhombic dodecahedron	58.3	71.1	39.1
Distortion degree			
High	16.7	18.4	14.1
Weak	58.3	13.2	10.9
Isometric	16.7	7.9	1.6
Etching degree			
Total	33.3	71.1	18.8
Intense	8.3	13.2	14.1
Caverns	16.7	18.4	17.2
Trigons	0.0	2.6	21.9

one yellow translucent cubic habit crystal has been described; all the rest of the specimens are classified as octahedron—dodecahedroid series. The prevalence of severely distorted crystals over isometric ones is noteworthy in terms of distortion degree (Table 2.6).

Crystals of the Kepina cluster contain no sharp-edged octahedrons, and are dramatically dominated by dodecahedroids versus the combined habit crystals (see Fig. 2.25).

The weight of combined habit crystals with octahedral faces does not exceed 0.01 ct. The surfaces (111) are mainly smooth, sometimes with triangular etch pits. Dodecahedral surfaces have sheaf like or splintery striations without shagreen. Rhombic and dodecahedral crystals exhibit parallel or sheaflike striations and shagreen, or less commonly splintery striations.

High share of crystals with etch marks is a distinction of the Kepina cluster diamonds, whereby severely etched individuals are few as compared to other weakly diamondiferous pipes.

Diamond coloration. The color palette is strongly dominated by colorless and epigenetically fancy colored brown crystals, and light-brown crystals are minor in number (Table 2.7). Diamonds of the Kepina cluster differ for higher concentration of transparent individuals.

Table 2.7 Color and presence of internal defects for diamond crystals from from Kepina field pipes

Diamond characteristic	Quantity of crystals from the same cluster, %		
	Cluster		
	Shocha	Kepina	Pachuga
Color			
Colorless	66.7	63.2	50.0
Yellow	8.3	10.5	1.6
Brown	8.3	23.7	15.6
Light-brown	16.7	2.6	31.3
Gray tint	0.0	0.0	1.6
Inclusions			
None	41.7	42.1	43.8
Black and dark	16.7	55.3	46.9
Transparent	16.7	2.6	1.6
Red	8.3	2.6	0.0
Brown feathers	0.0	0.0	3.1
Fracturing			
None	66.7	57.9	51.6
In bulk	25.0	23.7	18.8
Peripheral zone	8.3	10.5	17.2
Core zone	0.0	7.9	3.1

More than 40% of diamond crystals have fractures as internal defects; in 8% of specimens the fractures are only in the core zone, and in 11%—only in the periphery. Inclusions are found in more than half of the crystals.

Pachuga cluster diamonds are best studied; the diamond collection was relatively big and crystals from the collection were studied to the fullest extent. By integrity degree the ratio for the Pachuga cluster overall matches the ratio for the Kepina field, although the concentration of splinters is slightly higher (58%). The cleavages are natural and protomagmatic, the cleavage surfaces are exposed to dissolution.

All diamonds from pipes of the Pachuga cluster (64 pieces) are represented by crystals with layered (tangential) growth mechanism, with prevalence of dodecahedral habit individuals (39%).

Diamonds of the Pachuga cluster are characterized by elevated concentration of octahedral and combined habit crystals—14% and 29%, respectively (Table 2.6). Octahedrons reveals both sharp-edged smooth-faced individuals (pipes 693 and 695), and crystals with coarse-laminar surface structures (111), with triangular etch pyramids (Stepnaya pipe—688). Fine parallel or sheaf like striations or shagreen are more typical of the surfaces which match the rhombic dodecahedroid layout, both in dodecahedroids and in combinatory crystals. Splintery striations are less frequent.

The Pachuga cluster diamonds reveal lower percentage of etched crystals (below 20%), but the higher etching degree (Fig. 2.26, Table 2.6).

Viewing the distortion degree ratio, it is worth noting that the Pachuga cluster diamonds are dominated by isometric and weakly distorted crystals, contrary to the Kepina cluster diamonds.

Micromorphology of diamond crystal surface. Most typical forms of crystals from the Kepina field bodies are given in photos Figs. 2.26, 2.27, 2.28 and 2.29. Dodecahedroids and combinatory crystals with octahedral faces and dodecahedral surfaces are most typical of diamonds from the Kepina field pipes. The microrelief of faces and cleavage surfaces of crystals is characterized by widely occurring corrosion and dissolution processes, primarily due to development of the tessellated-blocky and drop-like microsculpture. Skeleton diamond crystals occur, although seldom. Apart from that, the crystal surfaces have regular pits and etch channels. The blocky

Fig. 2.26 Typical dodecahedroids of the Kepina field with varying striations

Fig. 2.27 Laminar dodecahedroids of the Kepina field with varying relief manifestations

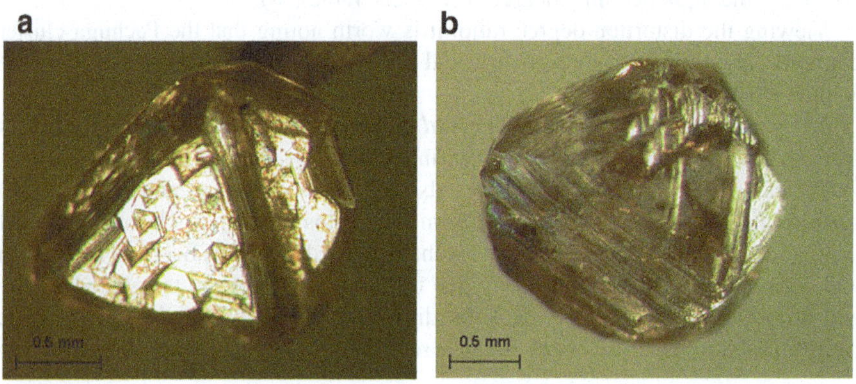
Fig. 2.28 Photo image: Combined habit crystals with octahedral faces and rhombic dodecahedral surfaces

sculpture of crystals is highlighted by presence of re-entrant angles, and a drastic face misalignment, which dictates emergence of a peculiar fir-shaped striations. In addition to drop-like blocky structure the curve-faced crystal surfaces also have coarse striations that translate into nonuniform furrows. Fine furrows cross the entire diamond surface, refracting at face seams, i.e. their shaped arises from the shape of dissolution layers. Apart from distorted curve-faced crystals and their irregular-shaped chips the collection also has diamond octahedrons and spinel twins. Sculptural accessories generated during diamond dissolution are often present on the surface of octahedral individuals. In case of major defects (fractures, microtwins) diagonal

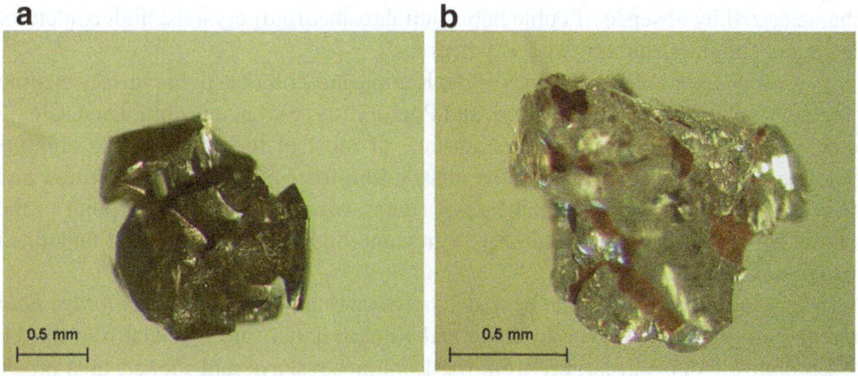

Fig. 2.29 Crystals of undefined shape, heavily etched

convex triangular layers are split into separate sections and create the polycentric lamination observed in diamond crystals from the Stepnaya pipe (AN–688). Crystal chips with etch channels on their surface frequently occur. Pure octahedrons were found among microcrystals.

Transparency and coloration. Much higher content of translucent crystals than on average in the Kepina field (more 55%) is the distinction of diamonds. In contrast to diamonds of the Kepina cluster, smoky-brown or brown-tinted crystals are present along with the colorless and brown crystals. Yellow crystals are virtually absent.

Internal defects. The presence of fractures in the crystal body is more typical of peripheral zones (19%) than for core ones (3%). More than 50% of crystals have black inclusions.

Thus, results of studying morphological and other features of crystals from low- and poor-diamondiferous bodies, and comparison of properties of diamonds from ADR deposits reveals that by size (weight), habit, surface nature, color and integrity degree diamonds are subdivided into three morphogenetic groups: 1. Diamonds from the V. Grib kimberlite pipe (deposit); 2. Diamonds from kimberlite pipes of the M.V. Lomonosov deposit and non-commercial kimberlite pipes of the Zolotitsa field (Snegurochka, Pervomayskaya, Koltsovskaya), 3. Diamonds from low- and poor diamondiferous pipes and bodies of the Kepina, Verkhotina, and Izhmozero fields of kimberlites and olivine melilitites.

2.3.3 Typomorphic Characteristics of Diamonds from Low- and Poor-Diamondiferous Kimberlites

Diamonds of weakly diamondiferous pipes and bodies of the Kepina and Zolotitsa fields drastically differ by morphology and other external signs. In view of this, diamonds of the Kepina field could be placed into a separate morphological group characterized by absence of cubic habit (tetrahexahedroid) crystals, high content of chips and splinters, and low integrity degree.

In weakly diamondiferous pipes of the Kepina and Zolotitsa fields surface etching processes are extensively developed, and Pomorye-type channels dominated. Cleavages exhibit protomagmatic nature: surface of most of the splinters and chips is characterized by epigenetic oxidative dissolution traces; often the dissolution and deep etch channels result in total loss of shape, which may indirectly point to the natural (not man-made) low integrity degree and higher concentration of undefined habit crystals.

The closest properties could be traced between diamonds from the ***Zolotitsa field pipes***—Snegurochka, Pervomayskaya and Koltsovskaya, and crystals from the M. V. Lomonosov deposit, largely from the Lomonosovskaya and Pionerskaya pipes, that form a separate morphological group, pursuant to (Zakharchenko et al. 2002; Kudryavtseva et al. 2005). Besides, weakly diamondiferous bodies of the Zolotitsa field are located in the same ore-hosting fault with the deposit, so are similar by geochemistry and mineralogy of rocks with the deposit pipes (Bogatikov et al. 1999). In view of this, and due to close properties of diamonds from the compared bodies, the studied diamonds could be placed into the same morphogenetical group with the M. V. Lomonosov deposit. Diamonds of that group are characterized by harsh prevalence of rounded rhombic dodecahedrons (dodecahedroids) (up to 74%), low concentrations of octahedral habit crystals ($\leq$15%) and elevated concentration of tetrahexahedroids (up to 17%), versus the V. Grib pipe. High and average diamond quality indicators are noted.

Results of studying diamonds from the Kepina field point to high share of small grade crystals; low percentage of octahedral habit individuals and a rise in their share in small grades; prevalence of rounded form diamonds, whose origin is linked to intense diamond dissolution. Cubes and tetrahexahedroids occur quite seldom, and were not found in the collection, by described based on reference data (Bogatikov et al. 1999). A skeleton crystal is also present among the studied specimens.

The Kepina field diamonds characterized by low integrity. More than a half of diamonds are represented as irregularly shaped ships of distorted curve-faced crystals. In small grades (<1 mm) the ratio of whole crystals rises, including dodecahedral habit crystals.

Noteworthy is wider development of deformation structures in the diamonds, versus the Zolotitsa field crystals.

Itching traces (caverns, itch channels and regular-shaped pits) are observed in the surface of a third of crystals. The ratio of crystals with etching traces to the total number of specimens differ for each cluster. The share of heavily etched crystals also varies.

Most of the diamonds are colorless, and the colored specimens are dominated by fancy brown individuals (8–23%). Notable is nearly complete absence of light-brown diamonds among crystals of the Kepina group, whereas the quantity of saturated smoky-brown reaches 24%. The amount of yellow diamonds does not exceed 5%, and a higher concentration of yellow-colored diamonds (10%) is solely noted for crystals of the Kepina cluster.

The major part, 95% of all crystals, contain black inclusions, unevenly distributed in the body. Some 3% of crystals contain transparent colorless inclusions, and 2%—red ones. Two crystals exhibited brown feathered inclusions, apparently of epigenetic origin.

The determined lateral zoning of kimberlites and associated rocks of the ADR (Bogatikov et al. 1999) manifested in alteration of the mineral composition of rocks and properties of the main accessories of diamonds also exhibits a consistent change in the diamond typomorphic properties, both for studied crystals of weakly and poorly diamondiferous bodies, and for diamonds from the province deposits. The diamond-bearing potential, integrity and transparency of crystals declines, the ratio of habit types alters, and oxidation dissolution (corrosion) and deformation degree rises in transition from the V. Grib kimberlite pipe to the kimberlite pipes of the Zolotitsa field (including the M. V. Lomonosov deposit and non-commercial Snegurochka, Pervomayskaya, and Koltsovskaya pipes), and further to kimberlite and olivine magmatites of the Kepina, Verkhotina, and Izhmozero fields.

References

Afanasiev, V.P., Zinchuk, N.N., Pokhilenko, N.P.: Exploration Mineralogy of Diamonds (650 p). Novosibirsk (2010)

Bogatikov, O.A., Garanin, V.K., Kononova, V.A., et al.: Arkhangelsk Diamondiferous Province (522 p). MSU Publishing (1999)

Garanin, K.V.: Alkaline ultramafic magmatites of Zimny Bereg: Their Potential Diamond-Bearing Capacity and Commercial Development Prospects (50 p). Ph.D. in geology and mineralogy thesis. MSU (2004)

Kopchikov, M.B.: Typomorphic Peculiarities of Diamond from Arkhangelsk Diamondiferous Province (17 p). Abstract of Ph.D. in geology and mineralogy thesis. MSU (2009)

Krasnova, N.I., Petrov, T.G.: Genesis of Mineral Individuals and Aggregates (228 p). SPb. Nevsky Kurier Publ. (1997)

Kudryavtseva, G.P., Posukhova, T.V., Verzhak, V.V., Verichev, E.M., Garanin, V.K., Golovin, N.N., Zuev, V.M.: Atlas: Morphogenesis of Diamonds and Their Mineral-Satellites from the Kimberlites and Other Relative Rocks from the Arkhangelsk Diamondiferous Province (1st ed) (624 p). Polyarny Krug Publ. (2005)

Makhin, A.I., Bartoshinsky, Z.V. Bekesha, S.N., Vinnichenko, T.G., Voloshinovsky, A.S., Vasiliv, V.V.: Technical report: to study of main typomorphic features of diamonds from Zolotitsa field kimberlites for the purposes of classification, to develop and implement a methodology for integrated study of diamonds in terms of the expedition. Archangelsk (1990)

Punin, Y.O.: Cleavage of crystals. In: Proceedings of the RMS, P. 110, Issue 6, pp. 666–686 (1981).

Vins, V.G., Yelisseyev, A.P., Sarin, V.A.: Physical fundamentals of contemporary approaches to treatment of natural diamonds and polished diamonds—section 21st Century Technologies—"Precious metals and precious stones" journal. No. **12**(180), 155–164 (2008)

Vins, V.G.: Optically Active Defects in Diamond—Regularities of Origination and Mutual Transformation. Abstract of DSc in physics and mathematics thesis (40 p). Altay State Technical University, Barnaul (2011).

Zakharchenko, O.D., Makhin, A.I., Khachatryan, G.K.: Atlas of Typomorphic Properties of Diamond from the East European Craton (Lomonosovskaya deposit) (104 p). TsNIGRI (2002)

Zinchuk, N.N., Koptil, V.I. Typomorphism of diamonds of the Siberian Platform (603 p). Nedra-Biznestsentr OOO (2003)

Chapter 3
Defect-Impurity Composition, Optical and Spectrometric Properties of Diamond from ADR Deposits

Defect-impurity composition of diamond is a crucial genetic characteristic of this natural material. By example of a wide range of minerals we are well familiar with impact that impurities have on the crystallizing of crystals, how they change crystallization parameters, properties and other characteristics of minerals. Therefore, we need to thoroughly consider defect-impurity composition of diamond from ADR deposits, and in this respect also focus on optical and spectrometric properties of diamond to conceive and substantiate any given concept of diamond crystallization in mantle conditions and correctly interpret a complex pattern of diamond evolution, up to its delivery to the Earth's surface.

3.1 Luminescence

When studying diamond photoluminescence, we can visually observe a structure homogeneity degree expressed by luminescence homogeneity/zoning, and determine specific defects manifested by color and intensity of crystal luminescence. Photoluminescence color is determined for diamonds of all samples (500 crystals in each studied sample), and spectra were acquired for most typical ones.

Photoluminescence spectra were recorded with Fluorolog 3 spectrofluorimeter at room temperature, excitation 350–360 nm. With a specimen inside the cell holder, the spectra were recorded at room temperature, excitation monochromator slit width 5 nm, unless otherwise indicated. The standard width of the emission monochromator slit was 1 nm, step 0.5 nm, accumulation time 0.1 s. The luminescence spectra were additionally recorded at excitation 450 nm, 500–520 nm. For a portion of crystals, the spectra were recorded under the microscope at 77 K with luminescence excitation by diode lasers 405 and 450 nm with an objective lens 10 × /0.3. PL of some specimens was also performed at excitation 785 by diode laser using In Via Renishaw spectrometer at 77°K with an objective lens 5 × /0.12.

V. Garanin et al., *Diamonds from the Arkhangelsk Province, NW Russia*, Springer Mineralogy, https://doi.org/10.1007/978-3-030-35717-7_3

The following emission systems previously described (Beskrovanov 2000; Yelisseyev and Kanda 2007; Khokhryakov and Palyanov 1990) were found in the spectra: N3—with a zero-phonon line 415–416 nm, H3—with a zero-phonon line.

Zonal and zonal-sectorial diamonds tending to transition from the younger layered-octahedral forms to cuboidal forms with fibrous structures are typical for the Lomonosov deposit. Besides, repeated zonality often turns up, which represents a typomorphic peculiarity of diamonds from the ADR.

For crystals from the Central Yakutia pipes an inverse type of zonality is more typical: cuboid or cuboctahedron → octahedron (Beskrovanov 2000). For diamonds from pipes of Srednemarhinsky and Malobotuobinsky districts of Yakutia, correlation of ADR diamonds study by the color cathodoluminescence method in conjunction with reference data (Zinchuk and Koptil 2003) shows more complex polyphase and oscillatory growth of crystals at the Lomonosov deposit.

Examination of diamonds from Zolotitsa field by the Fluorolog FL3 spectrophotometer (Horiba) at room temperature (luminescence excitation at 250–600, recording at 200–800 nm) discovered two groups of pipes dominated by yellow component of luminescence glow and pink component (Table 3.1) (Kudryavtseva et al. 2005; Kriulina et al. 2011).

Diamonds from the Karpinskogo-1 and Arkhangelskaya pipes are very similar by color distribution in diamonds of comparable rough samples. The luminescence characteristics of diamond are particularly noteworthy: due to a high fraction of crystals with black inclusions there is no luminescence in more than a half of the batch (68–69%). Approximately 7% have yellow and green glow (these are mostly yellow, orange and brown cubes), 10–14% are characterized by yellow and orange glow caused by stresses and dislocations in crystals and only 9% of crystals show pale-blue glow of varying intensity, typical for most of diamonds around the world.

Mineralogical examination of higher grade diamond ($-4 + 2$ mm, average weight 0.25–0.35 ct) from the Karpinskogo-1 and Arkhangelskaya pipes showed a difference between large-volume testing batches and lower grades ($-2 + 1$ mm) from earlier samples. Common patterns of typomorphic properties manifestation are preserved. An increase in the fraction of octahedron-dodecahedron row diamonds, a decrease in the fraction of luminescent gems (luminescence quenching due to inclusions) as crystal size grows are typical for all kimberlite pipes from the Lomonosov deposit. Distribution of crystals over crystal morphological forms, color, glow nature in the UV band inside differently sized varieties changes insignificantly, i.e. the main genetic tendency remains.

The Karpinskogo-1 pipe is characterized by predominance of crystals with weak featureless glow (luminescence is quenched due to a heap of impurities). About 50% of diamonds from the Karpinskogo-1 pipe do not show luminescent glow while being viewed under a gemological UV lamp with 365 nm wavelength.

Individuals with green-colored luminescence (H3, H4 centers) are widespread for 11% of crystals (of all crystals in the collection), 4% have yellow luminescence caused by presence of S2, S1, S3 centers. The crystals with blue and pale-blue glow (N3 center) comprise 9% (Table 3.2, Fig. 3.1).

Table 3.1 Distribution of diamonds from the pipes of ADR deposits by photoluminescence color

Photoluminescence color	Number of diamonds, %						
	Lomonosov	Pionerskaya	Pomorskaya	Arkhangelskaya	Karpinskogo-1	Karpinskogo-2	V. Grib
Pale-blue	38.7	40.1	27.5	8.9	7.1	22.5	62.5
Pink-violet	10.2	18.4	16.4	18.7	25.4	14.8	0.0
Yellow-green	30	20	27	37	41	36	0.0
Yellow	2.2	2.3	6.8	1.4	5.1	7	12.5
Zonal yellow-orange	2.2	0.0	2.9	2.3	2.1	0.9	2.5
Red-orange	6.6	4.6	4.8	2.3	3.1	0.9	0.0
Green	1.5	1.2	12.1	19.1	12.2	7.8	12.5
Green coat, blue core	8.8	13.8	2.3	10.3	4.1	10.4	10.0

Note The data for the Lomonosov, Pionerskaya, Pomorskaya, and Karpinskogo-2 pipes are quoting (Kudryavtseva et al. 2005)

Table 3.2 Occurrence frequency of diamonds from the Karpinskogo-1 pipe with various luminescence colors

Pipe	Rocks	Occurrence of diamonds with various luminescence colors, %							
		Pale-blue	Yellow	Yellow-green	Orange	Pink	Pale, undefined	None	Total
Karpinskogo-1	AKB + TR	8.8	10.2	6.0	0.7	0.0	5.1	69.1	100.0
	AKB	16.2	23.9	23.9	2.1	0	2.11	31.69	100.0

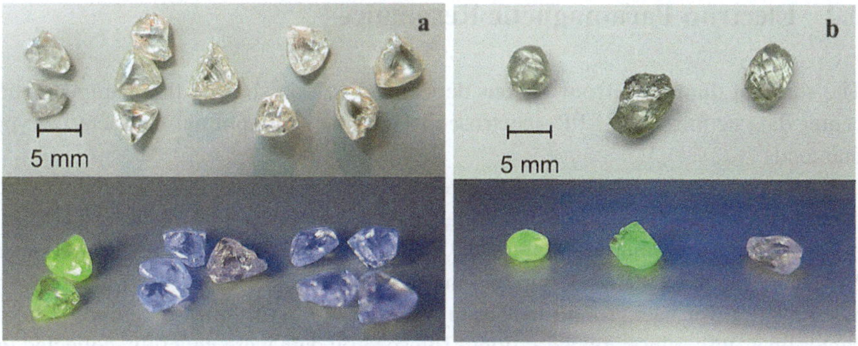

Fig. 3.1 Photo images of gem-quality diamond crystals from the Arkhangelskaya pipe luminescent under UV-radiation: **a** with predominant pale-blue luminescence among spinel twins, mackles; **b** yellow-green luminescence dodecahedroid and tetrahexaedroids, chips

The individuals with yellow and yellow-green luminescence, types of spectra with S1, S3 luminescence centers, prevail in the Arkhangelskaya and Karpinskogo-1 pipes (Bogatikov et al., 1999). This set of lines is atypical of diamonds from YDP. The 523.3 nm baseline that shows presence of an S2 impurity center (yellow-green area) and is missing in the spectra is the second difference of the spectra of crystals from the Lomonosov deposit.

A higher concentration of dodecahedral diamonds with blue–pale blue and pale-blue luminescence (Fig. 3.1) with spotted and zonal glowing (luminescence center N3) is typical of the Lomonosov and Pionerskaya pipes. The number of diamonds with yellow and yellow–green luminescence is small. Diamonds with pink luminescence probably indicate the potential (initial) commercial diamondiferous capacity of pipes (which is subsequently decreased due to dissolution processes) as this glowing is usually typical of highly productive bodies from the Yakutian Diamondiferous Province.

Diamonds from the Snegurochka pipe have pale-blue, pink-yellow, yellow luminescence. The prevailing yellow glow is usually connected to the presence of plastic deformation centers in crystals; in general, it is typical of diamonds from the Lomonosov deposit. Diamonds with pink luminescence probably indicate the potential commercial diamondiferous capacity of the pipe.

The V. Grib pipe stands out among other ADP pipes for the highest relative portion of diamonds with blue-pale-blue, and violet luminescence(~62%), 10 and 12% of crystals have yellow and green luminescence, respectively. Approximately 20% of diamonds do not luminesce. The V. Grib pipe diamonds are most similar by photoluminescence features to diamonds from the Pionerskaya (Table 3.1) and Lomonosov pipes (Kudryavtseva et al. 2005), and resemble diamonds from the Mirninskoe field of YDP (Zinchuk and Koptil 2003).

3.2 Electron Paramagnetic Resonance

The study of diamonds from various deposits of Russia showed that paramagnetic center data acquired by EPR spectroscopy provide an important characteristic of diamonds.

The paramagnetic center spectra in examined diamond crystal specimens were registered by EPR spectrometers Varian E-115 and CMS-8400 in X-band width (~9.4 GHz) at modulation amplitude 0.1 mT, modulation frequency 100 kHz and MW radiation power from 0.2 up to 5 mW. The spectra survey was carried out with accumulation for background reduction (duration up to 50 min) to determine faint center lines with minor concentrations. The spectral data were generally collected in a position of crystals wherein the quartic order centerline was parallel to the external magnetic field intensity vector ($H \| L_4$).

In ***kimberlites of the Lomonosov deposit*** crystals of all five groups of diamonds were recognized by the presence of the main paramagnetic centers related to nitrogen defects and plastic dislocations (Mineeva et al. 1996). These groups were recognized and established first for crystals of the Yakutia kimberlite diamonds: (1) with dominating P1 centers; (2) with dominating P2 centers; (3) with dominating N2 centers; (4) with simultaneously dominating P2 and N2 centers; and (5) without paramagnetic centers (Mineeva et al. 1996, 2009).

The observed set of defects in the studied crystals of the Lomonosov deposit and the ratios between their concentrations appeared to be typical of such groups of diamonds from other deposits. One or several types of centers could be identified in an isolated crystal (Table 3.3).

The P1 center is the most studied and simplest center in the diamonds. It results from an isolated N atom replacing a carbon atom and contributing a yellow component to diamond color. The concentration of main P1 centers in crystals of group I is comparable with such concentrations in diamonds from the YDP: in crystals of octahedral-dodecahedral type the concentration is about 0.1–0.4 ppm. The group of combinative yellow-colored crystals with P1 10–25 ppm can be identified, with up to 350 ppm in the cubic habit specimen sand in coated diamonds.

Patterns with dominating P1 center are most widespread among crystals from the Lomonosov deposit. Crystals with dominating P2 and N2 centers are significantly less common. P2 is a center that characterizes high degree of nitrogen aggregation in the structure; the model comprises 3 nitrogen atoms and a vacancy. The presence

Table 3.3 Paramagnetic centers in diamonds from various pipes of the Lomonosov deposit

Pipe	Number of crystals in the studied collection with certain paramagnetic centers, pcs							Studied crystals, total, pcs
	P1	P2	N1	N2	W7	W21	OK1	
Arkhangelskaya	42	40	–	8	8	2		74
Karpinskogo-1	36	10	–	–	2	2	6	40
Pionerskaya	8	16	–	4	–			40

of N2 centers is linked to plastic deformation of diamond crystals under gliding mechanism at elevated temperatures, above 1200 °C (Mineeva et al. 2009).

The original EPR studies of diamonds from the *Arkhangelskaya pipe* demonstrated predominance of diamonds with single P1 center. The diamonds from this pipe are distinct for availability of a multitude of crystals with centers arising in plastic deformation (N2 and W7), not only among diamonds of octahedral and dodecahedral forms, but also among cubes. However, a brown tint arising from impact of plastic deformation processes did not appear in the color of crystals.

The *Lomonosov and Pionerskaya pipes* are similar in their distributions of crystals with a particular set of paramagnetic centers. The Pionerskaya pipe contains crystals that correspond to subgroups with dominating P1 centers with a minimum degree of nitrogen defect aggregation. The higher degree of paramagnetic centers aggregation is typical of the Lomonosov pipe in comparison to diamonds from the Pionerskaya pipe.

The *Pomorskaya pipe* is characterized by total absence of N2 centers in crystals and a low degree of defect aggregation. Its peculiarities include an abnormally high number of diamond crystals with a low ratio between the concentrations of P2 and P1 centers that characterize the degree of nitrogen inclusion atom aggregation. This can be due to poorer developed post-crystallization structural defect transformation processes in the deposit.

Diamonds from the *Karpinskogo-1 pipe* are characterized by their moderate content of N2 centers, which is due to dislocations in crystals in the same manner as in diamonds from other deposits of the world. A high concentration of centers of isolated P1 nitrogen atoms is a feature of diamonds from the Karpinskogo-1 pipe, as well as the entire Lomonosov deposit (Mineeva et al. 1996).

The V. Grib pipe diamonds are characterized by low content (about 10%) of diamonds with dominating P1 centers. The presence of P2 centers of aggregated nitrogen atoms is typical of most of the diamonds, which is in good correlation with the high degree of perfection of the crystal structure (Mineeva et al., 1996). Note that the concentration of P2 centers here is lower by an order of magnitude than in such diamonds from the YDP and is less than 5×10^{17} centers/g. In groups of diamonds with dominating P2 and N2 centers the main center concentration corresponds to the same concentration in crystals from the YDP.

The high concentration of paramagnetic N2 centers connected to plastic deformations in crystals from the Grib pipe makes them different from the diamonds of the Lomonosov deposit. This is probably caused by a negligibly small concentration of xenoliths of catalyzed ultrabasites in the pipes of the Lomonosov deposit and the wide presence of these rocks in the Grib pipe. Types of spectra with luminescence S1 and S3 centers prevail among rounded crystals. A line at 523.8 nm that probably represents an individual center is observed in the spectrum of these crystals. This set of luminescence centers is unusual for crystals of diamonds from the YDP pipes.

Ni-containing centers are rather specific to crystals from Central Siberian subprovince. Ni-centers are absent in crystals from the Grib and Lomonosov deposits, which enables to conclude: this feature is typomorphic for the described deposits of the ADR (Bogatikov et al. 1999).

The diamond studies by EPR spectroscopy emphasize genetic division of diamonds from the southern and northern groups of pipes.

Diamonds of the Arkhangelskaya and Karpinskogo-1 pipe are characterized by high occurrence frequency and concentrations of isolated nitrogen atoms replacing carbons in the lattice (P1 center); this fact underscores that the Arkhangelskaya and Karpinskogo-1 pipe diamonds have high nitrogen concentrations. Furthermore, a low degree of defect aggregation indicates reduced temperatures of crystallization for most of crystals and a short time of post-crystallization annealing, so that even the C center did not translate into A (or some A centers were crushed to C (P1) centers). The published references on diamonds generally maintain that an A center (two nitrogen atoms) further evolves into a B center (4 nitrogen atoms and a vacancy), based on IR spectroscopy.

A P2 center with medium occurrence frequency was registered in the Arkhangelskaya pipe diamonds, which indicates higher nitrogen-vacancy defect aggregation in some crystals and, respectively, a longer stay at higher temperatures in the mantle.

The presence of isolated crystals with W7 centers and their high concentrations (as compared to specimens from other pipes of the Lomonosov deposit) in diamonds from the Arkhangelskaya pipe stands to note. These are complex nitrogen-vacancy centers that strain and appear in diamond structure when exposed to plastic deformation processes, which need to be factored in when characterizing diamond toughness properties.

The Pionerskaya pipe diamonds are slightly different from specimens of the Arkhangelskaya and Karpinskogo-1 pipes—the fraction of P1 centers is lower, whereas P2 and N2 shares are higher. The Pionerskaya pipe diamonds, and diamonds from other pipes of the deposit are mainly high-nitrogen, as indicated by EPR and IRS study results. Unlike in diamonds from the Arkhangelskaya and Karpinskogo-1 pipes, structural nitrogen is to a far greater degree aggregated, which proves its longer exposure to high temperatures that contributed to oxidation etching.

3.3 IR Spectroscopy

Lots of researchers have been studying mineral and structural defects in diamonds for half a century already (Argunov 2005; Beskrovanov 2000; Bokii et al. 1986; Sobolev 1978; Khachatryan 2003, 2010; Klyuev and Naletov 2008; Evans 1992; Taylor and Milledge 1995). Defined defects exceed 50 in number. This paper focuses on nitrogen defects represented by A, B centers, platelets and hydrogen, as the best known and indicative of temperature and chemical conditions of diamond genesis and evolution.

IR spectroscopy was applied to quantify defect-impurity composition of examined diamonds and total phase composition of trapped nanoscale inclusions. The IR absorption spectra in the 600–4500 cm^{-1} band were registered by two-dimensional mapping throughout the area of the manufactured plane-parallel diamond plates (the net with up to 1000 points) at intervals of 50–100 μm and resolution 2–4 cm^{-1} by FT spectrometer Tensor 27 (Bruker), with Hyperion 3000 microscope. Diamond intrinsic absorption was accepted as the internal standard (Zaitsev 2001).

For mixed-type diamond spectra analysis a full IR spectrum was resolved into individual A and B1 systems of absorption bands derived from natural diamonds of clear types Ib, IaA and IaB. Defect concentrations were determined using relationships offered in the papers (Sobolev and Lisoivan 1972; Boyd et al. 1994, 1995). Total nitrogen (Ntot, $N_A + N_B$) was defined by addition of identified concentration of this impurity in A- and B-forms. Hydrogen impurity content was estimated by absorption bands of C–H oscillations at the frequency 3107 cm^{-1} (De Weerdt et al., 2003), water (in a molecular form or as OH groups), carbonate and silicate phases—by maximum intensity of principal bands peculiar to them (Rossman 1988; McMillan and Hofmester 1988; Navon et al. 1988). In IR absorption spectra, apart from diamond intrinsic absorption bands, nitrogen and hydrogen impurity centers, we observed absorption bands of calcite and dolomite carbonate phases in microinclusions per 1400–1500 cm^{-1} and 860–880 cm^{-1} (McMillan and Hofmester 1988; Biellmann and Gillet 1992).

High nitrogen concentrations in diamond structure suggest its considerable content in diamond-forming environment. The key parameters of mineral-forming environment are temperature, pressure, oversaturation, oxidation–reduction potential. The experiments proved (Litvin and Bezrukov 1969) that compared with cubic growth sectors the octahedral ones are formed at higher temperatures and constant pressure, cubic sectors pinch out if temperature increases. Crystal morphology indicates to some oversaturation of the environment in formation of diamonds with higher nitrogen content in the structure. Gradual rise of carbon content in diamond-forming environment predicts different morphology of diamonds from plane-faced octahedrons with tangential growth to octahedrons with a step structure and polycentric growth, further to anti-skeleton forms (pseudorhombododecahedrons), highest over saturations result in formation of crystals with normal face growth mechanism (cubes and "split" crystals) (Krasnova and Petrov 1997; Punin, 1981).

The Yakutian diamonds, best known in literature (Kvaskov et al, 1997; Beskrovanov, 2000), tend to crystal morphology evolution in transition from simultaneous development of cubic and octahedral sectors solely to octahedral ones; in its turn, this indicates a temperature rise during diamond formation. The complex internal structure of the Arkhangelsk diamonds and a change of tangential growth to normal on the contrary indicates a temperature drop and/or its fluctuation by alternating zonal-sectorial and layered formation.

Altogether, the Lomonosov deposit is characterized by diamonds with high (3.8–8.5 cm^{-1}) relative concentrations of hydrogen defects (Table 3.4). Earlier studies showed (Blinova 1987; Beskrovanov 2000) that hydrogen impurity content in crystals relates to their growth mechanism. Thus, for example, there is a minor amount of hydrogen centers in crystals formed under the tangential growth (revealed in layered closed octahedral structure). Conversely, diamonds with normal and mixed growth mechanisms prevailing at the Lomonosov deposit are characterized by fibrous and sectorial internal structure, and differ for maximum high hydrogen concentrations.

Table 3.4 Concentration of defect-impurity centers of nitrogen, hydrogen and platelets in diamonds from ADR pipes

Pipe name	Number of specimens	Nitrogen concentration, at.ppm			%N_B	Hydrogen, cm^{-1}	Platelets, cm^{-1}
		N_A	N_B	N_{tot}			
Arkhangelskaya	306	524 2–1534	210 2–1775	735 19–2700	29 2.5–97	3.8 0–23	4.9 0–35
Karpinskogo-1	320	1010 0–2200	250 0- 990	1240 17- 2900	18 0–60	1.6 0–7.6	2.5 0- 20
Pionerskaya	280	560 0–1550	165 0–1000	732 0–2493	20 0–60	3,3 0–20	6,8 0–40
Snegurochka	55	370 25–1236	234 4–584	604 54–1612	40 3–74	4.2 0–21	9.8 0–28
Grib	116	482 0–2309	286 0–1857	743 0–2840	35 0–65	8.5 0–33	0.8 0–5.5

However, hydrogen impurity in thoroughly examined (Chap. 4) crystals of cubic habit with fibrous growth mechanism amounts to 0.5–1.5 cm^{-1}.

Occurrence of high-nitrogen diamond crystals with a low degree of B-form nitrogen aggregation (Kriulina 2012) is typical of the Lomonosov deposit in ADR. Diamonds with defect-impurity centers in A-form prevail among all studied collections, with their bimodal distribution in all the pipes. On average, the aggregated nitrogen proportion (%N_B) does not exceed 30% of total nitrogen impurity concentration ($N_A + N_B$) (Fig. 3.2).

Diamonds from the Zolotitsa field are characterized by a bimodal distribution of nitrogen impurities with the domination of A centers; bi- and trimodal distribution

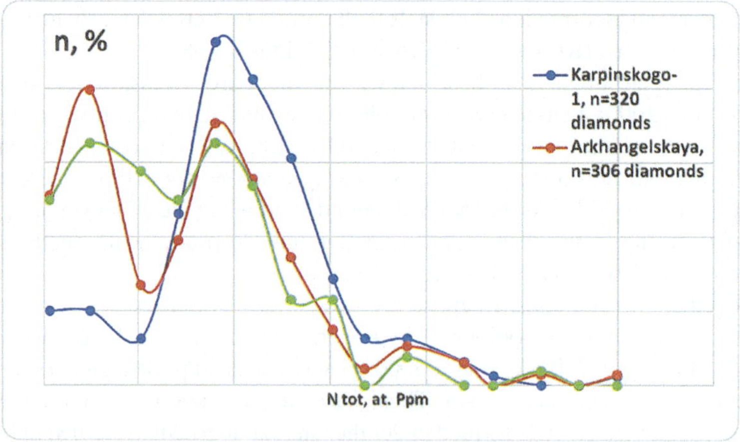

Fig. 3.2 Distribution of diamonds with different total concentration of nitrogen defect-impurity centers (Ntot) in the most productive pipes of M.V. Lomonosov deposit

of H and P impurities; reduced total nitrogen concentration in fine (<1 mm) crystals in contrast with macrocrystals.

Diamonds from the Karpinskogo-1 pipe show the highest content of total nitrogen ($N_A + N_B$), varying from 17 to 2900 at ppm—1240 cm^{-1} on average. Relative concentration of hydrogen in diamonds is 7.6, 1.6 cm^{-1} on average. The absorption coefficient in isolated crystals is directly proportional to hydrogen impurity content, and reaches 15 cm^{-1}. Relative platelets (P) concentration in diamonds is up to 20, 3.0 cm^{-1} on average. The proportion of aggregated nitrogen (%NB) is less than 30% (Khachatryan et al. 2008; Kriulina 2012) (Table 3.4). Altogether, diamonds of the Lomonosov deposit southern group pipes are very similar by nitrogen distribution and concentration in the structure.

Arkhangelskaya pipe. Type I diamonds are largely characterized by different (0–1534, average value 524 at.ppm) nitrogen concentrations in A-form and low (2–1775, average value 210 at. ppm) in B-form, i.e. nitrogen aggregation degree in a wide range (0–95%) is minor (av. value 29%). The hydrogen and platelets impurity relative concentrations are average (3.8 and 4.9 cm^{-1}, respectively).

Plane-face octahedrons and crystals with polycentric face growth belong to high-nitrogen diamonds of IaAB type with the nitrogen proportion in B-form at 15–60%; B2 band is registered in IR spectra. Some octahedrons are low-nitrogen. The average absorption coefficient of the CH group is 4.5 cm^{-1}, and reaches 16 cm^{-1}in individual crystals. In PL spectrum of these crystals at room temperature N3 or N3 and S3 system prevail, S2 is registered. Yellow color of these crystals is caused by N3 system.

Clear and yellow-colored dodecahedroids of the Ural type and tetrahexahedroids are classified as IaA type, nitrogen proportion in B-form is below 15%. About 20% of dodecahedroids are characterized by NB1 in the range of 20–97% (IaAB type), they are like group I crystals by PL characteristics. Several systems unmentioned in publications on diamonds from deposits of Russia are registered in IR absorption spectra of IaA type diamonds. The average absorption coefficient of CH group band is 4.7 cm^{-1}, and reaches 24 cm^{-1} in separate crystals.

Cubic and tetrahexahedral crystals vary by a set of defect-impurity centers and can be split into three diamond groups. The first group includes "classical" cubes with nitrogen defects in A- and C-forms and without B-center. Such crystals are rather rare. The second, most common group includes crystals containing nitrogen defect-impurity centers in A- and B-forms at very low concentrations of C-center, which is not registered by IRS (detected only by EPR), being an unusual phenomenon for crystals of cubic habit. The third group is represented by specimens with only C-center detected, such specimens make about 25% of the cube collection.

Low quantities of hydrogen (about 1 cm^{-1}) were registered in grey cubes with mixed growth mechanism. Total nitrogen concentration is up to 1000 at.ppm, nitrogen defect concentrations in B-form are up to 150 at.ppm.

Octahedral and rhombic dodecahedral habit diamonds of grey color or tint have maximum, of all studied, concentration of platelets (10–33 cm^{-1}), hydrogen (2–24 cm^{-1}) and nitrogen aggregation degrees % NB $\approx$ 70.

Diamonds from the Snegurochka pipe are represented by two groups of crystals combined by nitrogen center distribution parameters. The most common (about 50% of the collection) are medium- and high-nitrogen crystals dominating in grades $-4 + 3$, $-2 + 1$, with total nitrogen concentration Ntot from 400 to 1000 at.ppm. and a large proportion of nitrogen aggregated in B-form (40–60%). This group is unique, i.e. such a substantial number of diamonds with similar combination of parameters (defect concentrations) was detected for none of the studied pipes.

The Pionerskaya pipe offered bimodal distribution of diamonds by nitrogen center concentration. Group I: low-nitrogen diamonds (Ntot < 400 at. ppm) are characterized by a reduced proportion of nitrogen in B-form (% av. value = 14) and minimal occurrence of hydrogen centers (CH 0.48 cm^{-1}), but isolated crystals have the absorption coefficient CH (3107 cm^{-1}) 5–10 cm^{-1}.

Group II, the major population of diamonds from the Lomonosov deposit, is similar to diamonds from the Karpinskogo-1 and Arkhangelskaya pipes and is represented by high-nitrogen diamond crystals. Nitrogen total concentration is 800–1500 at.ppm. There are two apparent subgroups based on aggregation degree: the one with % B below 10% and high average concentration of hydrogen centers, and a subgroup with higher proportion of nitrogen in B-form (40–55%) and maximum hydrogen concentrations in diamond structure.

Nitrogen maximum concentration in diamonds from the Pionerskaya pipe reaches 2500 at.ppm.

In diamond crystals of Grib deposit the total nitrogen concentration ($N_A + N_B$) varies from 0 to 2840, 743 at.ppm on average (Palazhchenko et al. 2006; Khachatryan et al. 2006). Hydrogen relative concentration (CH) in diamonds up to 5.5, 0.8 cm^{-1} on average. Platelets relative concentration (P) in diamonds up to 32.5, 8.5 cm^{-1}on average (Table 3.4). The diamonds from the Grib pipe are characterized by higher P content and lower hydrogen content, the nitrogen aggregation degree ($\%N_B$) is considerably higher (2 times)—up to 60%, versus diamonds from the Lomonosov deposit pipes (Fig. 3.3). Low nitrogen concentration is typical of most crystals in the Grib pipe.

The study data analysis using IR spectroscopy enables to conclude that diamond-forming environment in the Lomonosov deposit was mostly saturated with nitrogen and hydrogen, versus other deposits of Russia; this resulted in formation of specific morphogenetic groups of diamonds. Maximum values were registered for diamonds from the Arkhangelskaya and Karpinskogo-1 pipes. By their thermodynamic parameters diamond-forming conditions of type I octahedral-dodecahedral crystals under classification by Yu. L. Orlov are similar to those of diamonds from low-titanium kimberlites of Yakutia.

Comparing data obtained by IR and UV visible spectrophotometry, the following common factors can be observed for studied diamond crystals from the Arkhangelskaya pipe (Fig. 3.4):

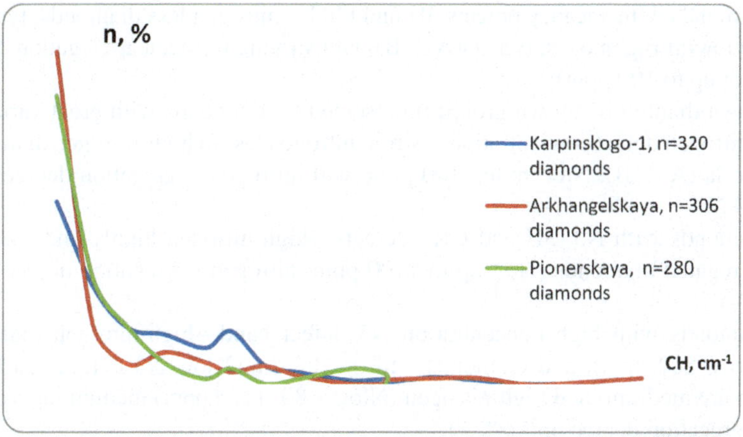

Fig. 3.3 Distribution of diamonds with different impurity nitrogen center absorption coefficient CH (3107 cm^{-1}) from the most productive pipes of the M.V. Lomonosov deposit

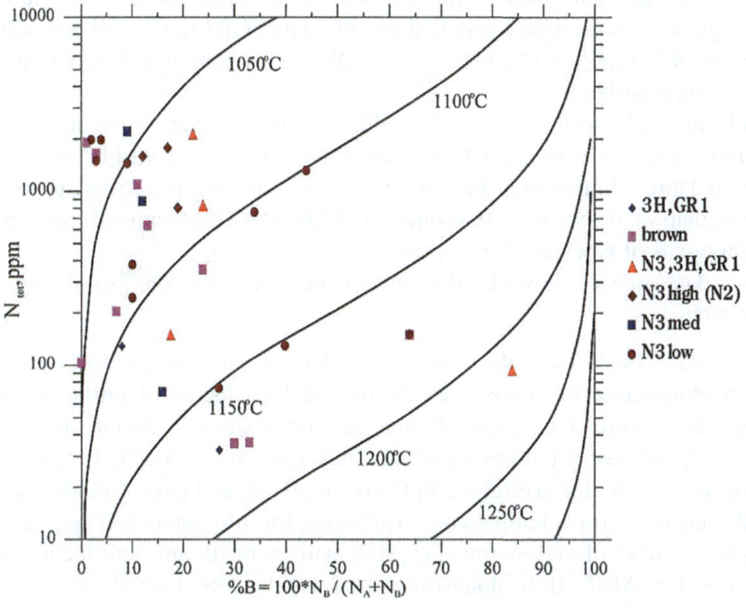

Fig. 3.4 Taylor V. R. diagram (1990) for type IaA + B diamonds. Estimated diamond age is 3 billion years. Imaging points stand for values of total and aggregated nitrogen concentration in diamond; imaging point color corresponds to groups discriminated based by optical spectra (by presence of nitrogen-vacancy defects)

- Diamonds with vacancy defects 3H and GR1—nitrogen less diamonds, type IIa, and low-nitrogen ones, type Ia(A + B),with varying nitrogen aggregation degree (Ntot up to 100 ppm);
- Brown diamonds (brown group) represented by diamonds with great variations of nitrogen defect concentrations—from nitrogen less to high-nitrogen diamonds, type Ia (A + B), Ntot up to 1100 ppm, with nitrogen aggregation degree up to 35%;
- Diamonds with N3, 3H and GR1 defects—high-nitrogen highly and medium-aggregated diamonds (Ntot up to 1100 ppm, nitrogen aggregation degree up to 85%);
- Diamonds with high concentration N3 defect band—high-nitrogen diamonds (above 800 ppm), and diamonds with medium and low N3 defect concentration are medium- and high-nitrogen (Ntot = 80–1100 ppm) medium-aggregated (aggregation degree up to 45%).

Data of nitrogen aggregation degree in studied specimens are given on the Taylor diagram, diamonds are divided into groups by optical spectra types.

Major nitrogen and vacancy defects were determined based on luminescence spectra, green laser excitation revealed the following: GR1 (LTPL 740 nm, red luminescence), N-V (LTPL 637 and 575 nm); LW UV excitation: H3 (LTPL 503 nm), N3 (415 nm, pale-blue).

GR1 red luminescence center, the radiation center, occurs in natural diamonds of all types under ionizing radiation of any kind. GR1 is responsible for both near-surface and three-dimensional dark-green color of diamonds occurring due to natural α- or γ-radiation, respectively (Gorobets et al. 2001). Green diamonds are typical of craterous areas of pipes in ADR deposits.

Study data analysis allowed to discriminate groups of diamonds similar by certain parameters:

- Diamond crystals from the Lomonosov deposit kimberlites are characterized by predominance of curve-faced forms and low degree of nitrogen aggregation, which correlates to data of spinelide composition evolution and relates to gradually increasing oxidation potential in stages prior to kimberlite pipe formation. As a result, fine crystals completely dissolved, and large crystals acquired a dodecahedral form at temperature insufficient for nitrogen defect aggregation.
- Higher number of yellow-tinted crystals with mineral and fluid inclusions was defined for ADR. It is noteworthy that bands with low absorption coefficients responding to carbonate impurity (microinclusions), and very occasionally to hydroxyl group and water, are sometimes registered in IRS spectra. Higher concentrations of nitrogen (50 < Ntot < 3000 at.ppm) and hydrogen defect-impurity centers adversely affect rough diamond quality. Octahedral-dodecahedral diamonds of type I according to Yu.L. Orlov's classification contain a reduced proportion of nitrogen in B-form (NB $\leq$ 30% on average) and have low platelets absorption coefficients (2 < Paver. < 5 cm^{-1}, with the exception of diamonds from the Snegurochka pipe Paver. = 8.5 cm^{-1}) when the IR spectrum absorption band crest shifts to the shorter wavelengths which, according to

published data (Blinova 1987; Vins 2011; Vins and Yelisseyev 2009, 2010) indicates a short duration of post-crystallization annealing at temperatures required for nitrogen defects transformation; relatively low absorption coefficients of the 1364–1370 cm^{-1} band suggest relatively lower temperatures of diamond crystal formation. Characteristics of diamond defect-impurity composition (av. value $N_B < 30\%$) evidence their evanescent high-temperature (above 900 °C) post-crystallization annealing in the magmatic substrate.

3.4 Raman Spectroscopy of Diamonds and X-Ray Spectrometry for Studying Inherent Mineral Inclusions

The typomorphic feature of diamonds from all pipes of the Arkhangelskaya diamondiferous province is about widespread occurrence of "diamond in diamond" type inclusions shaped as right octahedrons and triangular face fragments. This emphasizes intermittency of diamond formation nature (Garanin et al. 1991). Estimation of "diamond in diamond" type inclusion abundance requires specialized instrumental methods (for instance, color cathodoluminescence).

Black platelets graphite inclusions prevail in the Lomonosov deposit diamonds (67% of diamonds with inclusions); also observed are octahedral or slightly flattened black inclusions of chrome-spinelides (24%); pseudo-prismatic and tabular colorless inclusions (olivine, carbonates), often surrounded by graphitized feathers (15%); colorless idiomorphic and nonaligned in host mineral or fragmental "diamond in diamond" type inclusions. Rare inclusions of red-violet garnets have been registered in less than 1% of crystals.

Diamonds from the Grib pipe offer a variety of inclusions. The number of diamonds with graphite, sulfide inclusions (38%) is fewer than in pipes of the Lomonosov deposit.

As a rule, primary syngenetic inclusions in diamonds from the Grib pipe are represented by olivine and chrome-spinelides which chemically comply with inclusions from diamond and diamond-bearing magnesian peridotites. Secondary inclusions embrace mica, magnetite, monosulfide solid solution, pentlandite, millerite, pyrrhotite, silicate-sulfide mixture (altered olivine + pyrrhotite + pentlandite), serpentine, saponite, serpentized olivine, altered ortho- and clinopyroxenes, iron oxides.

Ilmenite and garnet (pyrope-almandine) inclusions prevalent in diamonds from the Lomonosov deposit pipes were not found among studied diamond inclusions from the Grib pipe.

A higher number of diamonds with inclusions of ultrabasic paragenesis was determined for all ADR pipes in tuff and xenotuff breccias. Coarse inclusions (more than 0.15–0.2 mm in size) are largely confined to peripheral regions of octahedral-dodecahedral crystals with blue and pale-blue photoluminescence.

To get authentic diagnostics results, inclusions were studied by Raman spectroscopy (spectrometer EnSpectr R532 with add-on microscope Olympys CX41) and electron probe microanalysis (energy-dispersive analyzer INCA-Energy 350).

Studying microinclusions requires custom selection of survey parameters for each inclusion. Peaks from the inclusion and the very diamond are initially registered, so the diamond initial spectrum must be "deducted" and diamond luminescence effect shall be taken into account. Diamond surface also matters (as a rule, irregular which impedes survey); spectrum registration through prepolished surface with inclusions in a short distance from it.

Raman scattering spectroscopy of inclusions in diamond crystals was made using Raman spectrometer (In Via Renishaw). Lasers with wavelength 325 nm (UV-laser), 352 nm (green laser) and 735 nm (red laser) and an appropriate range of lenses and gratings were applied. Spectra identification was made using RS microscope software standard library and KP Crystal Sleuth spectra software.

Composition analysis of key components for certain inclusions was performed using an energy dispersive spectrometer mounted on scanning electron microscope LEO 1430 VP. Near-surface microinclusions were identified in the electron backscattering mode; they were analyzed using a focused electron beam 15 keV, 10 nA. The obtained concentrations of key components were 100% standardized. The study was made in backscattered electrons, in two modes: surface and near-surface.

3.4.1 Graphite

Graphite presence is typical for most of studied diamonds with inclusions. Graphite is observed along feathers around colorless mineral inclusions or as a film covering fine crystals of trapped minerals (including diamonds), marks an intergrowth border in a diamond aggregate. In such cases graphite is formed after diamond crystallization, results in polymorphic transition of diamond under influence of stresses that cause fracturing of the diamond around inclusion. Subject to degree of feather wall graphitization, they look smoky-colored or absolutely black. The feather walls are in homogeneously graphitized in some cases, their color changing from black to faintly smoky, dark-grey, from an inclusion toward periphery. It was detected based on RS spectra: 1350 cm^{-1} (Zedgenizov 2011) and 1580 cm^{-1} bands are inherent in disordered amorphous carbon in graphite-like condition with conjugated sp^2-bonds (Khokhryakov 2000).

Another type of graphite is represented by {111} separate plates or accumulations of plate-like inclusions parallel to diamond's plane, presumably of syngenetic origin. It was detected by characteristic line 1455 cm^{-1} in RS spectra, (Figs. 3.5, 3.6). One diamond crystal can have both types of graphite (Fig. 3.7).

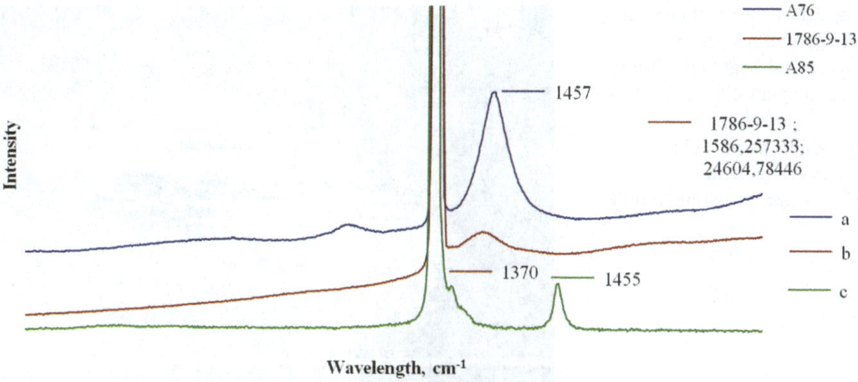

Fig. 3.5 Raman spectra of graphite: **a** plate-like graphite inclusion; **b** plate-like graphite inclusion orienting parallel to diamond's plane {111}; **c** graphite marks an intergrowth border in two-diamond aggregate

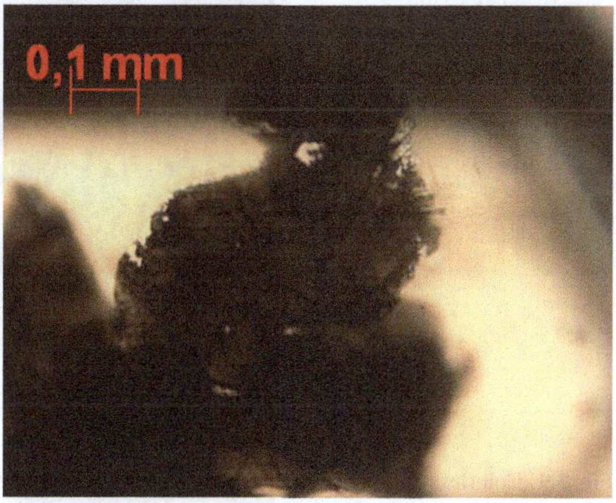

Fig. 3.6 Graphite plate-like inclusion in diamond (sp. A68). Optical microscope, transmitted light

3.4.2 *Olivine*

In all inclusions olivine is represented by colorless pseudo-prismatic and tabular crystals or crystal aggregates (Figs. 3.8, 3.9, 3.10, 3.11, 3.12 and 3.13). It was identified as forsterite (Fo > 90%) under RS spectra: characteristic peaks at frequencies 824 and 857 cm^{-1} (Fig. 3.14).

Diamonds of the Karpinskogo-1 pipe were defined to have three groups of olivine, by chemistry peculiarities. The first group is made up by medium-ferriferous olivine,

Fig. 3.7 Graphite marks an intergrowth border in two-crystal aggregate, forms separate plate-like inclusions oriented parallel to the diamond's plane {111} (sp. A85). Optical microscope, transmitted light

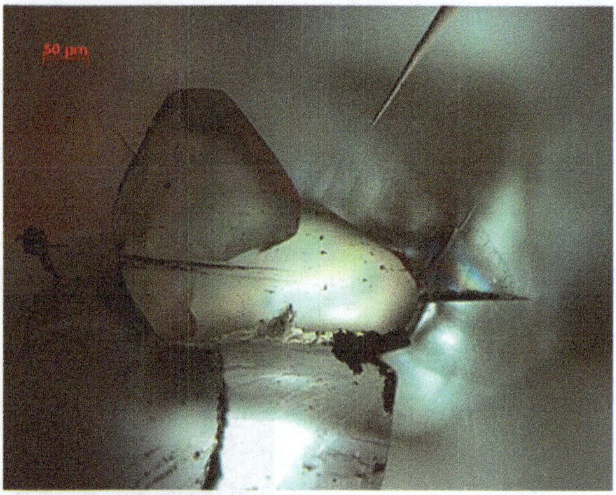

Fig. 3.8 The aggregate of two colorless forsterite crystals and plate-like graphite inclusions confined to them and evolving by cracks (sp. A57). Prepolished specimen; optical microscope, transmitted light

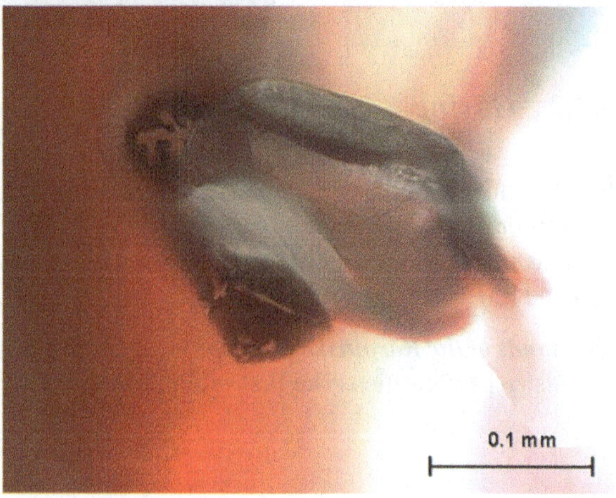

Fig. 3.9 Colorless tabular forsterite inclusion and graphitized cracks confined to it (sp. A85). Optical microscope, transmitted light

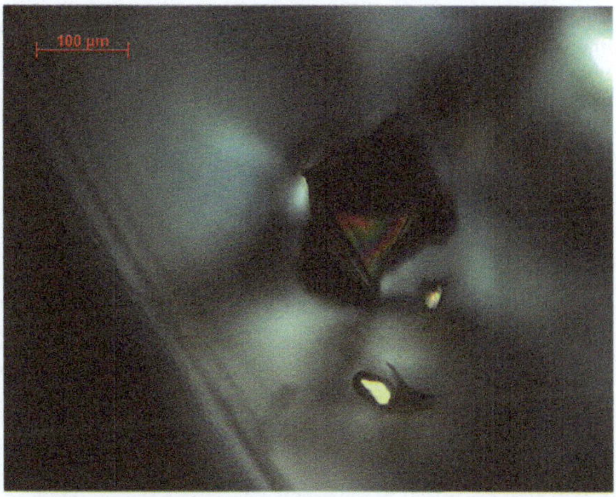

Fig. 3.10 Black chromite inclusion and colorless orthorhombic anisotropic crystals of forsterite in diamond (sp. A86). Optical microscope, transmitted light, crossed nicols

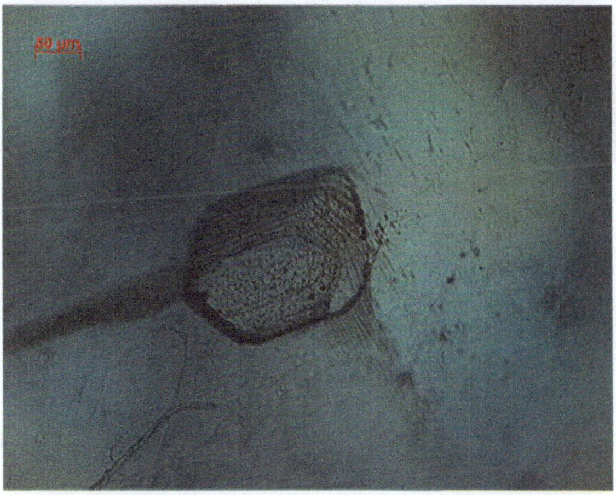

Fig. 3.11 Inductive synchronous growth surface of host diamond and captive mineral (Sp. A144). Prepolished crystal, optical microscope, reflected light

Fig. 3.12 Colorless orthorhombic forsterite crystal (sp. A150). Optical microscope, transmitted light

Fig. 3.13 Olivine inclusion in diamond (sp. 1786–24/1). Polished specimen, scanning electron micrographs, secondary electron micrographs

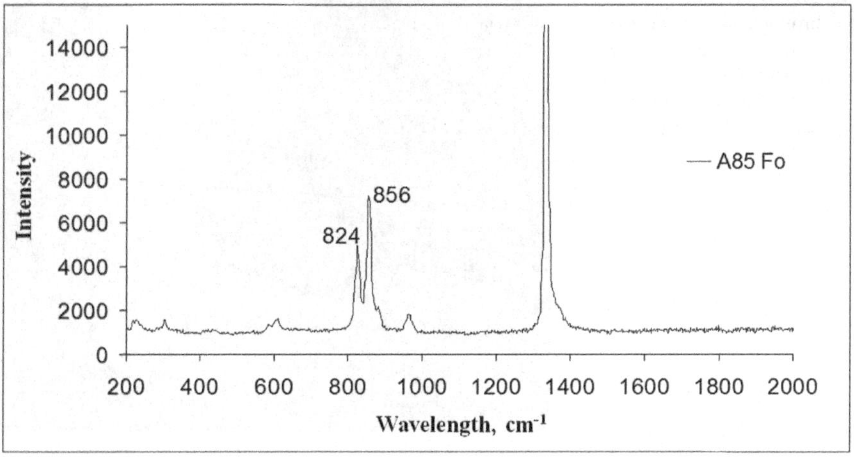

Fig. 3.14 Typical Raman spectra of forsterite inclusion in a diamond (sp. A85)

whose chemistry is similar to a mineral from diamond inclusions and diamond-bearing magnesian rocks—chemical-genetic groups (CGG) 2 and 3, under the classification (Garanin et al. 1991); the second group is represented by the only inclusion—high-ferriferous olivine (CGG 4) from diamond-bearing peridotites (Table 3.5).

Olivine inclusions of round or oval shape up to 1 mm in size. The chemistry is close to that of olivine from diamonds of other pipes in the Lomonosov deposit. Belongs to medium- and high-ferriferous varieties. FeO content in olivine from diamond is lower (6.1–8.1 wt%) than that in olivine from the Grib pipe. MnO content (0.00–0.1 wt%) is higher (Table 3.5).

Table 3.5 Chemical composition (wt%) of olivine inclusions in diamonds from the Karpinskogo-1 and V. Grib pipes

Chemical-genetic groups	Electron probe microanalysis of olivine inclusions composition in oxides							
	MgO	Cr_2O_3	MnO	FeO	SiO_2	CaO	NiO	Total
Olivine inclusions in diamond from the Karpinskogo-1 pipe								
2	51.40	0	0	6.62	41.54	0	0.27	99.84
2	50.98	0.45	0.10	6.65	41.71	0	0	99.89
2	50.42	0	0	6.36	41.47	0	0	98.25
2	50.42	0	0	6.36	41.47	0	0	98.25
2	51.40	0.04	0.08	6.26	41.40	0.03	0.38	99.59
2	51.60	0.01	0.10	6.53	41.50	0.01	0.35	100.10
2	51.20	0.07	0.11	7.08	41.60	0.02	0.32	100.40
2	51.40	0.06	0.11	6.53	41.70	0.02	0.32	100.14
2	50.90	0.03	0.08	6.77	42.20	0.02	0.35	100.35
2	51.30	0.04	0.10	6.9	41.10	0.02	0.35	199.81
2	51.70	0.05	0.09	6.85	41.60	0.01	0.36	100.66
2	51.30	0.01	0.08	6.7	41.60	0.03	0.33	100.05
2	51.40	0.01	0.10	6.62	41.90	0.03	0.32	100.38
2	51.00	0.05	0.10	7.13	42.10	0.01	0.31	100.70
2	51.50	0.05	0.11	6.85	41.50	0.02	0.33	100.36
2	51.20	0.08	0.10	6.92	41.10	0.02	0.32	99.74
3	50.53	0	0.11	7.71	40.61	0	0.33	99.29
3	50.02	0	0	7.5	41.50	0	0	99.02
3	51.00	0.05	0.07	7.12	41.50	0.02	0.34	100.10
3	49.80	0.04	0.11	8.12	41.50	0.04	0.34	99.95
3	50.53	0.03	0.09	7.35	41.70	0.02	0.34	99.83
3	50.6	0.03	0.09	7.27	41.80	0.02	0.31	100.12
3	50.5	0.02	0.10	7.58	41.80	0.03	0.35	100.38
3	50.5	0.04	0.07	6.84	41.60	0.01	0.34	99.40
3	50.8	0.04	0.09	7.25	41.70	0.03	0.34	100.25
3	50.3	0.01	0.11	6.37	40.80	0.02	0.31	97.92
4	49.2	0.02	0.09	9.24	41.20	0.02	0.33	100.10
Olivine inclusions in diamond from the V. Grib pipe								
CGG	MgO	Cr_2O_3	MnO	FeO	SiO_2	CaO	NiO	Total
1	51.33	0.05	0.05	7.13	41.10	0.02	0.18	99.86
1	50.45	0.06	0.04	7.49	40.68	0.02	0.23	98.97
1	51.11	0.76	0.04	7.46	40.14	0.02	0.21	99.74
2	51.84	0.10	0.00	7.52	40.12	0.03	0.38	99.99
2	50.09	0.05	0.14	8.62	40.25	0.03	0.27	99.45

Three olivine inclusions were observed to have stable elevated NiO content, 0.30–0.32 wt%, only three inclusions contained no NiO. Noteworthy is olivine inclusion in association with chrome-spinelide, which resulted in enriching olivine composition with Cr up to 0.45 wt%.

Diamonds from the V. Grib pipe have round or oval olivine inclusions, 120 × 80 μm in size. The chemistry is close to that of olivine from diamonds of the Lomonosov deposit, but FeO content in olivine from the Grib pipe diamonds is moderately higher (7.13–8.6 wt%) and MnO content (0.00–0.05 wt%) is lower (Table 3.5). Three olivine inclusions have lower nickel content (NiO 0.18–0.23 wt%).

Two olivine inclusions were observed to have higher nickel content (NiO 0.27–0.38 wt%) specific to olivine inclusions in diamond from the Lomonosov deposit pipes (Bartoshinsky 1992; Bogatikov et al. 1999; Kudryavtseva et al. 2005).

Two olivine groups can be identified based on chemistry features. The first group is represented by medium-ferriferous olivine, whose chemistry is similar to a mineral from diamond inclusions and diamond-bearing magnesian rocks—chemical-genetic groups (CGG) 2 and 3, under the classification (Beskrovanov and Spetsius 1991); the second group is represented by high-ferriferous olivine (CGG 4) from diamond-bearing peridotites (Table 3.5).

It was noted that magnesium-rich (Mg up to 52%) olivine inclusions are homogeneous, primary molten and normally positioned in the central or transitional region of diamond separately from other inclusions (Fig. 3.15). Iron-rich (Fe up to 42%) olivine frequently contains internal metal phase (solid solution breakdown structure). Chrome-rich (up to 0.65% Cr) olivine occurs in a crystal in association with chrome-spinellide or in aggregate with this mineral.

Two groups of olivine inclusions in diamond (Bartoshinsky 1992; Bogatikov et al. 1999) can be outlined based on calcium oxide / chrome oxide ratio, in view of the measurement error: 1) close by Cr_2O_3 (0.00–0.10) and CaO (0.01–0.04) content to olivine from inclusions in diamond of the ADR kimberlite pipes (Lomonosov, Karpinskogo-1), Mir pipe (Yakutia), Ebelyakh River placers; 2) olivine aggregated with chrome-spinelides with abnormally high Cr_2O_3 content (0.76 wt%) similar to such in aggregate with chrome-spinelide from the Lomonosov pipe Cr_2O_3 (0.45 wt%) (Bartoshinsky 1992; Bogatikov et al. 1999).

3.4.3 Garnet

Garnet inclusions in crystals of the Lomonosov deposit vary by color and belong to ultrabasic and eclogitic parageneses, while no garnet inclusions were detected in studied diamonds of the V. Grib pipe.

Garnet is represented in crystals from the Arkhangelskaya (Fig. 3.16) and Lomonosov pipes by red-violet isometric dodecahedral and flattened pseudo-prismatic crystals, 300–400 μm in size. The inclusions were identified by characteristic peaks of pyrope (364, 555, 642, 862, 916 cm^{-1}) in RS spectra (Fig. 3.17).

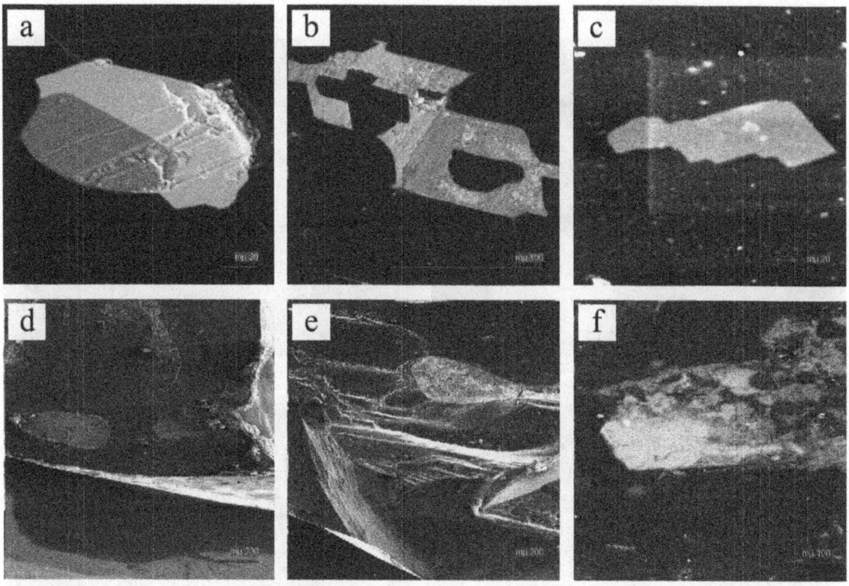

Fig. 3.15 Scanning electron micrographs of inclusions in diamond from the V.Grib pipe in back scattered (**a–c, f**) and secondary electrons (**d, e**): **a** olivine (left) and chrome-spinelide (right) aggregate 500^x; **b** slight intergrowth of silicate (olivine) and ore (mixture of sulfides) phase, 430^x; **c** olivine homogeneous inclusion, 150^x; **d, e** drop-like inclusions of chrome-spinelides, $20^\times$ and 50^x, respectively; **f** serpentinized olivine, 130^x

Electron probe microanalysis was used to determine quantitative composition of garnets (Table 3.6). It was defined that garnet inclusion in diamond No. A74 was represented by comparatively chrome-rich (16.8 wt% of Cr_2O_3) and calcium-rich (2.7 wt% of CaO) pyrope with knorringite minal at 45.28% of ultrabasic paragenesis; placed into the 1st chemical-genetic group—from diamond-bearing dunite-harzburgite and inclusions in diamonds (Sobolev, 1974) and to group G10 (Grutter et al, 2004).

Garnet inclusion composition analysis in diamond No. A193 showed that garnet was also represented by calcium-rich (7 wt% of CaO) pyrope; it was placed to the 29th chemical-genetic group of eclogitic paragenesis.

Two garnet inclusions from diamonds of the Karpinskogo-1 pipe were analyzed (Table 3.6). One of the garnets chemically matched with its peer from diamond-bearing dunites and harzburgites (CGG 1) and was magnesium-rich (22.4 wt% of MgO) with considerable content of Cr_2O_3 (10.2 wt%). The second garnet can be placed to CGG 15 and chemically matched with its peer from diamond-bearing magnesian and ferriferous eclogite. Characterized by reduced content of magnesium (8.5 wt% of MgO) and chrome (0.3 wt% of Cr_2O_3).

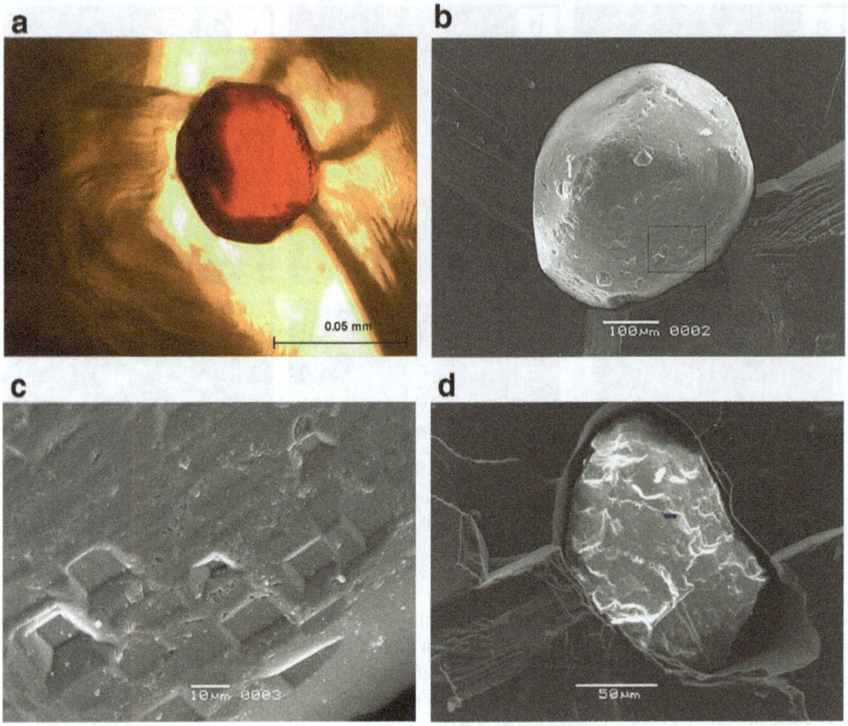

Fig. 3.16 Garnet inclusion in diamond from the Arkhangelskaya pipe, sp. A74: **a** optical microscope, transmitted light; field of vision width 0.25 mm; **b** electron microscope, secondary electrons; **c** garnet inclusion surface fragment, electron microscope, secondary electrons; **d** garnet of pyrope-almandine-grossular series, sp. A193, electron microscope, secondary electrons

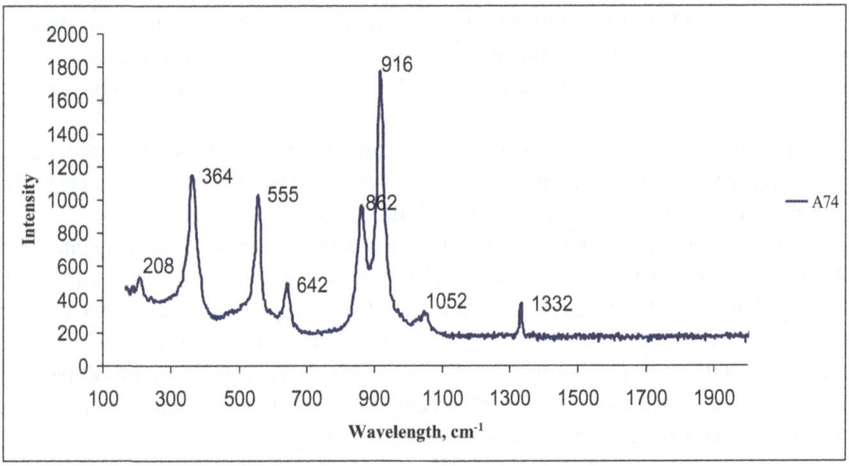

Fig. 3.17 Raman spectrum of garnet (pyrope) inclusion in diamond (sp. A74)

Table 3.6 Composition of garnet inclusions (wt%) in diamond from the Arkhangelskaya (specimens A) and Karpinskogo-1 pipes (specimens K)

Specimen No.	Chemical-genetic group	Electron probe microanalyses of garnet inclusion composition in oxides									
		SiO_2	TiO_2	Al_2O_3	Cr_2O_3	FeO	MnO	MgO	CaO	Na_2O	Total
A74	1	36.53	0	13.57	16.79	6.75	0.42	19.32	2.67	0.29	96.34
A193	29	41.29	0	23.48	0	12.34	0	13.1	7.02	0	97.23
K-1-1	1	41.4	0	15.6	10.2	6.66	0.34	22.4	2.4	0.02	99.06
K-1-2	15	39.4	0.56	21.5	0.03	18.0	0.35	8.16	11.2	0.28	99.48

3.4.4 Chromite

Chromite widely occurred in the studied diamond crystals. It can be situated in any region of host diamond; represented by single inclusions, 500 μm in size; inclusions are often brought onto the surface by etch channels and cleavages. Chromite inclusions can occur as numerous smaller crystals 10–150 μm in size (Fig. 3.18). There can be chromite and diamond aggregates. Most of chromite inclusions have unmistakable signs of being syngenetic to diamond: isometric octahedral inclusions or substantially flattened oriented in diamond (captive mineral faces are parallel to host diamond faces), with inductive surface of diamond and captive mineral intergrowth. Protogenetic inclusions have an idiomorphic shape, are situated without any crystallographic orientation in host diamond and moreover the diamond "grows around" a trapped inclusion.

The chemistry of chromite inclusions in diamonds was determined by electron probe microanalysis. Characteristic features of chromite in diamonds from the Arkhangelskaya pipe are given in Table 3.7.

By composition chrome-spinelides are placed to magnesian variety of chromopicotite. They are characterized by low iron content (8.57–16.62 wt% of FeO, 0–11.78 wt% of Fe_2O_3), low titanium content (0.06–2.37 wt% of TiO_2) and high chromium content (48.71–70.53 wt% of Cr_2O_3). Cr/(Cr + Al) and $Fe^{3+}/(Fe^{3+} + Fe^{2+})$ ratios evidence low fugitiveness of diamond-forming environment oxygen, and high chromium is likely to indicate a peculiar depth of mineral crystallization.

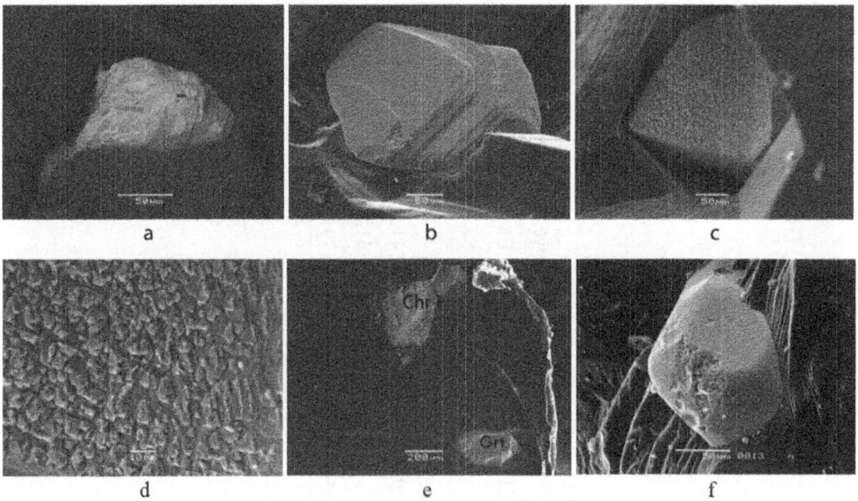

Fig. 3.18 Chromite inclusions in diamonds from the Arkhangelskaya pipe, images under SEM are secondary electrons: **a**–**c** cropping-out inclusion, sp. 157, sp. 1786-21-1, sp. A311; **d** chromite surface micro relief, shaped as second-generation octahedral crystals (sp. A311); **e** chromite and garnet inclusion (sp. A193), prepolished specimen; **f** cropping-out chromite inclusion, inductive growth line, sp. A13

Table 3.7 Electron probe microanalyses of chrome-spinelide inclusions in diamonds of the Arkhangelskaya, Karpinskogo-1 and Grib pipes

CGG	Electron probe microanalyses of chrome-spinelide inclusion content in oxides								
1	MgO	Al_2O_3	TiO_2	Cr_2O_3	MnO	FeO	Fe_2O_3	ZnO	Total
	Arkhangelskaya pipe								
	14.52	5.61	0.07	63.37	0.44	11.49	4.34	0.00	99.84
	12.59	5.42	0.06	65.23	0.35	14.1	2.48	0.00	100.23
	9.81	7.47	0.16	70.53	0.46	11.58	0.00	0.00	100.01
	12.19	5.91	0.08	66.82	0.46	13.62	0.06	0.00	99.14
	14.31	4.66	0.09	65.93	0.60	12.03	4.09	0.00	101.71
2	9.50	9.06	0.12	64.39	0.39	14.03	0.00	**3.64**	101.13
	15.65	5.65	0.29	61.00	1.17	8.57	5.36	**1.19**	98.88
	7.55	5.49	0.09	68.46	0.91	16.62	0.00	**2.67**	101.79
3	14.1	7.93	**2.37**	48.71	0.59	13.49	11.78	0.00	98.97
	11.87	7.12	**2.18**	52.26	1.02	14.7	7.96		97.11
Karpinskogo-1 pipe									
1	12	4.82	0.19	65.2	0.32	15.32	3.12	n/a	100.97
	13.6	6.71	0.16	64.1	0.29	13.35	3.17	n/a	101.38
	13	6.97	0.12	65	0.26	14.12	2.12	n/a	101.58
	12.66	7.2	0	63.52	0.26	13.79	1.69	n/a	99.12
	12.71	4.78	0	67.94	0.4	13.21	0.87	n/a	99.91
	12.24	4.03	0.11	68.13	0.24	14	1.13	n/a	99.88
	12.6	3.93	0.23	68.5	0.37	13.34	1.14	n/a	100.11
	12.97	4.95	0.1	67.89	0.36	13.2	1.21	n/a	100.68

(continued)

Table 3.7 (continued)

CGG	Electron probe microanalyses of chrome-spinelide inclusion content in oxides								
	12.67	4.89	0	67.34	0.36	13.77	1.6	n/a	100.63
	13.02	4.77	0.1	66.6	0.32	13.22	2.11	n/a	100.14
	12.73	4.92	0	67.07	0.3	13.78	1.78	n/a	100.58
	13.13	6.51	0	63.96	0.37	13.14	2.32	n/a	99.43
	12.17	4.34	0	66.74	0.45	14.36	2.11	n/a	100.17
	13.4	5.09	0.17	65.39	0.26	12.52	2.41	n/a	99.24
	13.6	4.27	0.04	65.9	0.23	12.25	2.84	n/a	99.13
	14.7	6.61	0.19	66	0.25	11.3	2.11	n/a	101.16
	13	4.48	0.06	65.3	0.22	13.07	2.7	n/a	98.83
	12.8	4.64	0.06	65.4	0.21	13.53	2.63	n/a	99.27
	14.2	6.04	0.07	66.1	0.23	11.18	1.35	n/a	99.18
	13.4	6.15	0.08	66.1	0.22	12.29	0.91	n/a	99.15
	12.5	4.36	0.3	66.4	0.23	13.78	2.03	n/a	99.59
	14.1	7.62	0.04	64	0.18	11.51	1.33	n/a	98.78
	13.7	7.28	0.01	65.2	0.2	11.77	0.48	n/a	98.64
	14.2	7.77	0.08	63.4	0.2	11.57	1.82	n/a	99.04
	13.9	7.85	0.07	63.1	0.19	12.2	2	n/a	99.31
	14.4	5.87	0.04	65.2	0.25	11.27	2.48	n/a	99.51
	14	7.47	0.02	63.5	0.19	11.52	1.53	n/a	98.23
	12	4	0.36	65.6	0.22	14.41	2.44	n/a	99.03
	12.7	3.78	0.34	65.9	0.24	13.32	2.65	n/a	98.93

(continued)

Table 3.7 (continued)

CGG	Electron probe microanalyses of chrome-spinelide inclusion content in oxides								
	13.9	6.66	0.18	64.2	0.23	11.87	2.03	n/a	99.07
	12.9	8.36	0.13	61.5	0.24	13.77	2.26	n/a	99.16
2	15.5	6.9	0	63.7	0.55	9.1	2.56	n/a	98.31
	14.88	7.4	0	63.73	0.47	10.51	2.4	n/a	99.39
	15.3	6.25	0.25	64.5	0.51	9.3	2.44	n/a	98.55
	16.3	2.98	0.11	67.5	0.16	10.55	6.39	n/a	103.99
V. Grib pipe									
3	11.89	3.74	64.82	0.33	0.31	14.37	3.40	n/a	98.86
3	11.81	3.81	65.10	0.31	0.36	14.37	2.95	n/a	98.71
3	11.83	3.82	65.63	0.32	0.36	14.13	2.87	n/a	98.96
2	11.28	4.17	64.40	0.34	0.30	15.70	3.42	n/a	99.61
2	11.18	4.23	64.60	0.36	0.32	15.62	3.05	n/a	99.36

Chrome-spinelide composition points have been plotted on the diagram proposed by N.V. Sobolev (1974), (Fig. 3.20).

Chromite inclusions were detected by Raman scattering method at frequencies within 685–720 (strong line), 565–590 (medium line) (Fig. 3.19).

Studied inclusions of spinelides are represented by high chromium picrochromite and medium chromium picroferrochromite. All inclusions can be divided into three groups by distribution of basic oxides.

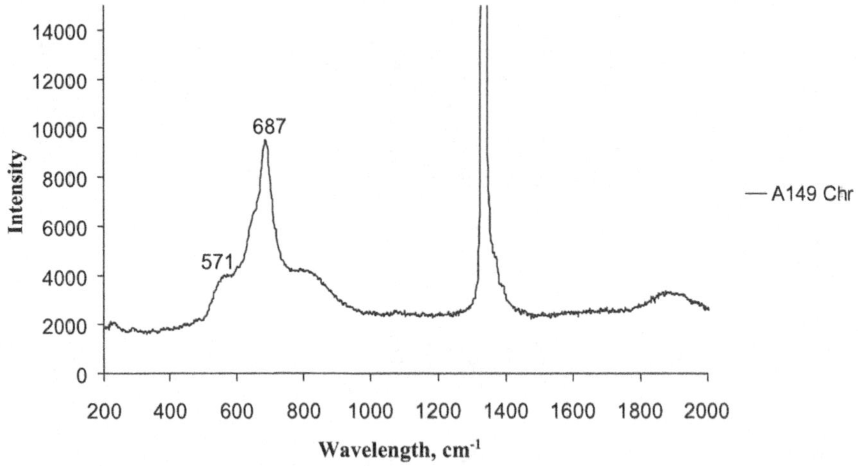

Fig. 3.19 Raman spectrum of chromite in diamond

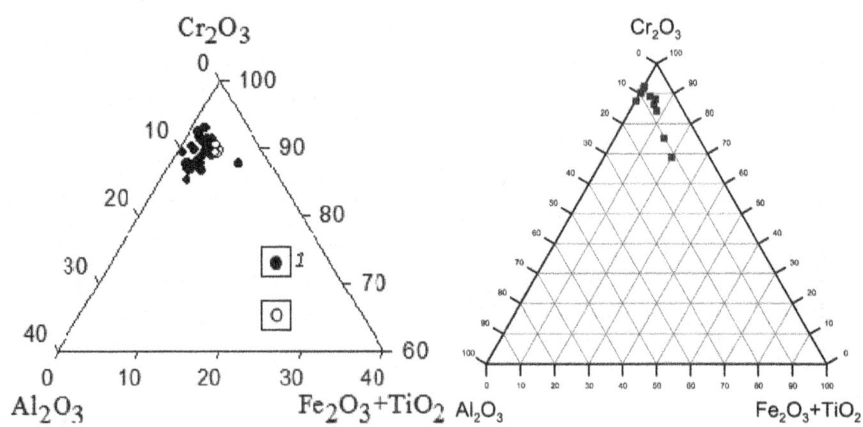

Fig. 3.20 Diagram of chrome-spinelide inclusion compositions (wt%) in coordinates Al_2O_3 – ($Fe_2O_3 + TiO_2$) – Cr_2O_3 in diamond from kimberlite pipes: V. Grib deposit (left); M.V. Lomonosov deposit: Arkhangelskaya pipe (right)

Group I combines 5 analyses (Table 3.7, An. 1–5) of high chromium picrochromites. Typical features of the group: high chromium content (63.37–70.53 wt% of Cr_2O_3), low alumina content (4.66–7.47 wt% of Al_2O_3) and high titanium content (0.06–0.16 wt% of TiO_2), and also prevalence of ferrous iron (11.49–14.1wt% of FeO) over ferric iron (0–4.34 wt% of Fe_2O_3).

Group II is represented by 3 analyses (Table 3.7, An. 6–8) of high chromium content picrochromite. Typical features of the group: reduced chrome content versus group I (61.00–68.46% wt% of Cr_2O_3), considerable magnesium content (7.55–15.65 wt% of MgO), alumina (5.49–9.06 wt% of Al_2O_3) and titanium content (0.09–0.29 wt% of TiO_2), considerable prevalence of ferrous iron (8.57–16.62 wt% of FeO) over ferric iron (0–5.36 wt% of Fe_2O_3). The distinctive characteristic of this group is elevated zinc content **(1.19–3.64 wt% of ZnO).** Chromites of this group under study are represented by syngenetic (specimen A192) and protogenetic (specimen A108) inclusions, brought onto the surface by cleavage, a chromite-diamond aggregate (specimen A192). Chromites were probably exposed to epigenetic transformations that resulted in zinc edging formation.

Group III is represented by 2 crystals of diamonds with medium chromium picro-ferrochromite (Table 3.7, An. 9 and 10). Typical features of the group: even lower chromium content (48.71–52.26 wt% of Cr_2O_3), **elevated titanium content (2.18–2.37 wt% of TiO_2)**, higher total ferruginosity with prevalence of design ferrous iron (13.49–14.70 wt% of FeO) over ferric iron (7.96–11.78 wt% of Fe_2O_3), plus reduced magnesium (11.87–14.1 wt% of MgO) and aluminum (7.12–7.93 wt% of Al_2O_3) content compared to spinelide of the first two groups.

Variation limits of the Al_2O_3, Cr_2O_3, TiO_2 contentin chromites of group I are typical of chrome-spinelides from the Arkhangelsk kimberlites. Elevated zinc content was observed in group II (1.19–3.64 wt% of ZnO). The over-baseline enrichment of ZnO chromites is probably stipulated by epigenetic (metamorphic, metasomatic) transformations. There was elevated titanium content in chromite (2.18–2.37 wt% of TiO_2) in group III. Subject to published data (Bogatikov et al. 1999; Garanin et al. 1991) high chromium (54–63 wt% of Cr_2O_3) and high titanium (up to 1.5 wt% of TiO_2) chrome-spinelides and grains with reduced alumina (8.5 wt% of Al_2O_3) content prevail among mantle associates of diamond.

Chrome-spinelide inclusions in diamond from the Karpinskogo-1 pipe are oval-shaped, average size up to 0.5 cm. Mineral chemistry is close to that from other pipes of the Lomonosov deposit and the Grib pipe. Chrome-spinelide is characterized by considerable content of Cr_2O_3 at 61.10–68.82 wt%, TiO_2 at 0.31–0.36 wt%; low content of FeO at 9.1–15.32 wt% and TiO_2 at 0–0.36 wt% (Table 3.7).

By chemistry properties it is similar to chrome-spinelide from diamond inclusions, high diamond-bearing dunites and harzburgites of CGG 1 (Garanin et al. 1991) (> 62.50% wt% of Cr_2O_3).

Chrome-spinelide inclusions meet the diamond potential criterion ($Cr_2O_3 > 62$, $Al_2O_3 < 8$, $TiO_2 < 0.7$ wt%), proposed by N.V. Sobolev (1974). Cr/(Cr + Al) high relationships are evidence of endogenous origin of diamond. $Fe^{3+}/(Fe^{3+} + Fe^{2+})$ low relationships are evidence of low fugitiveness of diamond-forming environment oxygen (Bogatikov et al. 1999).

Chrome-spinelide inclusions in diamonds from the V. Grib pipe are similar by their chemistry characteristics to chrome-spinelide from diamond inclusions, high diamond-bearing dunites and harzburgites (CGG 1, > 62.50 wt% of Cr_2O_3) (Garanin et al. 1991).

Chrome-spinelide inclusion in diamond is oval and pear-shaped, 160 × 100 μm in size. The mineral chemistry is close to that of chrome-spinelide from the Lomonosov pipe diamond. Chrome-spinelide is characterized by substantial content of Cr_2O_3 at 64.10–64.82 wt%, TiO_2 at 0.31–0.36 wt%; low content of Al_2O_3 at 3.74–4.23 wt%, MgO at 11.18–11.89 wt% (Table 3.7) as compared to chemistry of this mineral in inclusions from the Lomonosov pipe diamond.

Chrome-spinelide inclusions in diamonds from the Grib pipe meet the diamond potential criterion $Cr_2O_3 > 62$, $Al_2O_3 < 8$, $TiO_2 < 0.7$ wt%) proposed by N.V. Sobolev (1974). High Cr/(Cr + Al) ratios evidence the diamond endogenous origin (Fig. 3.20). Low $Fe^{3+}/(Fe^{3+} + Fe^{2+})$ ratios evidence low fugitiveness of diamond-forming environment oxygen (Bogatikov et al. 1999).

Two inclusion groups were found in one diamond crystal from the V. Grib pipe: homogeneous inclusion of high chromium (up to 65% Cr) chrome-spinelide situated in central region of diamond; association of magnesium-rich olivine (forsterite) and orthopyroxene (enstatite) inclusions which placed adjacently in the crystal transition region.

Aggregated inclusions of olivine and chrome-spinelides were found in three sawn crystals. Diamond-formation P&T parameters were determined using a thermobarometer (Nickel and Green 1985; O'Neill and Wall 1987) by this pair of minerals. Assuming that duration of diamond stay in the Earth mantle was 3.0 billion years, formation of dunite-harzburgite association diamond from the Grib pipe is supposed to had taken place in the depth of 150 km under pressure 45–46 kbar and at temperature about 1100–1250 °C.

3.4.5 Hematite

Epigenetic hematite was identified in some crystals of the Lomonosov deposit based on RS spectra: characteristic peaks 226, 292, 413 and 612 cm^{-1}, it appears as a black plate-like lining rimming diamond inclusion in diamond (Fig. 3.21).

3.4.6 Carbonate

Carbonate inclusions were found in a diamond from the Arkhangelskaya pipe. Carbonate is represented by colorless tabular crystals. According to Raman spectroscopy studies, the inclusions reveal a series of sharp peaks: 215, 327, 740 and 1093 cm^{-1} (Fig. 3.22), and peak shifting is observed in relation to pure magnesite

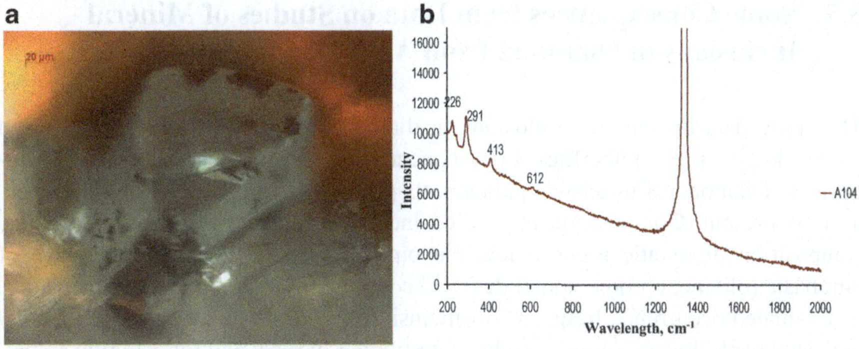

Fig. 3.21 Hematite inclusion in diamond: **a** photo image: black plate-like hematite lining rimming diamond-in-diamond inclusion (sp. 105), optical microscope, transmitted light; **b** Raman spectrum of hematite in diamond

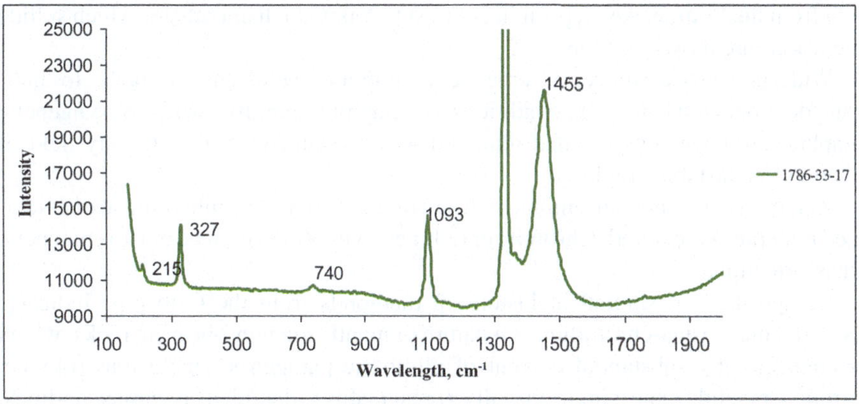

Fig. 3.22 Raman spectrum of magnesite inclusion and plate-like graphite inclusion in diamond

peaks; such shifting is attributed to isomorphous replacement of Mg for heavier Ca. Position of the lines suggests that carbonate is represented by magnesite.

Chromite and graphite inclusions became most widespread in studied specimens. Occurrence evaluation of inclusions in diamond showed the following sequence: chromite → olivine → garnet which matches occurrence evaluations of inclusions in diamonds of Yakutia.

In most cases inclusions are represented by single inclusions or several homonymous ones in one diamond crystal. Inclusions of different minerals were less often in one crystal. Moreover, the following mineral associations have been detected: olivine + graphite, olivine + carbonate + graphite, olivine + chromite + graphite, pyrope-almandine garnet + chromite, chromite + diamond.

3.5 Some Consequences from Data on Studies of Mineral Inclusions in Diamond from ADR Deposits

The study data on mineral inclusions in diamond from the Arkhangelskaya and Karpinskogo-1 pipe kimberlites suggest occurrence of diamonds based on two parageneses. Diamond of ultrabasic paragenesis prevails, diamond of eclogitic paragenesis is present. Composition of studied inclusions indicates magnesian-alumina composition of mantle rocks below the pipes, which is proven by presence of ultrabasic (olivine, chrome-spinelides) and eclogitic (pyrope-almandite) inclusions.

Evaluated occurrence frequency of inclusions in ultrabasic association diamond indicated the following sequence: chromite-olivine-pyrope; and for eclogitic association: carbonate-high-calcium pyrope. The same distribution is typical of inclusions in diamonds from Yakutia (Bulanova et al. 1990).

Sulfide inclusions were not found in the studied diamonds, probably due to low sulfur content in original protolith, which is confirmed by mantle xenolith composition from the Lomonosov deposit in contrast to Yakutian diamonds for which sulfide inclusions are most common.

Widespread occurrence of epigenetic graphite inclusions is likely to indicate diamond annealing in conditions of graphite stability, whereas syngenetic graphite presence suggests that diamond was crystallized in the stability field of diamond-metastable graphite.

Zinc-rich chromites, chromium-rich and titanium-rich chromites typical of associated minerals were found. Chromite enrichment with ZnO might be due to epigenetic transformations.

Composition of studied inclusions in diamonds from the Grib pipe indicates essential magnesian-chromium-rich nature of mantle diamond-bearing rocks, which is confirmed by substantial content of ultrabasic paragenesis inclusions (olivine, chrome-spinelide) and single visually (presumably) identified as garnet findings, and absence of ilmenite. Relying on P&T parameter values we may conclude that formation of the studied syngenetic mineral inclusions in the diamond occurred in the field of diamond thermodynamic stability in acute reducing conditions of diamond-forming environment (Bogatikov et al. 1999). Using olivine and chrome-spinel aggregates we determined formation of the diamond originated from dunite-harzburgite association at $P = 45$–46 kilobar, $T = 1100$–1250 °C.

Single sulfide inclusions of epigenetic origin were registered in diamonds from the Grib pipe: present in fine aggregates with silicate phases (olivine, altered olivine) generally rhomb-shaped, submitting to crystallographic orientations of diamond. They follow the diamond crystallization laws, by composition can be made by high-nickel phases (up to 60%), presumably millerite, and iron-nickel phases (iron content—20%, nickel content—36.0%), presumably pentlandite.

References

Argunov, K.P.: Diamonds of Yakutia – Novosibirsk, p. 402. SO RAN Publ, Geo Branch (2005)

Bartoshinsky, Z.V.: Crystal morphology of diamond from kimberlites of Arkhangelsk diamondiferous province. Miner. Sb. **2**(46), 64–73 (1992)

Beskrovanov, V.V.: Diamond Ontogeny, p. 264. Nauka, Novosibirsk (2000)

Beskrovanov, V.V., Spetsius, Z.V.: Morphology and physical properties of diamonds from mantle xenoliths. Miner. J. **13**(5), 31–41 (1991)

Biellmann, C., Gillet, Ph.: High-pressure and high-temperature behaviour of calcite, aragonite and dolomite: a Raman spectroscopic study. Eur. J. Miner. **4**, 389–393 (1992)

Blinova, G.K.: Structural impurities as growth mechanism indicators in natural diamond crystals. Rep. AS USSR **294**(4), 868–871 (1987)

Bogatikov, O.A., Garanin, V.K., Kononova, V.A., et al.: Arkhangelsk Diamondiferous Province. MSU Publishing (1999), 522p

Bokii, G.B., Bezrukov, G.N., Klyuev Yu.A., et al.: Natural and Synthetic Diamonds. Nauka (1986), 224p

Boyd, S.R., Kiflawi, I., Woods, G.S.: The relationship between infrared absorption and the A defect concentration in diamond. Phil. Mag. b. **69**, 1149–1153 (1994)

Boyd, S.R., Kiflawi, I., Woods, G.S.: Infrared absorption by the B nitrogen aggregate in diamond. Phil. Mag. b. **72**, 351–361 (1995)

Bulanova, G.P., Spetsyus, Z.V., Leskova, N.V.: Sulphides in Diamonds and Xenoliths of Yakutian Kimberlite Pipes, p. 120. Nauka, Novosibirsk (1990)

De Weerdt, F., Pal'yanov, Y.N., Collins, A.T.: Absorption spectra of hydrogen in 13C diamond produced by high-pressure, high-temperature synthesis. J. Phys. Condens. Matter **15**, 3163–3170 (2003)

Evans, T.: Aggregation of Nitrogen in Diamond. The Properties of Natural and Synthetic Diamond, pp. 259–290. Academic, London (1992)

Garanin, V.K., Kudryavtseva, G.P., Marfunin, A.S., Mikhailichenko, O.A.: Inclusions in Diamonds and Diamondiferous Rocks: Monograph. MSU Publishing, Moscow (1991), 240p

Gorobets, B.S., Rogozhin, A.A.: Luminescence Spectra of Minerals. Reference Book (2001), 312p.

Grutter, H.S., Gurney, J.J., Menzies, A.H., Winter, F.: An updated classification scheme for mantle-derived garnet, for use by diamond explorers. Lithos **77**, 841–857 (2004)

Khachatryan, G.K.: Classification of diamonds from kimberlites and lamproites according to distribution of the nitrogen centers in crystals. Ores Miner. **2**, 46–60 (2010)

Khachatryan, G.K.: Improved method of evaluating nitrogen concentrations in diamond and its practical application. In: Geological Aspects of Mineral Resources Base at ALROSA: State-of-the-art, Prospects, Solutions, pp. 319–322. Mirny (2003)

Khachatryan, G.K., Palazhchenko, O.V., Garanin, V.K., Ivannikov, P.V., Verichev, E.M.: Genesis of "Non-equlibrated" Diamonds Crystals from Karpinskogo-1 Pipe According to Cathode Luminescence and IR Spectroscopy Data, No. 2. pp. 38–45. Newsletter of Moscow University, Series 4. Geology (2008)

Khachatryan, G.K., Verichev, E.M., Garanin, V.K., Garanin, K.V., Kudryavtseva, G.P., Palazhchenko, O.V.: Distribution of Structural Defects in Diamonds from the V.P. Grib Pipe (Arkhangelsk Diamondiferous Province), No. 6, pp. 29–37. Newsletter of Moscow University, Series 4. Geology (2006)

Khokhryakov, A.F., Pal'yanov, Yu.N.: Morphology of diamond crystals dissolved in the aqueous silicate melts. Miner. J. **12**(1), 14–23 (1990)

Khokhryakov, A.F.: Experimental study of the formation of rounded diamond crystals. Newsletter of OGGGGN RAS. No. 5. Issue 15. vol. 1, pp. 80–88 (2000)

Klyuev, Yu.A., Naletov, A.M.: The effect of some color centers on the coloration of natural and synthetic diamonds. Superhard Mater. **4**, 61–66 (2008)

Krasnova, N.I., Petrov, T.G.: Genesis of Mineral Individuals and Aggregates. Nevsky Kurier Publishing (1997), 228p.

Kriulina, G.Yu.: Constitutional characteristics of diamond fields of the Arkhangelsk and Yakutian diamondiferous provinces. In: Abstract from PhD in Geology and Mineralogy Thesis, p. 24. MSU Publishing (2012)

Kriulina, G.Yu., Garanin, V.K., Rotman, A.Ya., Koval'chuk, O.E.: Peculiarities of Diamonds from the Commercial Deposits of Russia. No. 3, pp. 23–34. Newsletter of Moscow University, Series Geology (2011)

Kudryavtseva, G.P., Posukhova, T.V., Verzhak, V.V., Verichev, E.M., Garanin, V.K., Golovin, N.N., Zuev, V.M.: Atlas: Morphogenesis of Diamonds and Their Mineral-Satellites from the Kimberlites and Other Relative Rocks from the Arkhangelsk Diamondiferous Province, 1st ed. Polyarny Krug Publishing (2005), 624p

Kvaskov, V.B., Vecherin, P.P., Zhuravlyov, V.V.: Natural Diamonds of Russia, p. 230. Polaron (1997)

Litvin, Yu.A., Bezrukov, G.N., Butuzov, V.P.: Crystallization Mechanisms and Kinetics Minsk, p. 477 (1969)

McMillan, P.F., Hofmeister, A.M.: Infrared and Raman spectroscopy. In: Hawthorne, F.C. (ed.) Spectroscopic Methods in Mineralogy and Geology. Rev. Miner. Min. Soc. Am. **18**, 99–160 (1988)

Mineeva, R.M., Bershov, L.V., et al.: First announcement on peculiarities of paramagnetic centers in diamond crystals from kimberlites of Arkhangelsk Province. Rep. RAS **348**(5), 668–670 (1996)

Mineeva, R.M., Titkov, S.V., Speransky, A.V.: Structural defects in natural plastically deformed diamonds: evidence from EPR spectroscopy. Geol. Ore Depos. **51**(3), 261–271 (2009)

Navon, O., Hutcheon, I.D., Rossman, G.R., Wasserburg, G.J.: Mantle-derived fluids in diamond micro-inclusions. Nature **335**, 784–789 (1988)

Nickel, K.G., Green, D.H.: Empirical geothermobarometry for garnet peridotites and implications for the nature of the lithosphere, kimberlites and diamonds. Earth Planet. Sci. Lett. **73**, 158–170 (1985)

O'Neill, H.S.C., Wall, V.J.: The olivine-orthopyroxene-spinel oxygen geobarometer, the nickel precipitation curve, and the oxygen fugacity of the Earth's upper mantle. Petrol. J. **28**(6), 1169–1191 (1987)

Palazhchenko, O.V., Verichev, E.M., Garanin, V.K., Kudryavtseva, G.P.: Morphological and spectroscopic peculiarities of diamonds from V. Grib deposit from Arkhangelsk diamondiferous province. Article 1. Morphology of diamond crystals. In: Proceedings of IHLs. Series Geology and Exploration, No. 2, pp. 14–22 (2006)

Punin, Yu.O.: Cleavage of crystals. Proc. RMS **110**(6), 666–686 (1981)

Rossman, G.R.: Vibrational spectroscopy of hydrous components. In: Hawthorne, F.C. (ed.) Mineralogical Society of America Reviews in Mineralogy, vol. 18, pp. 193–206 (1988)

Sobolev, E.V., Lisoivan, V.I.: On the nature of intermediate type diamond properties. Rep. AS USSR, **204**(1), 291 (1972)

Sobolev, N.V.: Deep-Seated Inclusions in Kimberlites and the Problem of the Upper Mantle Composition, p. 264. Nauka, Novosibirsk (1974)

Sobolev, N.V.: Nitrogen centers and growth of natural diamond crystals. Problems of Earth Crust and Upper Mantle Petrology, pp. 245–255. Nauka, Novosibirsk (1978)

Taylor, W.R., Jaques, A.L., Ridd, M.: Nitrogen-defect aggregation characteristics of some Australasian diamonds: time-temperature constraints on the source regions of pipe and alluvial diamonds. Am. Miner. **75**, 1290–1310 (1990)

Taylor, W.R., Milledge, H.J.: Nitrogen aggregation character, thermal history and stable isotope composition of some xenolith-derived diamonds from Roberts Victor and Finch. In: Extended Abstract of the 6-th International Kimberlite Conference, Novosibirsk, pp. 620–622 (1995)

Vins, V.G., Yelisseyev, A.P.: Effects of growth conditions on spectral characteristics of type Ib synthetic diamonds. Promis. Mater. **6**, 36–42 (2009)

Vins, V.G., Yelisseyev, A.P.: Effects of HPHT annealing on the defect-impurity structure of natural diamonds. Promis. Mater. **1**, 49–58 (2010)

Vins, V.G.: Optically active defects in diamond—regularities of origination and mutual transformation. In: Abstract of DSc in Physics and Mathematics Thesis, p. 40. Altay State Technical University, Barnaul (2011)

Yelisseyev, A., Kanda, H.: Optical centers related to 3d transition metals in diamond. New Diamond Front. Carbon Technol. **17**(3), 127–178 (2007)

Zaitsev, A.M.: Optical Properties of Diamond: A Data Handbook, p. 502. Springer Verlag, Berlin (2001)

Zedgenizov, D.A.: Composition and evolution of crystallization media of fibrous diamonds of lithospheric mantle of the Siberian Craton. In: Abstract of DSc in Geology and Mineralogy Thesis, p. 416. Novosibirsk (2011)

Zinchuk, N.N., Koptil, V.I.: Typomorphism of Diamonds of the Siberian Platform, p. 603. Nedra-Biznestsentr OOO (2003)

Chapter 4
Recent Data on Diamond from the M. V. Lomonosov Deposit

In recent years, cubic habit diamonds have been attracting most of the attention. As turned out in detailed studies of their internal structure and constitutional characteristics, their formation passes several stages. Simultaneous complex transitions evolve on color parameters or morphological characteristics in one crystal. The detailed study of similar diamond crystals from the M. V. Lomonosov deposit allows to take note of newly revealed features of cubic diamond formation, which surely advances our data on diamond and its genesis.

4.1 Cubic Habit Diamond from M. V. Lomonosov Deposit

Widespread occurrence of cubic diamonds in kimberlite pipes is a typomorphic feature of the M. V. Lomonosov deposit (Kudryavtseva et al. 2005). Specimens from production sample drawn in the Arkhangelskaya pipe were studied in the paper; prior to that, we studied diamonds with similar characteristics from the Karpinskogo-1 pipe.

It should be noted that there is a wide diversity of cubic habit crystals in the Lomonosov deposit. Uniqueness of these diamonds is evident, since the very first mineralogical characterization disclosed these could be placed into none of the varieties identified by Yu. L. Orlov in the mineralogical classification (1984). He described 3 varieties where crystals can have cubic habit: II (visually homogenous transparent yellow cubes with fibrous structure), III (translucent gray crystals with zonal-sectorial internal structure), IV (crystals with a translucent colorless octahedral core formed in the layered growth, and non-transparent fibrous coat forming a cubic habit crystal in strong growth). This paper presents the results of studying internal structure in several samplings of diamond cubic crystals, whereby crystals make rows of transition from one sampling to another (either by color or morphological characteristics) in terms of their properties, and the sharpest distinctions are shown by end members of the rows.

V. Garanin et al., *Diamonds from the Arkhangelsk Province, NW Russia*, Springer Mineralogy, https://doi.org/10.1007/978-3-030-35717-7_4

Fig. 4.1 Photos of crystals typomorphic for southern group of the M. V. Lomonosov deposit bodies

Transparent visually homogenous yellow, orange-brown cubes of varying color intensity with smooth and saddle-like faces, combined crystals with cubic and tetrahexahedroid faces (Fig. 4.1) were selected for study from all the diversity of crystals. Under Yu. L. Orlov's classification (1984) accepted in Russia, these cubes belong to variety II. Weight of crystals amounts to 0.04–0.40 ct. The diamonds were ground a little more, parallel to the plane (100) by 1/2 of the crystal body to uncover their internal structure.

By internal structure and defect-impurity composition visually homogenous yellow-colored cubic habit crystals from the M. V. Lomonosov deposit are subdivided into three morphogenetic groups (Table 4.1).

Subsidiary absorption lines in the region 1350–1600 (Fig. 4.2) *are registered in many yellow-colored crystals of both cubic and dodecahedral habit in diamonds from the Arkhangelskaya and Karpinskogo-1 pipes.* In the sources the lines are interpreted as associated to hydrogen, but these assumptions yet have not been corroborated by experiments.

The diamonds are characterized by high concentrations of nitrogen in A, C, N3 forms but contain no nitrogen in complex aggregated B1 and B2 forms of absorption systems; typical crystals are given in the figure (Fig. 4.2).

The ***1st group of diamonds*** comprises cubes with smooth faces or combination of tetrahexahedroid and cube faces. The color is from yellow to orange-brown. There crystals have high content of nitrogen in the form of C-center (up to 275 at. ppm), with no A- and B-centers (or a C-center with insignificant evidence of A-centers). The group is represented by 10 specimens (crystals 1785-2-5, 1785-2-6, A186, A199, A94 and others). Luminescence in cathode rays is pale-blue and faint, photoluminescence—fancy yellow, yellow-green, orange. The spectrum contains 350, 396, 570 nm bands. The diamonds are placed into variety II by C-defect prevalence, under classification by Yu. L. Orlov. The 1st group diamonds are characterized by parallel bands

Table 4.1 CCL, IR spectroscopy and isotopic composition data of carbon in cubic habit diamond from the Arkhangelskaya pipe

Crystal group	Crystal No.	Crystal photo	CCL photo	Internal structure	Crystal zone	IRS data, at.ppm						Physical class.	$\delta^{13}C$,‰	PLS data	
						N_C	N_A	N_B	P	RzP,cm^{-1}	H			N3	Other peaks
I	A94; A1791; 1785-2-6		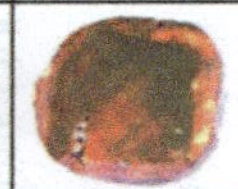	Homogenous with plastic deformation bands	Whole body	178	0	0	0	-	0	Ib	-4,9	N3	350, 396, 570
	Coat A186; A1786	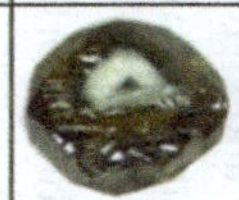	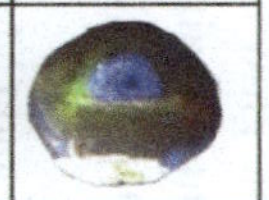		Core	Non-transparent							-4,9	N3,S3	
					Edge	275	0	0	0	-	1		-4,9	N3,S3	
II	1786-2-26-2; 1787-3-4; A19	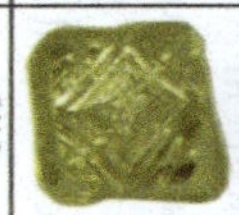		Homogenous	Core	80	360	0	-	-	1,6	Ia+Ib	-4,1	N3,H3	512, 503-570
					Edge	87	726	0	-	-	2,0		-4,4	N3,H3	
III	A26; 1791-5-2; 1787-26-2			Zoning, by cube	Core	0	1100	497	5	1378	4/27	IAB	-9,8	N3,H3 S3	313, 319,347*, 396-400, 420, 500, 649, 471*, 488*, 650*, 671*, 517, 536, 570 * only in crystal with H3
					Edge	0	785	265	5	1378	4/19		-8,4		
	A2; 1786-27	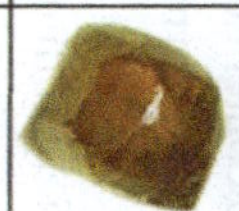		Cubic kernel	Core	0	754	212	4,7	1368	1,3	IAB	-5,4	N3	482, 517
					Edge	0	1000	0	-	-	1,5		-5,4	N3	
	A29			Octahedral kernel	Core	0	950	130	0,6	1362	3	IAB	-7,1	N3	317, 400, 482, 515
					Edge	0	1200	221	-	-	2,2		-5,3	N3	
				Multiple zoning		0	784	265	4	1378	1		-8,4	N3	430, 517, 522

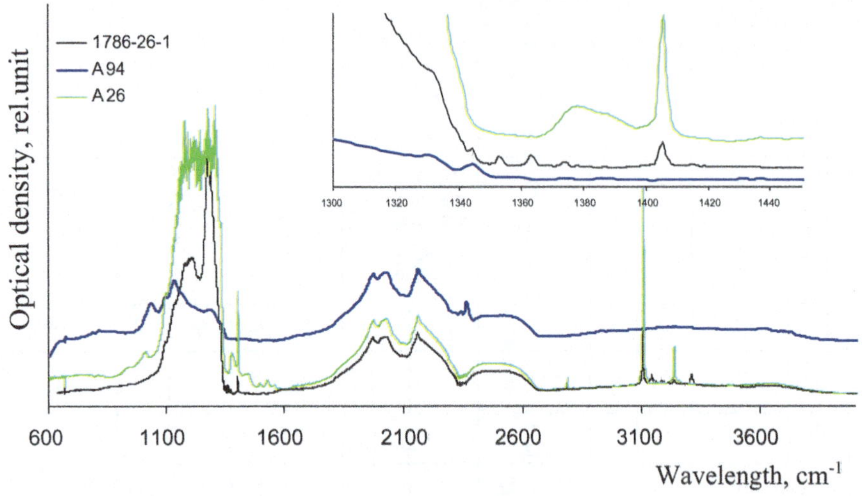

Fig. 4.2 Optical density spectra of diamonds with C-defects (A94, resolution 4 cm^{-1}), A- and C-defects (1786-26-1, resolution 1 cm^{-1}), with A-, B- and C-defects (A26, resolution 1 cm^{-1}). The 1300–1440 cm^{-1} band has been zoomed in the tab

(parallel-oriented){111} and red glow in CCL, related to plastic deformation marks registered in EPR spectra (Fig. 4.3) by N2, M2 centers (Makeev et al. 2011). The crystals have homogeneous fibrous structure or act as a thick coat for primary seed of cubic or octahedral shape. Carbon isotopic composition is 4.9‰, i.e. the cubes reveal a substantial share of deep mantle carbon (Galimov, 1984; Galimov et al, 1994).

The ***2nd group of diamonds*** comprises transparent yellow diamonds of varying color intensity and cubic habit with saddle-like faces (with large deep negative tetragonal step pyramids), less frequently with additional tetrahexahedroidal surfaces (crystals A19, 1786-26-1, 1787–3-4).The crystals have blue glow in CCL, are characterized by fibrillose homogeneous internal structure. Photoluminescence glow is weak and pale caused by N-3 defect (415 nm) and weak manifestation of H-3 defect in the 503–570 nm band.

They exhibit average content of nitrogen defects in A- and C-forms (300 < Ntot < 1000 at. ppm), whereby A-form nitrogen drastically dominates, B- and P-centers are missing.

These are typical variety II diamonds under classification by Orlov (1984). Apart from known absorption systems A, B, P, 3107 cm^{-1}, poorly studied bands appear at frequencies 1353, 1363, 1374 cm^{-1} in cubic crystal absorption spectra without P-system; moreover 1363 cm^{-1} band occasionally occurs individually (Vasiliev et al. 2011). The example of the spectrum with these bands is given in Fig. 4.4. The 1363 cm^{-1} band is not identical to the P-band as it has smaller width; moreover, the P-system is situated in the 1380–1370 cm^{-1} range in low absorption coefficients. The nature of these absorption bands requires further investigation. The series of

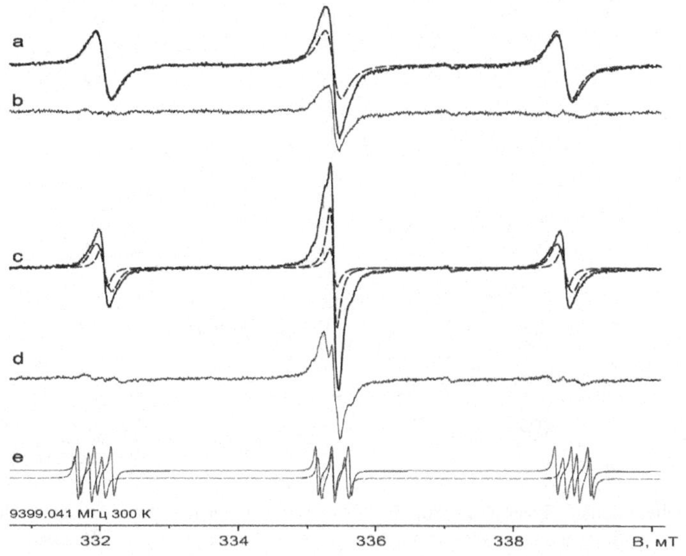

Fig. 4.3 Model components (dashed lines in a and c of ERP spectra) at B||[001] of defective diamonds A199 and A94 (**a**, **b**), sp. A186 (**c**, **d**); **c** and **d** are residual spectra after removal of P1-center signals, full lines are experimental spectra, dashed lines are P-1 center design spectra. For comparison, the data were calculated for B alignment ||[001], EPR spectra: M-2-(full line) and M3-centers (dashed line) of plastically deformed diamonds as reported by Mineeva et al. (2007)

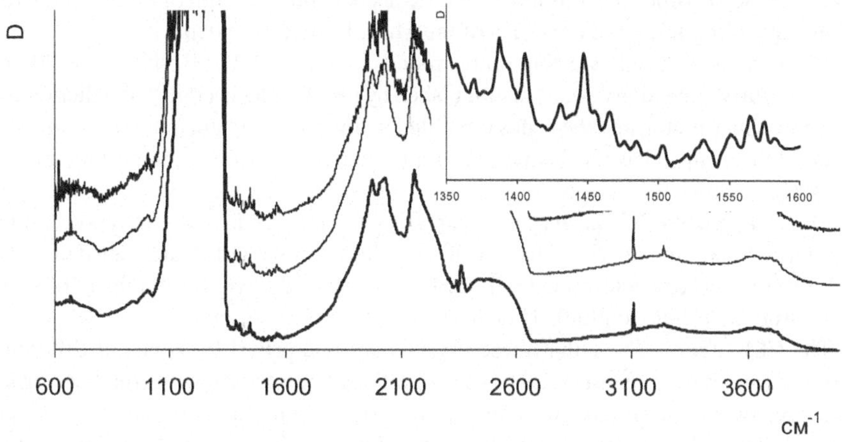

Fig. 4.4 IR absorption spectra of crystals with a set of absorption bands in the 1350–1600 band (in the insertion). Diamonds from the Karpinskogo-1 pipe, sp. К3–5, 3–20, 3–10

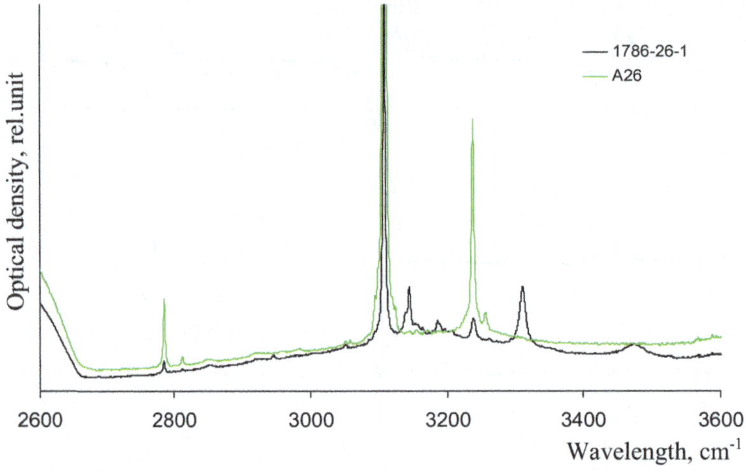

Fig. 4.5 Optical density spectra of diamond 1786-26-1 with a set of narrow bands in the range of CH-group fluctuations and normal absorption spectrum in this range, cryst. A26, resolution 1 cm^{-1}

narrow bands at frequencies 3310, 3188, 3144, 2945 cm^{-1} is registered in spectra of some cubic yellow crystals.

The typical absorption spectrum with indicated subsidiary peaks, and the absorption spectrum with the prevailing system 3107 cm^{-1} is given in Fig. 4.5. The presence of these absorption bands makes the IR-absorption spectrum of these diamonds unique, enabling to identify crystals of the M. V. Lomonosov deposit.

The sources (Orlov 1984; Natural Diamonds of Russia 1997; Beskrovanov 2000) contain data on subsidiary peaks distinguished by researchers in cubic habit diamonds from pipes in Yakutia, and the peaks identified in this paper remain unique. Diamonds of the 2nd group have the heaviest isotopic composition ($\delta^{13}C$ center/margin = 4.1/4.4‰).

The ***3rd group of diamonds*** is apparently characterized by the longest mantle evolution. Such cubic crystals have yellow, yellow–brown color, are transparent of translucent, a core is determined only by CLL expressed in various intensity of blue glow (from pale-blue to blue). They have complex zonal structure.

The CLL figures show that these crystals are comprised by zones of different geometry and have a seed-crystal (a very small octahedral core) with zonal-sectorial structure. In the worst case we can see multistage shape transformation: a small octahedral core has layered octahedral zoning; a central cubic zone was formed by mixed growth mechanism (with simultaneous development of cubic and octahedral sectors and gradual outcrop of the latter); it was followed by crystal homogeneous growing under normal mechanism with gradual nitrogen concentration drop; abrupt outcrop of cubic growth sectors and acquiring octahedral shape took place in the fourth stage, whereupon a break in diamond crystallization was apparently observed; in the final diamond crystallization stage a thick coat with normal growth mechanism was formed to give final cubic appearance to the diamond.

In the central zone there is a core of octahedral or cubic shape which contains nitrogen defects in the forms of A- and B-centers. The external zone of these crystals (coat) is formed in normal or mixed growth mechanisms. Under IRS data, all the group 3 crystals are high-nitrogen: 900 < Ntot < 1750 at. ppm (cryst. A26, A29, 1791-5-2, 1786-26-2, 1786-27, 1794-1, 1794-2, A141). The following pattern is noted: crystals with a coated cubic seed contain only A-defect; diamonds with multiple straight zoning by cube in the coat and in the core have A- and B defects; and specimens with an octahedral core in fibrous coat have defects in the form of A-, B-, C-centers. The 1362–1378 cm^{-1} band is registered in the IR spectrum of 30% of group 3 crystals. Zones with normal, tangential and mixed growth mechanism in some crystals recur 2–3 times.

Luminescence is very bright and varied in color. Structural complexity is reflected in the PL spectra that apart from the basic bands (N3, S3, H3) contain bands with maximums of 313, 347, 396, 411, 482, 500, 649, 671 nm. These crystals can be placed into variety III by PL and IR spectra under Yu. L. Orlov's classification, and to the variety IV by core presence. The lightest composition of carbon ($\delta^{13}C = -9.8$ ‰) is observed in cubes with normal growth mechanism and straight zoning by cube in CCL rays. Insignificant increase in carbon isotopic composition weight is noted in zonal crystals towards the margin ($\delta^{13}C$ center/margin =−7.1/−5.4‰ and −9.8/−8.4‰).

Cubic habit diamond crystals with thick non-transparent coat and deep encircling etch channels were additionally studied. These specimens have yellow, yellow-green, gray colors.

The diamond samplings viewed in this paper exhibit similarities to / differences from cubic habit crystals from non-placers and placers of Yakutia. Cubic habit diamonds are endemic for ore bodies in the Yakutian Diamondiferous Province (Afanasiev et al., 2011): these are absent in pipes of the Malobotuobinsky, Srednemarhinsky diamondiferous regions, occur in small quantities (less than 3%) in the bodies of the Daldyn-Alakit region (Udachnaya, Yubileynaya, Komsomolskaya pipes) and are typomorphic placers in the north-east of the Siberian craton (Titkov et al. 2006; Zinchuk and Koptil 2003). Best studied were the diamonds prevalent in placers of the Anabarsky district (Zedgenizov 2011; Titkov et al, 2015), with fancy orange-yellow, orange-brown color due to nitrogen presence only in the form of C-center, and being gem stones (Orlov 1984; Zinchuk and Koptil 2003; Serov and Shelementiev 2008).As reported by Zinchuk and Koptil (2003), these are characterized by light carbon isotopic composition (from −9.9 up to −16.8%), which differentiates them from the M. V. Lomonosov deposit cubes enriched with heavy carbon ($-4.9 < \delta^{13}C < -9.8$‰). Pursuant to (Galimov 1984, Galimov et al. 1994), these are forms of deep-seated carbon brought into the Earth crust during mantle degassing. The differences prove regional typomorphism of cubic habit diamonds and suggest that various sources of carbon engaged in crystal formation in the last stage of diamond crystallization exist in two regions. Such are indicators most typical of ultrabasic paragenesis I and cubes of variety III, under classification by Yu. L. Orlov, but differ in optic-spectrometric characteristics.

Group 1 cubic habit diamond with fibrous structure and high concentrations of C-defect exhibits growth lower temperatures and formation in the latest stage of diamond crystallization. Group 2 cubes with prevalence of A-centers over C were likely formed deeper, with high content of carbon heavy isotope (^{13}C) and nitrogen in the environment; since their concentrations are considerably higher, these crystals have been exposed to short-term annealing and a part of nitrogen passed into non-paramagnetic conditions. A minor increase in ^{12}C isotope content in group 3 diamonds with A- and B-forms of nitrogen defects suggests generation at a relatively earlier diamond crystallization stage, from the source of carbon close by its composition (probably caused by difference of upper mantle isotopic composition), and subsequent short-term annealing that contributed to nitrogen defects in A- and B-forms, and S3 defect occurrence in PL spectra with the preservation of the C-center. By growth conditions, these crystals can be placed between octahedral, dodecahedral diamonds of variety I and cubic ones, described in groups 1 and 2, by the presence of A- and B-centers and high concentrations of nitrogen in A-form.

Formation of outwardly uniform cubic crystals in the Arkhangelskaya and Karpinskogo-1 pipes evidences highly oversaturated conditions in the latest stages of diamond crystallization. High diversity of diamond internal structure forms suggests a complex multi-stage history of their crystallization in conditions multiply changed chemistry of the environment. Abrupt alteration of growth mechanisms, changes of defect concentrations and their transformations (B and P formation) suggest possible breaks and short-term annealing of the crystals (Taylor et al. 1990; Taylor and Milledge 1995).

The following conclusions could be derived based on cubic habit diamond spectroscopy and internal structure studies

- *Visually homogenous yellow-colored cubic habit crystals from the M. V. Lomonosov deposit are characterized by varying internal structure, composition and distribution of structural-impurity defects, which explains their division into three morphogenetical groups: (1) cubes with homogeneous structure contain defects in the form of single nitrogen atoms and deformation centers; 2)crystals with homogeneous internal structure, nitrogen defects in A- and C-forms and subsidiary peaks in ranges 1374-1354 u 3310-2945 cm^{-1}; (3) cubes with a central zone formed under a tangential or normal growth mechanism, with various distribution of nitrogen in C-, A-, B-forms.*
- *Orange-brown color of combined crystals with cube and tetrahexahedron faces is caused by structural defects: high concentrations of individual nitrogens and plastic deformation.*
- *Dominated combined growth mechanism proves high over saturations with hydrogen and hydrocarbons in the diamond crystallization environment, at temperature sufficient to form octahedron growth sectors. The crystal morphological evolution trend (outcrop of the latter and changing octahedral shapes to cubic) proves a decrease in temperature during growth of diamonds in the*

M. V. Lomonosov deposit, potential impact of pressure which results in plastic deformation of crystals.

- *Microinclusions in cubic habit crystals are represented by silicate-carbonate phases with prevalence of essentially carbonate phases.*
- *Formation of outwardly uniform cubic crystals in the Arkhangelskaya pipe proves high carbon oversaturation in the latest stage of diamond crystallization and, relying upon the data on carbon isotopic composition, a mantle source with carbon isotopic composition ($-4.9 < \delta^{13}C < -9.8$‰). A minor increase in weight of carbon isotopic composition from the central zone toward the periphery of the diamond is observed.*
- *Visually same-type (by morphology and color) cubic crystals have different internal structure, which disables using for their discrimination the conventional mineralogical classification by Orlov (*1984*), put into place based on the studies of Yakutia's diamond crystals.*
- *An important feature of some cubic habit zonal crystals is presence of both a monocrystals core and an intermediate polycrystalline porous zone, the latter identified when studying their internal morphology by SEM, CCL, EPR methods. It proves the abrupt alteration of physical and chemical conditions and probable breaks in diamond crystallization.*
- *The results of this study corroborate that complex-zoned and plastically deformed cubes of the M. V. Lomonosov deposit are one-of-a-kind specimens among diamonds from the kimberlite bodies, such specimens are not described in diamonds from utilized pipes of Yakutia.*
- *Cubic, tetrahexahedral and dodecahedral habit diamonds (cubes exposed to dissolution processes) bear maximum resemblance to such crystals from placers of north-eastern Yakutia by mineralogical and optic-spectral characteristics. This enables to predict both similarity in petrochemical composition and affinity in conditions of their ore body formation.*

4.2 Microinclusions in Diamonds

Crystals with reduced transparency (as compared to octahedrons) are present among diamonds from the M. V. Lomonosov deposit in dodecahedral, tetrahexahedral and transitional in-between forms; along with the surface micromorphology this drastically complicates gemological evaluation of color and degree of crystal flaws. The studies indicated that a drop in the diamond transparency is caused by presence of microinclusions scattered in the crystal body.

Photo images in the high-resolution optical microscope were obtained for crystals with natural plane faces, and for crystals with polished surfaces. Dark and light subvisible pin-point inclusions reducing transparency level of crystals were observed in the diamonds. Typical representatives of these crystals are tetrahexahedrons saturated with inclusions (Fig. 4.6).

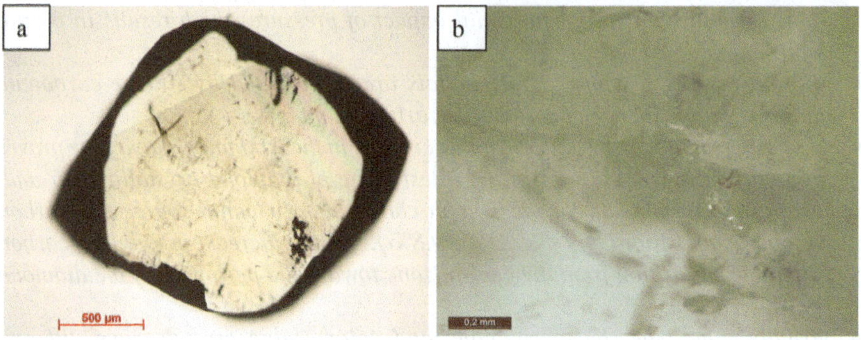

Fig. 4.6 The image of cubic habit diamond crystal, plane-parallel plate as per (110): **a** main view; **b** fragment of crystal with scattered microinclusions

A slight yellow-tinted diamond crystal of dodecahedral habit with a high transparency degree from the Arkhangelskaya pipe may be used as an example. The left part of the photo (Fig. 4.7a) shows a registered macroinclusion in the form of disk-shaped feather with graphite, and in the central zone the surface is smoothest and most suitable to observe microinclusions.

The high-magnification photo shows multiple whitish pin-point inclusions in varying depth inside the crystal body (Fig. 4.7b); the screen display connected to the microscope we can see finer whitish inclusions (Fig. 4.7c).

In the study of diamonds with microinclusions, peculiar features of inclusion chemical phase composition have been correlated to external morphology and major nitrogen center content (Tables 4.2 and 4.3). The diversity of mineralogical forms and internal structure peculiarities of studied crystals are given in Fig. 4.8.

X-ray spectral microprobe analysis with 17–20 microinclusions has been carried out for each crystal prepared in the form of plane-parallel plate. As a rule, these are spectra of inclusions similar in composition and located in the same zone of crystal. The values (Table 4.4) have been averaged to obtain representative composition of inclusions characterizing the crystal.

Octahedron-dodecahedroid series diamonds. Diamonds of the M. V. Lomonosov deposit contain no discernible cloud-like inclusions but are characterized by presence of microinclusions scattered in the body of crystals and reducing their transparency (Fig. 4.7). Based on IR-spectroscopy data, such crystals belong to IaA type under the physical classification. High-nitrogen diamonds (Ntot 1290–2450 ppm), with a medium degree of nitrogen aggregation in B-form (about 20–50%). The crystals stand out for high absorption coefficients on hydrogen peaks (3107 cm^{-1}) up to 25 cm^{-1}. Meanwhile, low-intensity peaks of silicate absorption and 1650 cm^{-1} peaks responding to inclusions of water molecules are registered in crystals' spectra. The spectra have no peaks corresponding to OH forms.

As evidenced by the X-ray spectral analysis, combined habit octahedron-dodecahedroid diamonds are characterized by carbonate–silicate phase composition with higher K-Na component and ferruginosity.

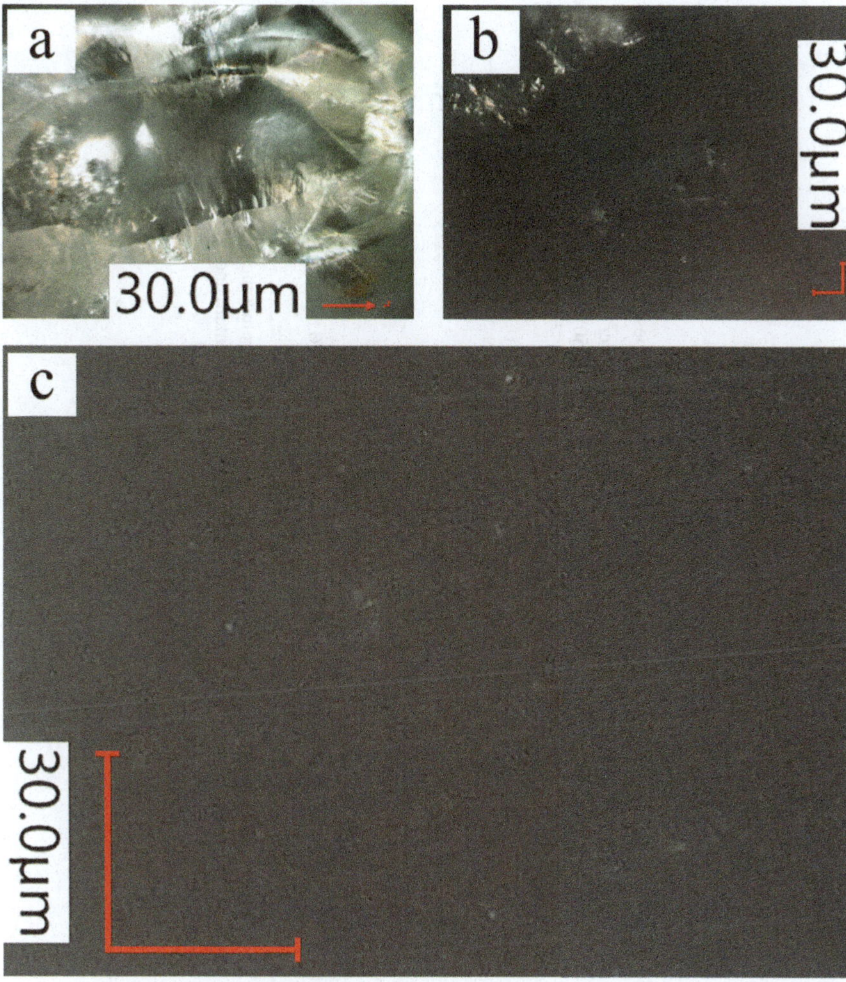

Fig. 4.7 Crystal from the Arkhangelskaya pipe (sp. 1A-4-17) with microinclusions, a photo under optical microscope with different magnification: **a** crystal main view; **b** a zone with microinclusions; **c** magnified fragment with pin-point whitish inclusions 0.5 μm in size

Tetrahexahedral Habit Diamonds Visually homogeneous crystals: from totally transparent to individuals whose transparency might be reduced due to microinclusions scattered in the bodies. The IR absorption spectra analysis established (Table 4.2) that impurity nitrogen total concentration in cubic diamonds varies in a wide range from 170 up to 2100 at. ppm, with an average degree of nitrogen aggregation (%B = 34) and high hydrogen absorption coefficients up to 16 cm^{-1} and carbonate phases. Such combination of structural-impurity defects is dictated by zonal-sectorial structure, which is shown in abnormal double refraction figures.

Table 4.2 Defect-impurity composition of diamonds with nanosized inclusions, Arkhangelskaya and Karpinskogo-1 pipes

Diamond crystal shape	Nitrogen centers						Hydrogen CH (cm^{-1})	Water (1650) (cm^{-1})	Silicates (800), cm^{-1}	Carbonates (1430) (cm^{-1})
	Ntot (ppm)	A (ppm)	B1 (ppm)	%B	P, cm^{-1}	C, cm^{-1}				
Tetrahexahedroid	2143	1403	740	34	–	–	16	–	+	1.5
Octahedron	1800/1290–2450	920/625–1210	950/660–1240	30/20–50	–	–	20/15–25	+	+	–
Coated diamond, thin coat	450	350	98.8	23	5.9	–	0.7	2.3	–	2.5
Coated diamond, thick coat	1040/770–1300	815/700–970	195/64–340	17/9–21	–	+	3.2/0.5–6.6	–	–	1.6/1–5.5
Compact enamel cube	850/710–1000	700/550–800	160/100–190	19/14–25	–	–	0.8/0.4–1.2	1.5/0.8–2.5	1.7/1.3–2.1	0

Note "–" absorption bands are not registered; " + "—weak absorption bands are registered. The numerator—an average value; the denominator—the minimum and maximum values*Note* SO_3, Br—elements are absent. The numerator—an average value; the denominator—the minimum and maximum values

Table 4.3 Average compositions of microinclusions in diamond crystals of different habits, Arkhangelskaya and Karpinskogo-1 pipes

Diamond crystal shape	Average compositions of microinclusions (wt%)											
	Na_2O	K_2O	Cl	Al_2O_3	SiO_2	P_2O_5	MgO	CaO	TiO_2	MnO	FeO	BaO
Tetrahexahedroid	0.0	15.0	0.3	3.4	39.6	9.9	4.8	2.4	8.8	0.0	9.0	6.8
Octahedron	0–4.2	25/10–40	1.2/0–2.5	3.0/0–6	8.8/7–12	1.7/0–4.7	17/10–30	17/2–28	2/2.4–3.3	0–0.2	22/5–40	0.7
Coated diamond, thin coat	4.0	8.6	0.2	6.2	36.5	3.4	12.0	16.2	2.4	0.0	0.0	10.5
Coated diamond, thick coat	5.2/4–5	7.0/5–10.5	1.6/1.2–2.0	1.5/0.7–2.5	11/9.2–12	4.4/4–4.7	27/10–35	23/20–28	3.7/3.3–4.2	0.1	13.5/9–21	1.7/0.7–2.8
Compact enamel cube	1.3/ 0.3–1.6	9/6.8–10	0.5/0–1	9.5/6.5–15	52/49–61	1.7/0.2–2.7	4.5/2.6–5.7	8/5–9.4	2.8/1.2–3.3	0	11/ 10–25	0

Note SO_3, Br—elements are absent. The numerator—an average value; the denominator—the minimum and maximum values

Fig. 4.8 Diamonds of octahedral, tetrahexahedroid and cubic habit with microinclusions, Arkhangelskaya pipe, crystal weights 0.10–0.20 ct

Table 4.4 Representative compositions of microinclusions in diamond crystals, Arkhangelskaya and Karpinskogo-1 pipes

Spec. No	O	Na	Mg	Al	Si	P	S	Cl	K	Ca	Ti	Mn	Fe	Ba	K + Na	Si + Al	Ca + Mg + Fe
Variety II crystals																	
A209	63.4	3.5	15.5	0.2	2.8	1.0	0.0	1.1	2.0	7.1	0.8	0.0	2.3	0.3	**5.5**	**3.0**	**24.8**
1787-3-2	63.2	2.7	5.3	1.1	4.3	1.3	0.0	0.7	4.5	10.0	0.8	0.0	6.0	0.1	**7.1**	**5.3**	**21.3**
1785-4	63.0	3.4	15.1	0.5	3.4	1.1	0.0	0.9	2.3	6.8	0.9	0.0	2.5	0.2	**5.7**	**3.8**	**24.3**
1787-3-1	69.6	1.6	3.5	1.4	8.2	0.2	0.0	0.2	2.8	7.9	0.6	0.0	4.0	0.0	**4.4**	**9.6**	**15.4**
A-207	66.9	3.6	4.7	0.9	5.8	0.4	0.0	0.6	4.8	7.2	0.7	0.0	4.4	0.0	**8.4**	**6.7**	**16.3**
1785-3	59.2	1.0	3.3	3.4	18.7	1.0	0.0	0.5	4.7	4.2	0.9	0.0	3.1	0.0	**5.7**	**22.2**	**10.5**
A-210	71.1	0.2	1.5	2.9	16.7	0.1	0.0	0.0	3.4	2.9	1.0	0.0	6.8	0.0	**3.6**	**19.6**	**11.2**
Variety IV crystals																	
A-213 center	65.1	0.6	1.6	4.8	21.7	0.0	0.0	0.0	4.4	2.2	1.1	0.0	1.8	0.0	**5.0**	**26.5**	**5.6**
A-213 margin	69.3	0.9	1.7	4.9	17.7	0.2	0.0	0.1	3.4	1.2	0.4	0.0	0.4	0.0	**4.2**	**22.6**	**3.3**
1783-313 margin	77.3	0.5	0.3	2.5	16.9	0.0	0.2	0.1	2.6	0.6	0.3	0.0	6.7	0.0	**3.1**	**19.4**	**7.6**
1787-313 center	73.3	1.1	0.6	3.8	18.3	0.0	0.0	0.1	3.6	0.7	2.4	0.0	4.1	0.0	**4.7**	**22.1**	**5.4**
1784-33-5ear1	70.1	2.2	4.6	2.0	9.8	0.7	0.0	0.1	3.1	5.1	0.5	0.0	2.3	0.0	**5.3**	**11.8**	**12.0**
1784-33-5 center	73.0	1.4	3.3	2.2	12.5	0.3	0.0	0.6	4.2	2.9	1.2	0.0	1.0	0.0	**5.6**	**14.7**	**7.1**
1784-33-5 ear2	77.3	0.0	4.1	1.3	6.8	0.1	0.0	0.0	2.4	4.8	0.3	0.0	2.8	0.0	**2.4**	**8.1**	**11.8**
1784-32-2 rim	80.8	0.0	6.9	1.4	1.4	0.0	0.0	0.0	5.1	3.3	0.3	0.0	0.9	0.0	**5.1**	**2.8**	**11.0**
1784-25-1	73.9	0.0	1.6	1.1	10.7	2.3	0.0	0.2	5.1	0.6	2.1	0.0	2.0	2.4	**5.1**	**11.8**	**4.2**
1784-34-7	73.4	0.7	6.5	0.3	4.0	0.4	0.4	1.2	13.5	0.8	0.0	0.0	13.4	1.2	**14.2**	**4.3**	**20.7**

Aluminosilicate phases prevail in microinclusions, with high titanium content (up to 8.8%)—2 to 3 times higher than in crystals of other shapes (Tables 4.3 and 4.4).

Cubic crystals, yellow and non-transparent, so-called "enamel cubes" with a thick coat, are medium-nitrogen (Ntot 710–1000 at. ppm), and the amount of nitrogen defects is somewhat reduced from the core towards the diamond margin. Some crystals of the group are characterized by presence of a minor octahedral core. A-center is the dominating form of nitrogen in the main body and margin zones of diamond.

Cubes are characterized by a low degree of nitrogen aggregation into B-form (14–25%). In IR absorption spectra, the registered hydrogen peaks are minimum by intensity among the studied groups of crystals (0.4–1.2 cm^{-1}), but silicate phase peaks inclusions are specific to them. Prevalence (maximum values among crystals of various shapes) of aluminosilicate microinclusionsis confirmed by the X-ray spectral analysis data: SiO_2 49–61 wt%, Al_2O_3—6.5 to 15 wt%, carbonate phases are noted in negligent quantities.

The study has not identified microinclusions in yellow-colored transparent cubes of near-gem quality.

Mapping of plates by hydrogen defect, carbonate and silicate phase presence indicated local irregular burst of their content. Distribution of these microinclusions does not correlate to the figure of nitrogen and hydrogen impurity distribution in diamond.

Coated diamonds of variety IV are unique by inclusion presence. We shall view the crystals in more detail.

The studied diamond with thin coat fragments has low nitrogen content (Ntot = 450 at. ppm). The crystal core is built under layered growth mechanism, a cubic seed is noted in the center, identified in UV spectrum. Relatively high concentrations of nitrogen centers and platelet presence (P = 5.9 cm^{-1}) are observed in the core. The hydrogen absorption coefficient is 0.7 cm^{-1}. Microinclusion composition meets the carbonate–silicate mixed phases.

Diamonds with thick non-transparent coats from the Arkhangelskaya and Karpinskogo-1 pipes are distinguished by a wide range of hydrogen center absorption (0.5–6.6 cm^{-1}). The highest coefficients of 1.0–5.5 cm^{-1}carbonate phase absorption (at 1430 cm^{-1}) are registered in IR absorption spectra.

By content of major nitrogen centers the diamonds placed into a morphogenetic group of moderately nitrogen low-aggregated crystals: total nitrogen 700–1300 at. ppm, % B from 9 to 21. Minor quantities of C-defects are noted in some crystals (Fig. 4.8).

Arrangement of points on the triangular diagram of microinclusion compositions according to the microprobe analysis results enables to conclude that essentially carbonate trend of microinclusion compositions for cuboids and tetrahexahedroids, and aluminosilicate, silicate-carbonate trend for coated crystals (Fig. 4.9) are specific to the diamonds from the Arkhangelskaya and Karpinskogo-1 pipes.

Variety I diamonds with cloud-like inclusions characterized by growth mechanism change and prevalence of octahedral growth sectors in the final stage of diamond

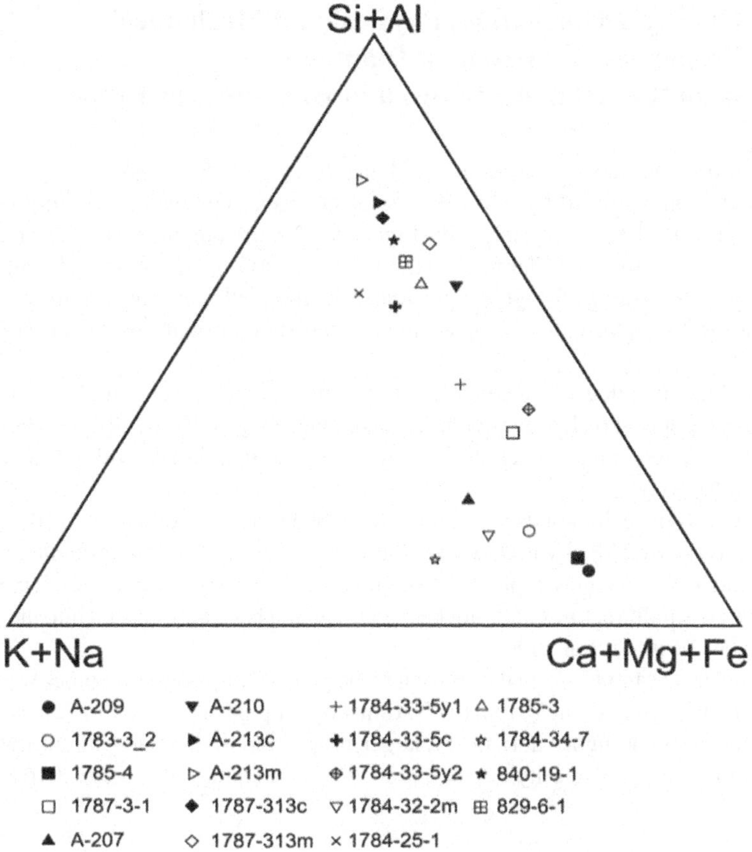

Fig. 4.9 Special characteristics of melt/fluid nanosized microinclusion composition in diamonds of the M. V. Lomonosov deposit

crystallization have silicate, aluminosilicate trend of microinclusions with essential alkalinity.

Tetrahexahedral and cubic habit crystals in the M. V. Lomonosov deposit originated in deep mantle carbon (δC^{13}—4 to 5%) (Kriulina 2012*), have micro-fibrillose structure and aluminosilicate composition of microinclusions.*

Both carbonates and aluminosilicates, i.e. carbonate–silicate trend of nanosized inclusions, have been found in variety IV diamond coats.

4.3 Quality, Deformation, Physical and Mechanical Properties Estimation of Diamond from the Arkhangelskaya and Karpinskogo-1 Pipes

Most of diamond crystals from the Arkhangelskaya and Karpinskogo-1 pipe specimens are characterized by a low degree of nitrogen aggregation according to the IRS data; nonetheless, crystals with higher % B exist among diamonds from the Karpinskogo-1pipe. It is known (Naletov et al., 2007), that diamond toughness properties improve with high proportion of aggregated nitrogen in B-form, i.e. it is possible to suggest that complex planar defects contribute to crystal structure hardening.

The diamond integrity percentage in both studied pipes is rather high—50 to 70% of diamonds have no cleavages or external damage (Fig. 4.10, Table 4.5). The major part of damaged crystals is comprised by chips, i.e. individuals without more than 30% of the body.

Craterous facie diamonds are characterized by higher fracturing: up to 10–15% of diamonds are overfilled with fractures ("broken"), often filled with graphite; from a quarter to half of crystal samples have singular medium and large fractures. Diamonds with single small fractures have not been marked and have been placed into the group of crystals "without fractures".

Comparing samples from the Arkhangelskaya and Karpinskogo-1 pipes we determined that rough diamonds from the Karpinskogo-1 pipe have better toughness characteristics even at higher horizons: a larger proportion of crystals without fractures (up to 60%) and by integrity degree (up to 70% of undamaged). In the 2016 sample

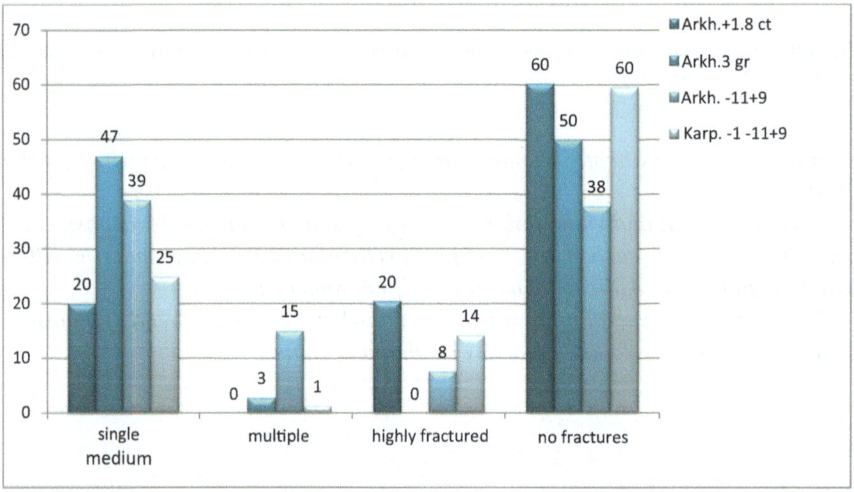

Fig. 4.10 Occurrence frequency of diamonds with varying fracturing degree in studied samples from the Karpinskogo-1 and Arkhangelskaya pipes

Table 4.5 Occurrence frequency of diamonds with varying fracturing degree in studied samples from the Karpinskogo-1 and Arkhangelskaya pipes

Pipe		Arkhangelskaya						Karpinskogo-1	
Size grade		+1.8 ct		3 gr		−11+9		−11+9	
Number of crystals		pcs	%	pcs	%	pcs	%	pcs	%
Fractures	Single medium	13	15.7	11	7.6	70	14.9	27	6.0
	Transverse	3	3.6	56	39.4	115	24.5	59	13.2
	Multiple	0	0.0	4	2.8	71	15.1	6	1.3
	Highly-fractured, "broken"	17	20.5	0	0.0	36	7.7	64	14.3
	Ferritization along fractures	0	0.0	0	0.0	0	0.0	25	5,6
No fractures		50	60.2	71	50.0	178	37.9	267	59.6
Total		83	100	142	100	470	100	448	100

from the Arkhangelskaya pipe, diamonds with numerous fractures filled with graphite comprise a third of studied mass, whereas the 2008 samplings contain a half of such crystals, i.e. the quality of graphitized crystals is deteriorating as the pit mining depth rises.

Most of fractured diamonds in the Karpinskogo-1 pipe sample contain graphite inclusions and epigenetic inclusions of iron oxides and hydroxides: ferritization along fractures was registered in 6% of crystals (Table 4.5).

In morphology of crystals plastic deformation marks have been registered in less than 4% of specimens from the Arkhangelskaya and in 7% of specimens from the Karpinskogo-1 pipes. Nevertheless, crystals with sharp plastic deformation bands (planes) and creating surface blocky relief irregularity prevail among deformed crystals.

Plastic deformation processes were weakly revealed in diamond crystals of the Arkhangelskaya and Karpinskogo-1 pipes (Table 4.6), and are mainly detected only by spectral methods.

The completed studies lead to the conclusions

1. Higher brittleness of diamonds and crystal lattice stress condition are caused by high concentrations of impurity-structural nitrogen defects, which are small in size, worsen strength of carbon atomic bonding and disable formation of inherent bonds between nitrogen atoms, typical of planar defects.
2. Diamond crystal lattice stresses in many crystals are relaxed due to fracturing.
3. Higher brittleness of the M. V. Lomonosov deposit diamonds is registered for dodecahedron and combined crystals with zonal-sectorial and microblocky internal structure.

Mineralogical and gemological characteristics of the Karpinskogo-1 pipe diamonds indicate its higher industrial and commercial potential versus the Arkhangelskaya pipe. Crystals of the Karpinskogo-1 pipe differ from the Arkhangelskaya pipe diamonds for better color and integrity characteristics even at upper mining horizons.

The quantity of gem quality and near-gem quality crystals is improving while the pit mining depth increases, at the expense of reducing the proportion of individuals with graphitization marks, chips and splinters.

4.4 Investigation of Diamonds Inert ToX-Ray Excitation

The collection of "non-luminescent" diamonds was specifically selected and created at the Mining and Processing Plant of Severalmaz (ALROSA company division) from tailings of X-ray luminescence separation, from the batch of heavy mineral concentrate weighing 5 tons.

The crystals were studied using optical and electron microscopy, infrared spectroscopy, ultraviolet spectroscopy, electron paramagnetic resonance.

Table 4.6 Occurrence frequency of diamonds with plastic deformation (PD) marks in studied samples from the Karpinskogo-1 and Arkhangelskaya pipes

Pipe		Arkhangelskaya						Karpinskogo-1	
Size grade		+1.8 ct		3 gr		−11+9		−11+9	
Number of crystals		pcs.	%	pcs.	%	pcs.	%	pcs.	%
Magnitude of PD	Without marks of PD	83	100	142	100	460	97,87	419	93.53
	Sharp bands of PD	Without marks of PD				6	1.28	5	1.06
	Bands of PD					4	0.85	3	0.70
	Shagreen surface					2	0.43	14	3.17
	Glide lines					–	–	7	1.41
Total						470	100.00	284	448

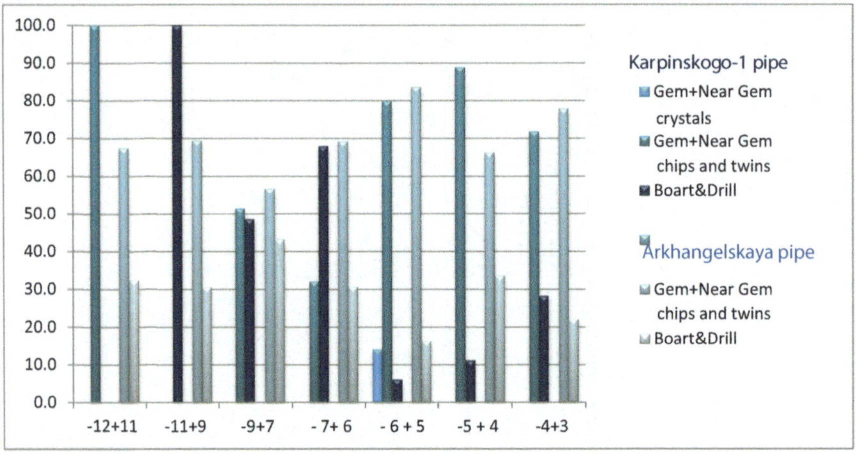

Fig. 4.11 The distribution bar chart of "non-luminous" diamonds under code list (Gem, Near Gem, Industrial) in different size-weight groups, the Karpinskogo-1 and Arkhangelskaya pipes

4.4.1 Qualitative Evaluation of Diamonds Inert to X-ray Excitation

Initial qualitative evaluation of rough diamonds from the X-ray luminescence separation tailings was carried out to acquire data regarding the need for their extraction. The larger crystals and their chips found in the X-ray luminescence separation tailings are generally mined during mineralogical examination of tailings; such specimens were not delivered for investigation. The largest crystals are occasional and weigh 0.69 and 1.29 ct., correspond to Boart and Rejections grades by quality.

But the quantity of individuals in enlarged 'Gem + Near Gem chips and twins' and 'Boart & Drilling' classes among non-luminescent diamonds of combined − 9+5 grade in both pipes varies from sharp prevalence of near gem positions to virtually total absence of such (Fig. 4.11), due to low representativeness of specimens. Nevertheless, there are gem quality crystals among non-luminescent diamonds.

4.4.2 Defect-Impurity Composition of Diamond Crystals Non-luminescent in X-rays

Analysis of summarized results from Table 4.7 shows that diamonds in the Arkhangelskaya and Karpinskogo-1 pipes are similar by total nitrogen concentration but aggregation degree (%B) is higher in the Karpinskogo-1 pipe, as well as absorption of B2 system.

The spectra of the Karpinskogo-1 pipe luminescent crystals with high aggregation degree are shown in Fig. 4.12; such diamonds have intense luminescence in UV-light

Table 4.7 Average values of parameters defined by IRS for diamonds from the Karpinskogo-1 and Arkhangelskaya pipes

Pipe	Quantity (<150 ppm,)	Ntot (ppm)	B1 (%)	ν_{B2} (cm^{-1})/n/a B2 (%)	B2 (cm^{-1})	CH (cm^{-1})
Karpinskogo-1	79 (7)	973	33	1368.7/2	12.2	4.56
	Low luminescent 33 (0)	922	17	1372.1/35	3.4	6.2
Arkhangelskaya	38 (0)	1063	10	1372.7/35	4.7	6.4
	Low luminescent 169 (28)	876	3	1375.1/84	9.3	2

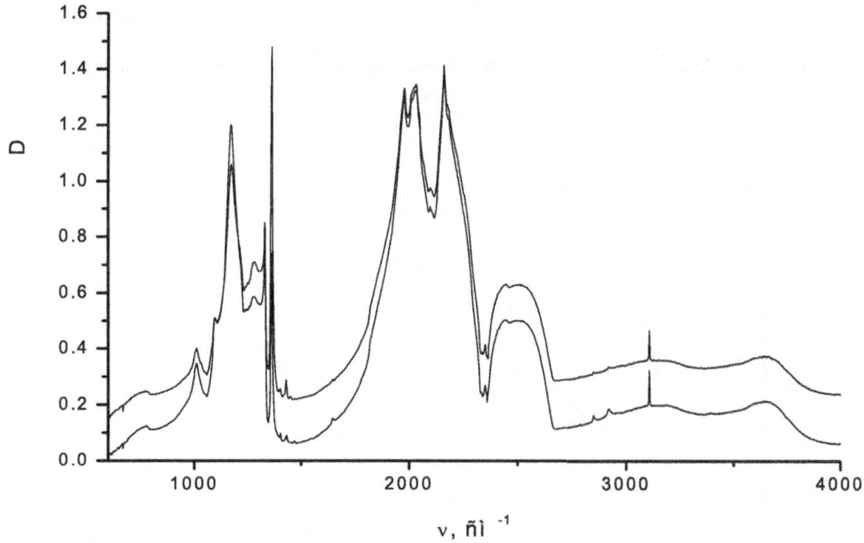

Fig. 4.12 Typical spectra of luminescent crystals K-6-11, K-6-10 with high degree of nitrogen aggregation (%B = 97), the Karpinskogo-1 pipe

and X-rays. Also observed is nearly straight-line correlation between increasing concentration of B2 defects along with growing proportion of nitrogen in B1-form (Fig. 4.13).

B2 band is registered in 15% of specimens, its average value is 9.3 cm^{-1}. Subsidiary CH bands are not registered in the ranges of 1300–1600 and 3000–3500 cm^{-1} in crystals with B2 system. Diamonds inert to X-ray exposure are characterized by total absence of B2 band. The IaAB-type spectra of crystals (specimens A3-8-4 and A3-8-14) from X-ray luminescence separation tailings are given in Fig. 4.14, where B2 band presence is controversial, nitrogen concentration is 1700 ppm, aggregation degree amounts to 5%.

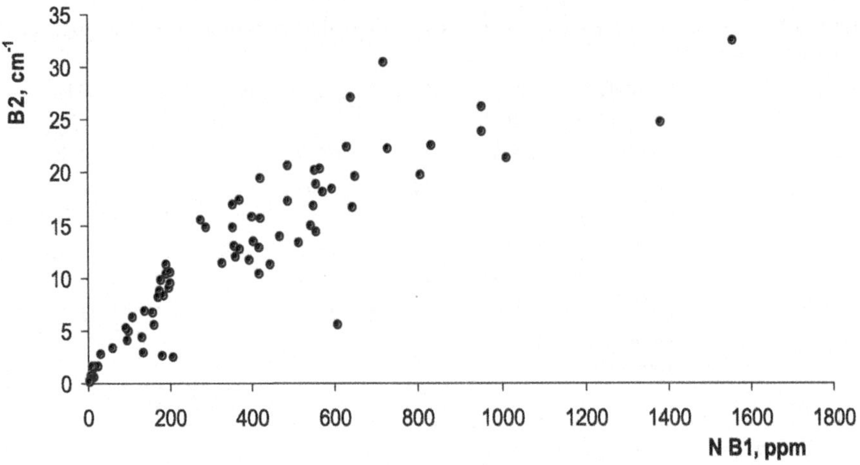

Fig. 4.13 Dependence of B2 band intensity from concentration of nitrogen in the form of B1 defects in the sampling from the Karpinskogo-1 pipe

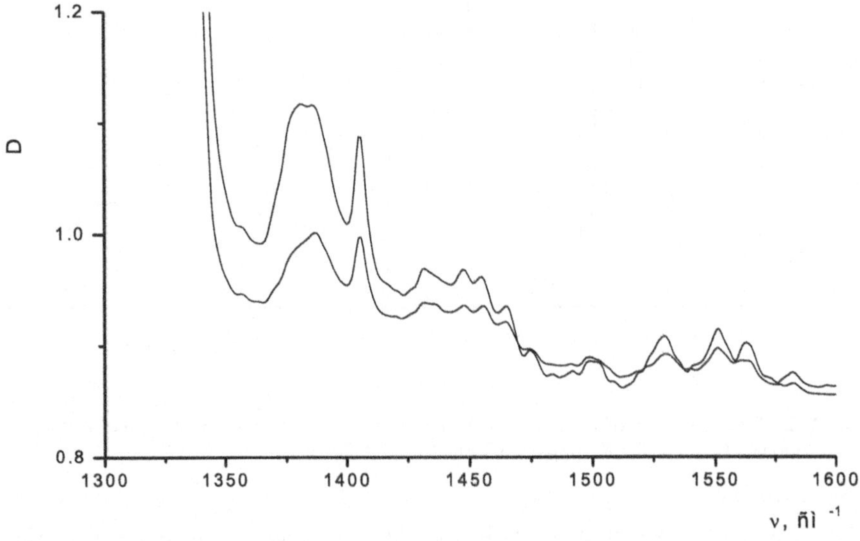

Fig. 4.14 The spectra of crystals A3-8-4 and A3-8-14 in the range 1300–1600 cm^{-1}

After analyzing the diagrams (Figs. 4.15 and 4.16) showing relationship of total nitrogen concentration (Ntot) and aggregated B-form nitrogen proportion in studied diamond collections from the Arkhangelskaya and Karpinskogo-1 pipes it can be concluded that diamonds that are not luminescing under X-rays are characterized by low degree of nitrogen aggregation in B-form, which proves their short-term post-crystallization annealing.

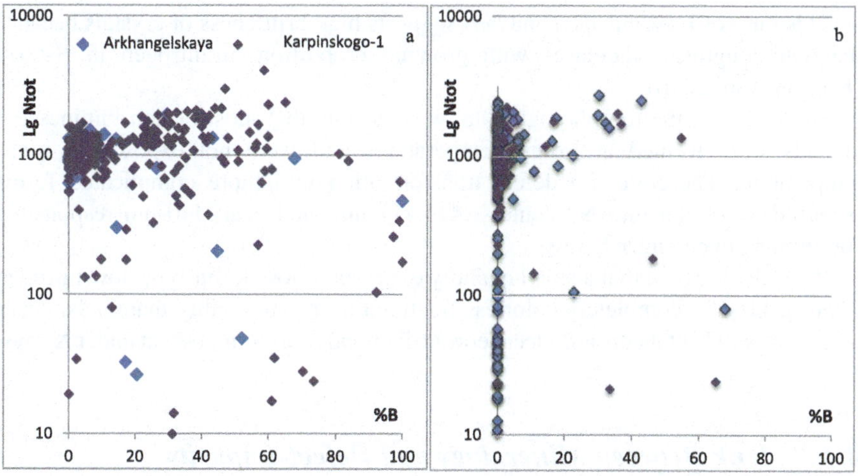

Fig. 4.15 Diagram of relationship between total nitrogen concentration (Ntot) and aggregated nitrogen proportionin the B-form in studied diamond collections from the Arkhangelskaya and Karpinskogo-1 pipes: **a** collection from industrial mining (luminescent in X-rays, 295 crystals); **b** from X-ray luminescence separation tailings (non-luminescent in X-rays, 281 crystals)

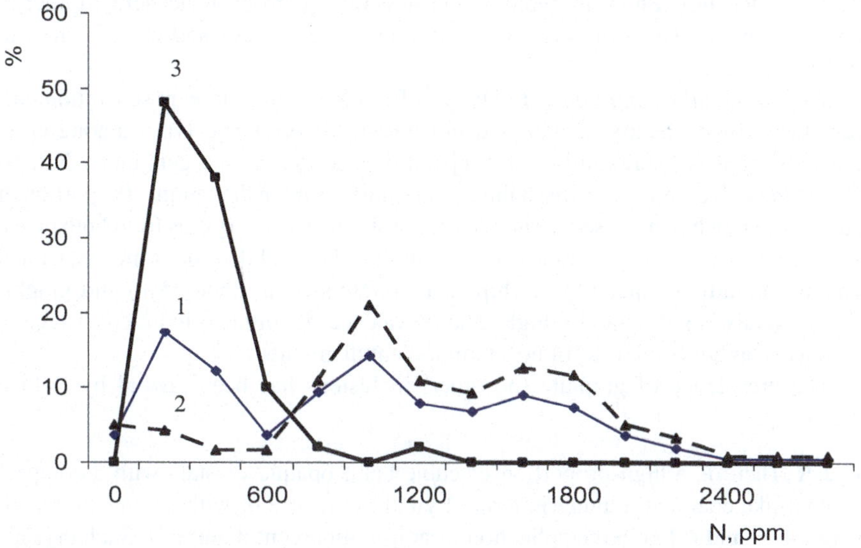

Fig. 4.16 Distribution of non-luminescent crystals of the Arkhangelskaya pipe by total ($N_A + N_{B1} + N_C$) nitrogen (1) concentration, concentration in crystals without C-defects (2) and with C-defects (3)

This ratio of total nitrogen and %B suggests high brittleness of crystals because diamond toughness increases with growing proportion of nitrogen in B-form (Khachatryan, 2010).

Summarizing the morphology data, we can state that most of non-luminescent crystals were formed at lowest temperatures and were not exposed to high-temperature. Therefore, no defect transformation into more complicated forms occurred, inter alia into N3 centers (415.417 nm) and peak 440 nm responsible for luminescence under X-rays.

It should be noted that a small quantity of crystals close to IIa-type, low-nitrogen ("nitrogenless"), completely colorless fractured near-gem quality diamonds, occur among variety I octahedron-dodecahedroid diamonds non-luminescent under X-rays.

4.4.3 Link Between Mineralogy and Defect-Impurity Composition of Diamond Crystals in the Collection Mined from X-ray Luminescent Separation Tailings

The diamonds have been studied by IRS and spectral photoluminescence (analyzed by E. A. Vasiliev, Ph.D. in geology and mineralogy) to evaluate defect-impurity composition similarity of crystals luminescent under X-rays and crystals inert to X-ray exposure.

Mineralogical examination of diamonds from X-ray luminescent separation tailings has shown identity of diamond mineralogical forms and types among crystals mined under X-ray luminescent separation and non-luminescent under X-rays. Dodecahedral, combined, cubic habit crystals are present in the sample. Proportion of cubes is somewhat increased in samples of non-luminescent crystals from both pipes, versus industrial mining samples. While tetrahexahedroid diamonds are occasional among the tailings, quantity of chip sand splinters (more than 55%) and graphitized crystals is substantially high. The prevalence of splinters indirectly suggests increased higher brittleness of non-luminescent diamonds.

The prevalence of graphite (no other) inclusions has been proved by Raman scattering.

Cubes There is a high proportion of cubic habit opaque crystals with a compact enamel-like coat and minute (in term of yield ratio) core or without any core at all (Figs. 4.17 and 4.18) in the collection of non-luminescent diamonds. Such crystals are of little use in monocrystal industrial application and are only applied in abrasive powder manufacturing.

In terms of mineralogy, these crystals are crucial as containing a multitude of microinclusions.

Crystals of this group do not luminesce under a gemological UV-lamp (365 nm) and either glow at exciting by a more intense laser source or do not glow, or else give faint yellow photoluminescence. Unfortunately, many individuals turned out to

Fig. 4.17 Chips of cubic habit crystals (variety II) from the X-ray luminescent separation tailing collection

Fig. 4.18 The crystals of various habits from the X-ray luminescent separation tailing collection: red-brown cube, yellow cube and curve-faced crystal splinter

be opaque to be studied by IRS; we only managed to register absorption only for several specimens in their fine cleavages. Sampling of crystals (non-luminescent and low-luminescent under X-ray) from the Arkhangelskaya pipe is notable for a high proportion of crystals with C-defects (26%).

Carbonate and water absorption bands are registered in coatings of cubic crystals; the crystals have green tint on basic yellow, yellow-brown color. Water peaks are not registered in variety II crystals, but such crystals have very high concentration of nitrogen in C-form, up to 350 ppm (150 at. ppm on average). The crystals can be classified as low-nitrogen by content of nitrogen in A- and B-forms: about a half of crystals contain only 10–50 at. ppm, the others contain up to 250–400 at. ppm.

The spectra of crystals with C-defects non-luminescent under X-rays often contain carbonate absorption bands with maximum sat 1400, 780 nm, and molecular water with the maximums around 1650 (deformation vibrations) and 3000–3500 (stretching vibrations).

Overall, 50 out of 289 non-luminescent crystals in the sampling C defects identified by IR absorption. 10% of crystals contain C-defects as the only impurity form of nitrogen, and 80%—as dominating together with A-defects (Table 4.8). No crystals with bands A, B2 and C in their IR absorption spectra were found. This regularity

Table 4.8 Data on defect-impurity composition of diamond non-luminescent under X-rays for crystals of cubic and tetrahexahedral habit

Habit, color	Value	CH (cm^{-1})	A (ppm)	B1 (ppm)	Ntot (ppm)	B1 (%)	C (ppm)	Syst. 1500	1430 cm^{-1}	H_2O (cm^{-1})
Cube, coated yellow	Av. value	0.9	513	14	526	0	–	–	Carb	1650
	Min–max	0.2–2	250–900	0–40	250–930	0–3				
Cube black	Av. value	0.96	653	11.2	664.2	13.6	84.5	-	carb	1650
	Min–max	0.1–1.9	30–1300	0–55	81–1308	0	92			
Cube yellow	Av. value	0.09	16.5	0	16.5	0	148.4	1500	Carb rarely	-
	Min–max	0–0.2	0–40	0	0–40	0	98–230			
Tetra-hexa-hedron	Av. value	0.09	16.5	0	16.5	0	–	1500	–	3500
	Min–max	10–20	800–1600	0–30	811–1630	0–2				

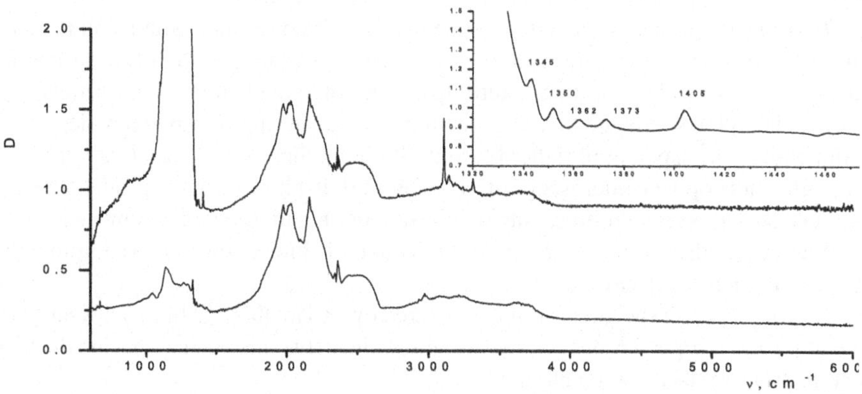

Fig. 4.19 The optical density spectra of crystals from the Arkhangelskaya pipe, with C-defects (3A-16-12) and A-defects (3A-16-26) prevailing in the spectrum

is proved by experimental studies (Vins 2011), as when C-defects gradually transform into A and further aggregate in B1 and B2 defects, nitrogen almost completely translates into more complicated forms.

The spectra of diamonds from the Arkhangelskaya pipe that contain nitrogen in C-defect form are given in Fig. 4.19. Identifying C-defects at low concentrations is possible not only by a 1345 cm^{-1} peak, but by a shoulder in the 1130 cm^{-1} range as well.

In contrast to the data of the paper (Hainschwang et al. 2012), in which narrow bands were found in Ib-type diamonds with C-defects, in the studied collection the bands in the 1400–1600 range occur in IaA-type diamonds, colorless, i.e. not containing C-defects in concentration detectable by infrared and by optical spectroscopy, and bands in the 3000–3400 range occur only in IaA-type crystals. In Ib-type crystals and crystals changing to Iba-type the classical hydrogen 3107 cm^{-1} band has the absorption coefficient at 0.1–0.2 cm^{-1}.

C-defects occur only in single crystals with low fluorescence, in IR absorption spectra of diamonds mined under X-ray luminescent separation from the Arkhangelskaya and Karpinskogo-1 pipes. This confirms the assumption that complex defects whose configuration is close to N3 and B2 centers are responsible for luminescence under X-rays.

Therefore, X-ray fluorescence is absent in these crystals due to two reasons:

1. *X-ray fluorescence quenching owing to a large number of carbonate and water microinclusions.*
2. *Absence (entirely low concentrations) of centers with the absorption peak at 400–440 nm, which causes luminescence. Nitrogen in diamond structure is mainly present as single atoms; the nitrogen aggregation degree in more complicated forms including N3 and 40–440 nm is very low, which is insufficient for luminescence occurrence even under the most sensitive X-rays*

Tetrahexahedroids are mainly represented by whole crystals with a small share of chips with protomagmatic nature of cleavage. Crystals are slightly transparent due to drop-like and hillocky sculpture (from pattern to finely hummocky relief); the prepolished plates indicate that the specimens have varying transparency degree—from entirely transparent to rather blurred. Studying these specimens under a high-magnification optical microscope it was found that blurs in the crystal body are associated with scattered tiniest inclusions (about one micron and less in size).

Tetrahexahedroids of this group have faint and visible yellow tint, predominantly 1–2 large fractures (cleavage).

C-center in these specimens is not registered by IRS method (Table 4.7). Nearly all the impurity nitrogen is concentrated as an A-center (two nitrogen atoms in adjacent tetrahedrons replace carbon atoms).

A-center concentration is high (800–1550 at. ppm), but at the same time a proportion of nitrogen in B-form only amounts to 1–2%, platelets are totally absent, i.e. complex aggregated defects are absent or only traces of their presence are observed. This also proves the absence of optically active centers that cause luminescence under X-ray and UV radiation.

Noteworthy are extremely high concentrations of CH (3107 cm^{-1}) hydrogen unstructured impurities, with absorption coefficient from 10 to 20 cm^{-1}. Subsidiary peaks of molecular water (the absorption peak of 3500 cm^{-1}) are registered in IR spectra as well.

Octahedron-dodecahedroid crystals The collection of non-luminescent crystals includes I variety diamonds, under the classification by Yu. L. Orlov (Fig. 4.20). Colorless and near-colorless transparent curved-faced chips and splinters with protomagmatic and mixed nature of cleavage are present in equal proportions. The share of chips with man-caused cleavage is much higher than in the batches from industrial mining, i.e. luminescent crystals.

Absorption systems with a row of narrow bands in the 1350–1600 cm^{-1} range also appear in absorption spectra of IaA, IaA + IIb–type diamonds from the ADR, i.e.: 1575, 1564 1550 1503 1465 1456, 1448, 1431, 1406, 1387 cm^{-1}, with the absorption coefficient up to 0.9 cm^{-1}, and width at a half of height being 4 cm^{-1} at 2 cm^{-1} resolution, (Figs. 4.21, 4.22, 4.23 and 4.24).

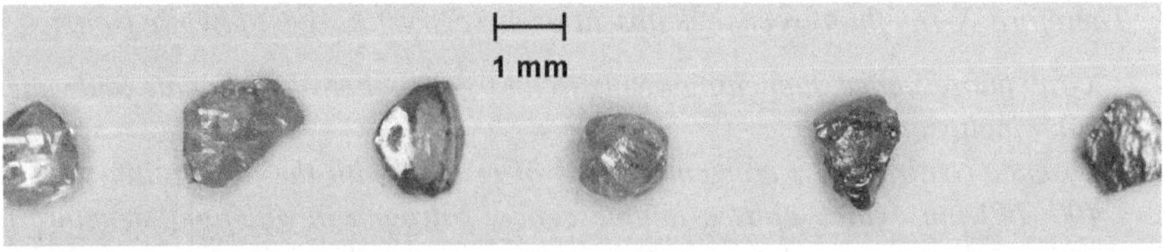

Fig. 4.20 Chips and splinters of curve-faced crystals from the X-ray luminescent separation tailing collection

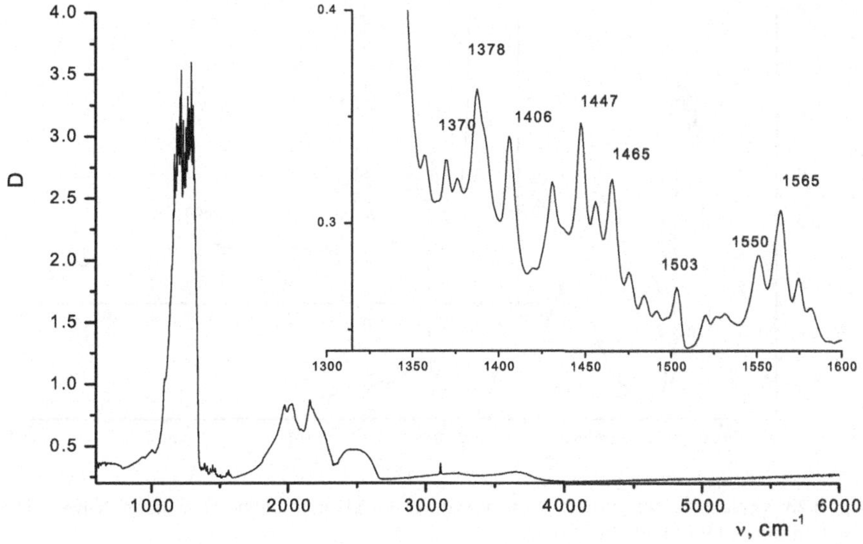

Fig. 4.21 Optical density spectrum of luminescent crystal A11-105. The insertion shows the 1300–1600 cm^{-1} range

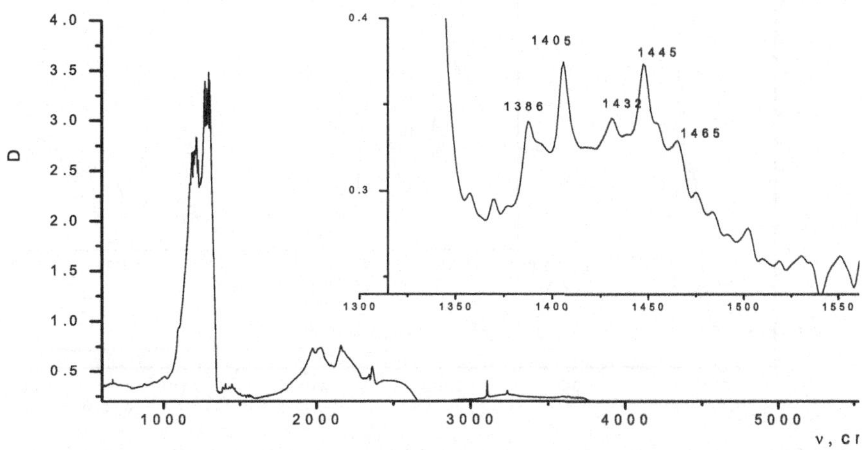

Fig. 4.22 Optical density spectrum of luminescent crystal A11-33. The insertion shows 1300–1600 cm^{-1} range

The second system bands lie in the 3000-3400 cm^{-1} range and have maximums at 3311, 3189, 3154, 3144, 3050 cm^{-1}, the absorption coefficient up to 2 cm^{-1}, and width at a half of height being 4 cm^{-1} (Fig. 4.25).

Bands of these systems are present in the spectrum along with the known lines 3107, 3237, 2786, 1405 cm^{-1}, and apparently belong to group CH of specific low-temperature configuration (Kriulina 2012). The set of these bands is not permanent;

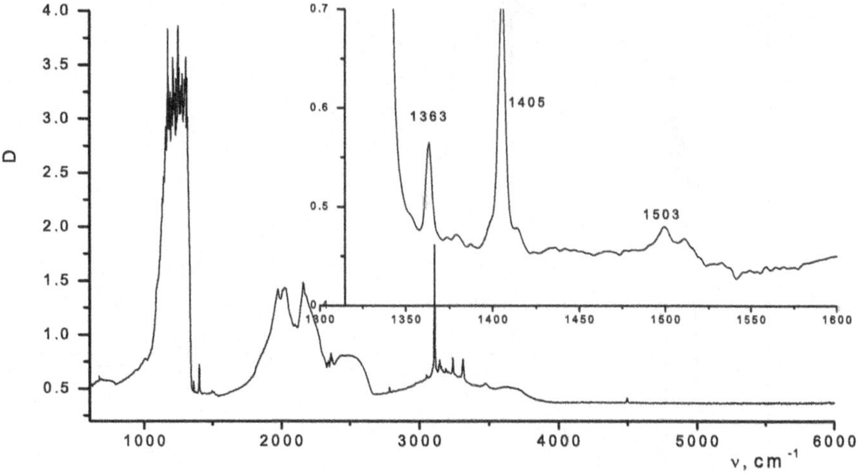

Fig. 4.23 Optical density spectrum of crystal A3-11-54, non-luminescent under X-rays. The insertion shows 1300–1600 cm^{-1} range

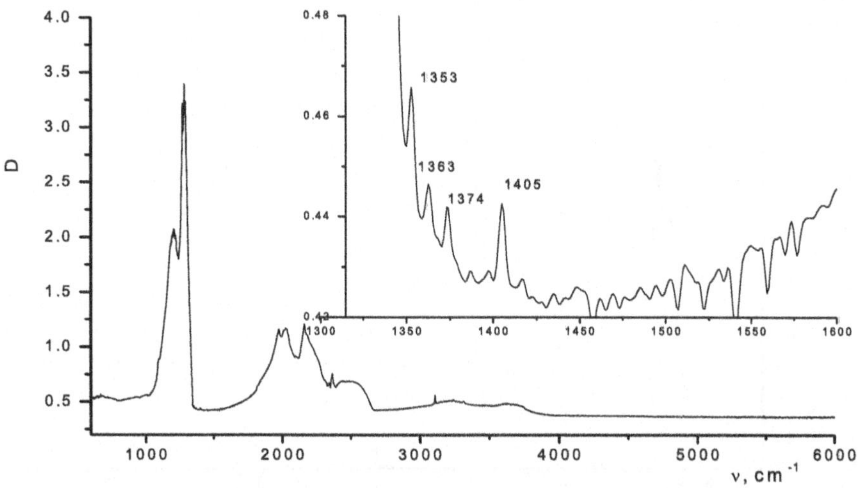

Fig. 4.24 Optical density spectrum of crystal A3-11-37, non-luminescent under X-rays. The insertion shows 1300–1600 cm^{-1} range

1363 band occurs individually, 1353, 1361, 1374 bands are registered together, just the same as the set of 1386, 1432, 1445, 1465 cm^{-1}.

A few plastically deformed crystals are present in the collection of non-luminescent diamonds. The band is registered at 4160—an "amber" center occurring due to plastic deformation (Massi et al. 2005) in one-of-a-kind man-caused pink-brown chip of dodecahedroid with sharp plastic deformation bands (sp. A3-11-72, Fig. 4.26) in IR absorption spectrum.

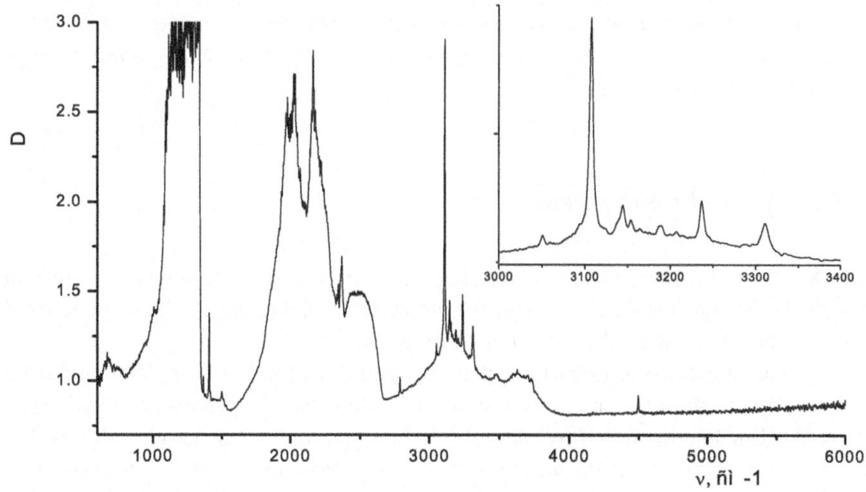

Fig. 4.25 Optical density spectrum of crystal K-10-4. The insertion shows 3000–3400 cm^{-1} range

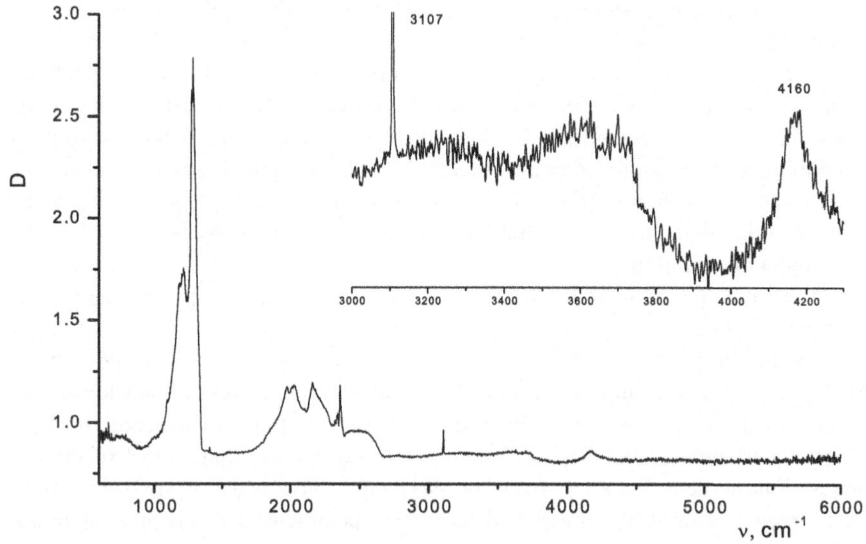

Fig. 4.26 Spectrum of crustal A3-11-72, the insertion contains 3000–4300 cm^{-1} range

By physical type, "non-luminescent" octahedron-dodecahedroid diamonds belong to Ia, IaAB-types, differ from luminescent ones by existence of subsidiary absorption bands associated with microinclusions, which are likely to affect X-ray fluorescence intensity and accumulation. It can be assumed based on similarity of IR spectral types and presence of weak N3 band in PL spectrum that these crystals have low X-ray luminescence, but being predominantly splinters (plane specimens) they were just shielded from the X-ray luminescent separation source/detector by other minerals or never got a chance to accumulate the sufficient X-ray luminescence charge to be registered.

4.4.4 X-ray Luminescence

The X-ray luminescence spectra were registered by the X-ray luminescence unit in which the X-ray tube 5BXV-7 with Re anticathode and water coolinzg were used (analyzed by V. A. Rassulov, Ph.D. in geology and mineralogy).

Diamond ***X-ray luminescence*** is characterized by a broad structureless band from 330 to 600 nm with the maximum about 480 nm (visually perceived as pale-blue glow, A-glow band) excited by radiation with quantum energy 5.5 eV—X-ray and UV-radiation, protons, electrons (Dean and Male 1964; Dean 1965; Rassulov et al. 2009). The spectra of crystals that luminesce and do not luminesce under X-rays are given in Fig. 4.27.

According to research findings it is apparent that diamonds from the collection of non-luminescent specimens did not "react" to excitation in the process of X-ray spectral analysis.

The X-ray luminescence A-band, a broad structureless band from 330 to 600 nm with the maximum about 480 nm, prevails almost in all luminescent diamonds. However, there are individuals in the industrial mining series which absorption maximum is not strongly marked, orange cubes of Ib-type and tetrahexahedroids of IaA-type, i.e. among diamonds luminescent under X-rays (checked by XLS). Possibly it is related to the fact that luminescence intensity is by far lower than the instrument detection limit.

Further the diamonds from two samplings were examined by spectral photoluminescence method.

Photoluminescence spectra were registered by the microspectrophotometer MSFU-312 at room temperature in the integral mode at resolution not lower than 3 nm in the range 390–850 nm. Under excitation by the pulsing nitrogen laser LGI-505 with wavelength 0.337 μm. According to research findings most of crystals among luminescent diamonds from the Arkhangelskaya and Karpinskogo-1 pipes have green and light-blue glow and there are specimens with green, orange and yellow-green glow (Fig. 4.28) among non-luminescent crystals.

Data on presence and occurrence frequency of separate photoluminescence centers (Table 4.9) show distribution of N3-centers (with the zero-phonon line 415–416 nm), both in luminescent and non-luminescent crystals. Meanwhile, H3 absorption systems with the zero-phonon line 503.2 nm responsible for green glow, S3 with the zero-phonon line 496.7 nm responsible for orange glow and also other bands are oftener registered in non-luminescent crystals.

Luminescence color in most of crystals with excitation is within the 300–380 nm range using N3 and S2 systems, or these systems are registered together (Fig. 4.29).

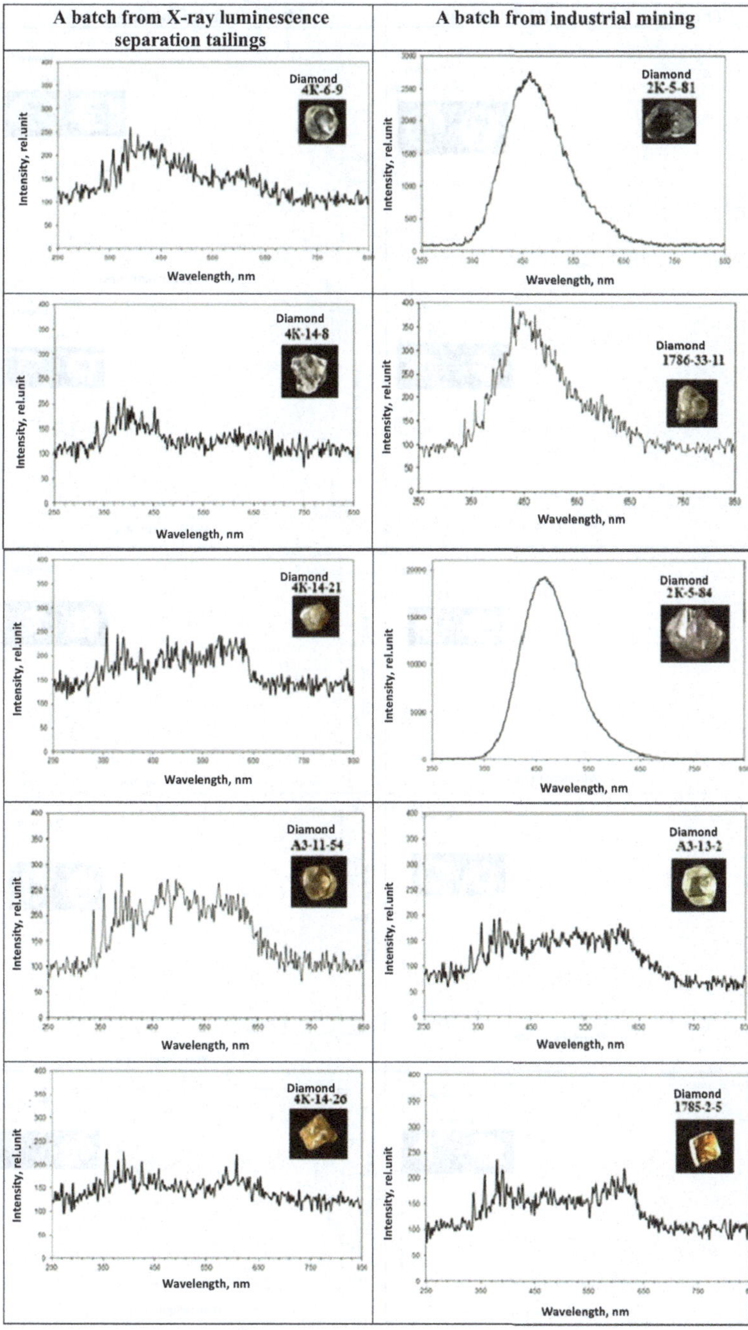

Fig. 4.27 X-ray spectra of luminescent and non-luminescent diamond crystals from the Arkhangelskaya and Karpinskogo-1 pipes, the spectra of diamonds from the XLS tailing collection similar in morphology (no X-ray luminescence) to the spectra of the industrial mining series diamonds (with X-ray luminescence) are given as an example

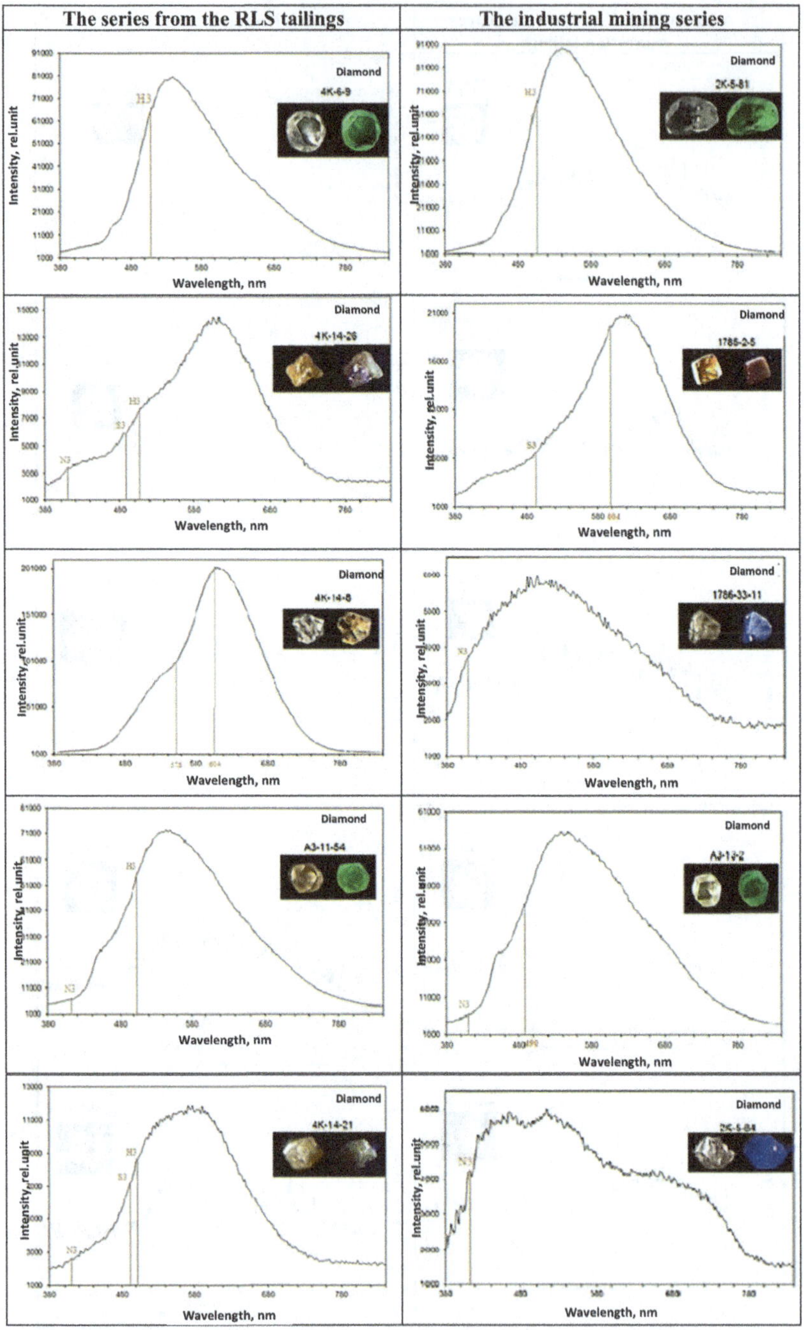

Fig. 4.28 The diamond photoluminescence spectrum at room temperature

Table 4.9 A fragment of the table with crystal-by-crystal registration of photoluminescence center distribution in luminescent and non-luminescent diamonds

Coll	№	Specimen	N3	490 nm	578 nm	H3	S2	S3	604 nm
NRL	1	4K-6-9				+			
	2	4K-14-8			+				+
	3	4K-14-21	+			+		+	
	4	4K-14-26	+			+		+	
	5	A3-11-15						+	+
	6	A3-11-54	+			+			
RL	7	2K-5-81				+			
	8	2K-5-84	+						
	9	A3-13-2	+	+					
	10	1785-2/5						+	+
	11	1786-33-11	+						

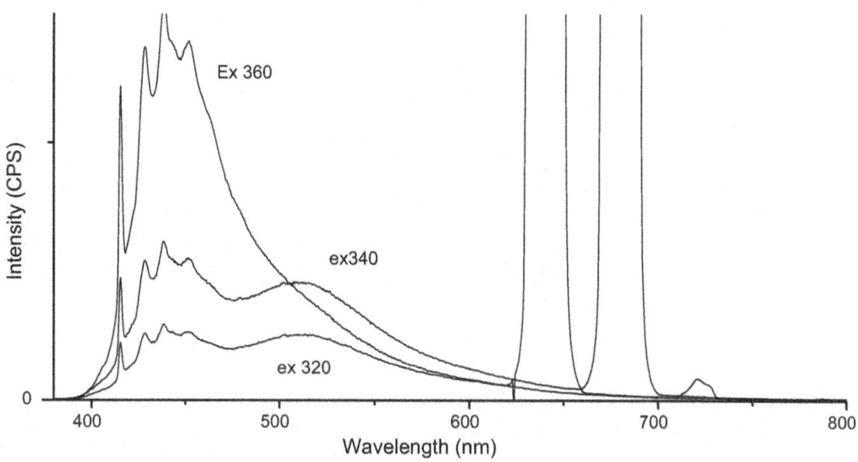

Fig. 4.29 PL spectrum of crystal K-6-13 at 23 °C, excitation 320, 340, 360 nm in the 620–700 range of the second order excitation line

Occurrence frequencies of visible luminescence bands in the studied samplings are given in Table 4.10.

In some cases, pink luminescence is caused by N3 system and a broad structureless band with the maximum of some 600 nm. S1 and S2 systems exhibit in a broad band with the maximum of 520 nm, if excited in the range of 300–380 nm.

S1 and S2 bands are specific to cubic crystals and cube growth pyramids combined shapes and octahedrons. N3 defects appear while B1 defects are being formed, so they occur in crystals with B1 and B2 systems. The N3 system rarely occurs in low-temperature crystals with C-defects and can indicate a complicated internal

Table 4.10 Occurrence frequency of luminescence systems in diamonds from the Arkhangelskaya and Karpinskogo-1 pipes

Pipe	Occurrence frequency of luminescence systems (%)						
	N3, registered/dominates	478 nm	490 nm	503 nm	S3 registered/dominates	585 nm	655 nm
Karpinskogo-1	90/81	7	32	14	27/7	30	20
Karpinskogo-1/low luminescent	49/29		6	0	82/69	3	23
Arkhangelskaya	68/47		37	0	68/58	3	
Arkhangelskaya/low luminescent	26/3		20	25	60/34	10	

structure—a high-temperature central part of the crystal and low-temperature coat—these are "coated" crystals.

The bands with maximums of 473, 489, 491, 496, 536, 560, 573, 583, 604, 693, 700, 710, 740, 787, 818 nm appear in the spectra under excitation of 405 and 450 nm at 77 K. Moreover, lines 787, 718 nm that accompany S3 band are registered under excitation of 450, and 793 nm line is registered under excitation of 350 nm. Under excitation of 787 nm the bands with maximums of 804, 814, 903, 910, 927, 933, 945, 948, 962, 985, 1020 nm (Figs. 4.30, 4.31, 4.32, 4.33, 4.34 and 4.35) are registered in the spectra.

The band with the maximum of 986 nm—H2 (a nitrogen-vacancy center) was registered in the Karpinskogo-1 pipe crystal spectra.

The bands with maximums of 804, 823 nm are registered in crystals of all types. Pursuant to the published data, absorption of the N1 system, which is inactive in luminescence (Zaitsev 2001), begins in this range (826 nm).

The band with the maximum of 933 nm is specific to crystals with a high nitrogen aggregation degree and is accompanied by an accessory at 912 nm (Fig. 4.30) in the case of high intensity. There is no information on these bands in diamond luminescence reference guides (Dishler 2012; Zaitsev 2001).

925, 948 nm bands oftener occur in crystals with low degree of nitrogen aggregation and are registered in spectra of most of crystals (Fig. 4.32). During spectra registration the laser power was changed from 0.1 to 100%, depending on luminescence intensity. It is not possible to directly associate some or other luminescence

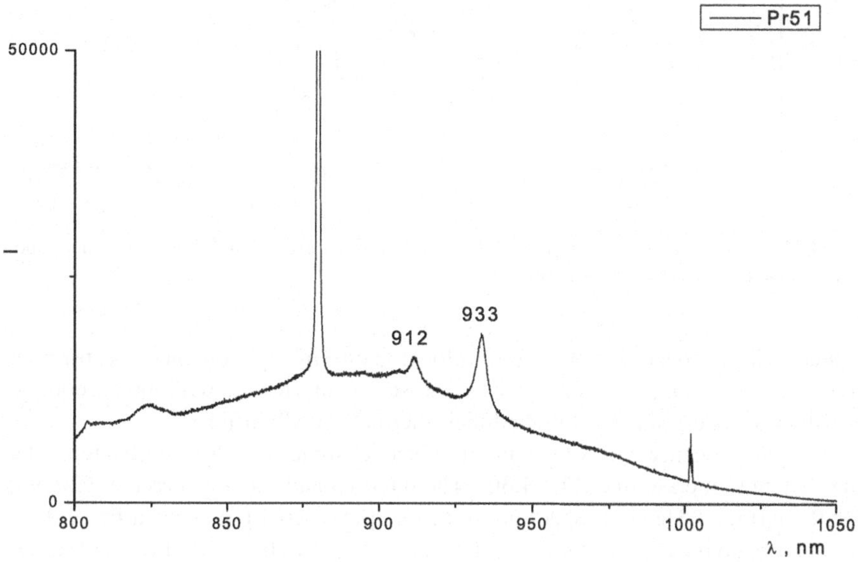

Fig. 4.30 Luminescence spectra of crystal Pr51 at 77K, excitation 785 nm, power 10%

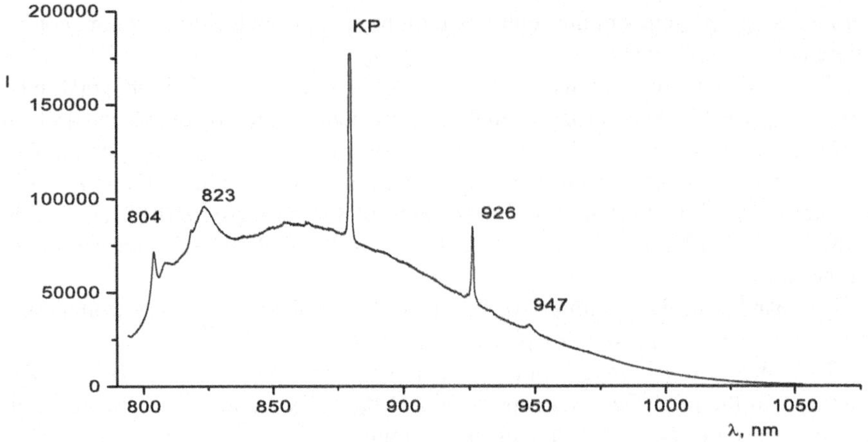

Fig. 4.31 Luminescence spectra of crystal A100-4, at 77K, excitation 785 nm (typical spectrum)

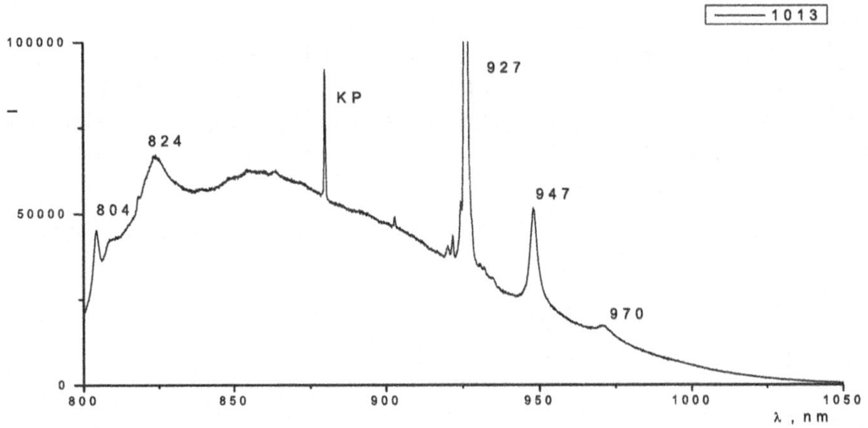

Fig. 4.32 Luminescence spectra of crystal A101-3, at 77K, excitation 785 nm (927 much more intense than Raman scattering line (RS))

centers with nitrogen defects due to a local nature of luminescence spectra registration under laser excitation by the microscope, in view of widespread complex zonal-sectorial crystal structure and their thermal-growth history.

The luminescence band of 926 nm in spectra of some crystals is more intense than the Raman scattering line (Fig. 4.30). The intense bands are registered at 920, 961, 1020 nm (Fig. 4.31) in the spectrum of the Karpinskogo-1 pipe crystal Pr 20–1.

There is no other band (luminescence intensity is less 0.1% RS, Fig. 4.32) except the RS line in individual crystals at 100% laser power.

Crystal 107-1 from the Arkhangelskaya pipe gave very intensive 903 and 927 nm bands and weak 945, 861 nm bands (Fig. 4.35).

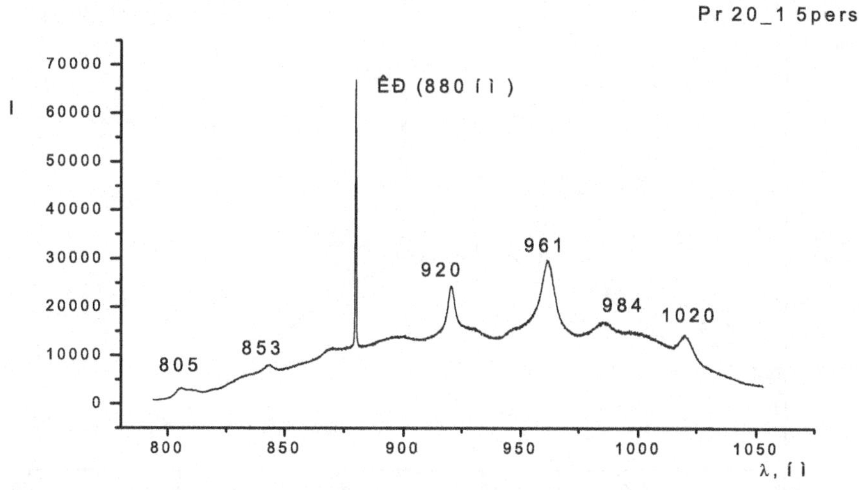

Fig. 4.33 Luminescence spectra of crystal Pr20-1, at 77K, excitation 785 nm, power 5%

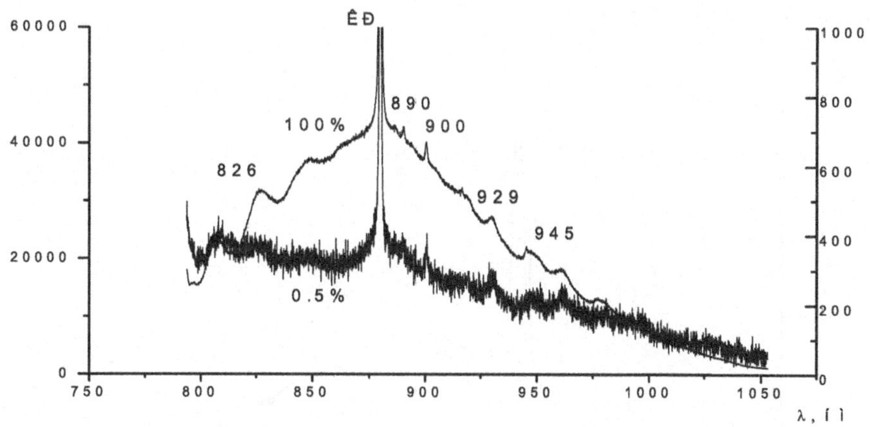

Fig. 4.34 Luminescence spectra of crystal K-6–9, at 77K, excitation 785 nm, power 0.5% and 100%

The double with maximums of 883, 885 nm, given in references as the 884 nm band, attributed to nickel atom Ni^+ (Yelisseyev and Kanda 2007), has been registered in luminescence spectrum of crystal Pr. 26-4. Intensity of this system is by an order of magnitude higher than RS intensity; 800 and 869 nm bands, not registered in the spectra of other specimens (Fig. 4.36), are also registered in the spectrum.

Luminescence intensity of all systems is reduced in the sampling of low luminescent crystals from the Arkhangelskaya and Karpinskogo-1 pipes. N3 system occurs in registered photoluminescence spectra, it was detected in crystals with the B2 band

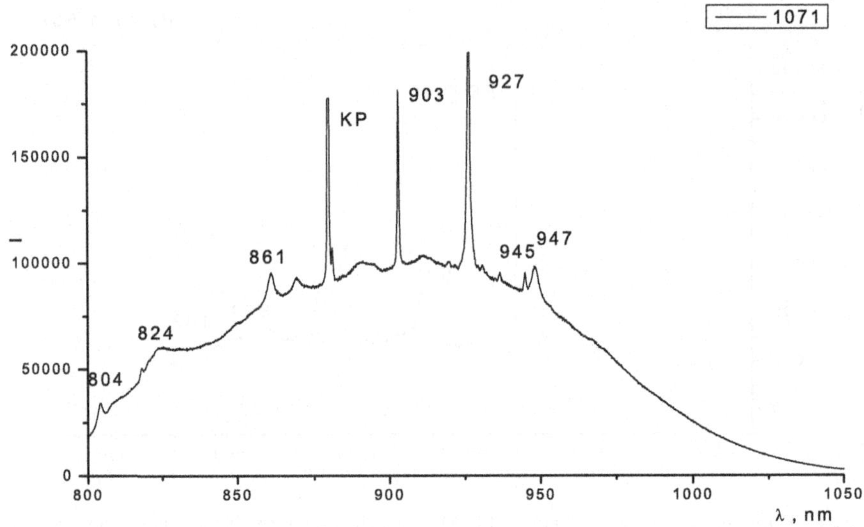

Fig. 4.35 Luminescence spectra of crystal from the Arkhangelskaya pipe (sp. 107-1) at 77K, excitation 785 nm, power 50%

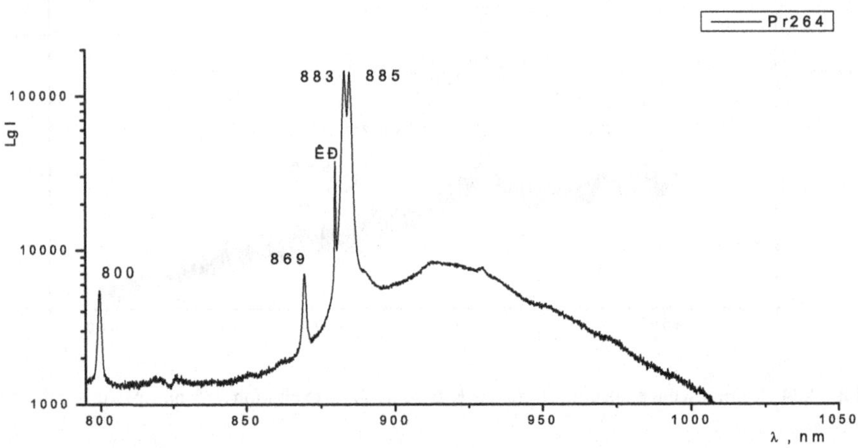

Fig. 4.36 Luminescence spectra of crystal K-6-9, at 77K, excitation 785 nm, power 5%

in the absorption spectra (specimens A3-11-72, 61,71, 104) or in low-nitrogen or nitrogenless crystals (specimens A3-11-66, 101, 96).

4.4.5 *Paramagnetic Centers in X-Ray-Luminescent and Non-Luminescent Diamond Crystals (EPR Spectroscopy)*

Four samplings of specimens were examined by EPR to study composition and concentrations of paramagnetic centers in diamond crystals that are non-luminescent and luminescent under X-ray radiation. The sampling of crystals from the Arkhangelskaya pipe comprised 76 non-luminescent specimen sand 39 luminescent crystals (comparison group). The sampling of crystals from the Karpinskogo-1 pipe included 11 non-luminescent and 23 luminescent crystals, respectively. 149 specimens were studied in total.

The crystals from the Arkhangelskaya and Karpinskogo-1 pipes are characterized by high content of nitrogen centers in P1 form ("C" center according to the classification accepted for IR spectroscopy); generally they do not contain nitrogen in other forms (including in A and B forms detected using IRS methods). Average concentrations of P1 paramagnetic centers in studied samplings of diamond crystals are given in the bar chart (Fig. 4.37).

P1 center is undoubtedly dominating, both by average concentrations and by occurrence frequency in diamonds of studied samplings. Concentrations of other centers are in general lower (Fig. 3.27).

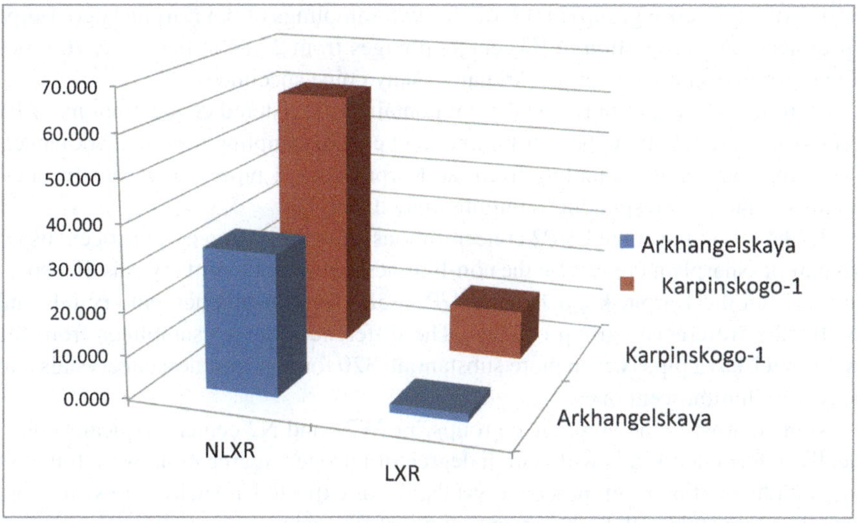

Fig. 4.37 Distribution bar chart of P1 center average concentrations (in ppm) in the studied diamond samplings (NLXR—non-luminescent under X-rays, LXR—luminescent under X-rays)

4.4.6 Summary of Results and Comparative Analysis Based on EPR Data

Comparative concentrations of paramagnetic centers nThe diagrams with average concentrations of paramagnetic centers, in which the sector of circle corresponds to percentage of all detected centers, were built for each sampling of crystals and average concentrations (from the whole sampling) were denoted by figures (Fig. 4.38).

The analysis of paramagnetic center concentrations by samplings and also diagrams of composition enabled to determine differences between X-ray non-luminescent diamonds and diamonds of corresponding "comparison groups" presented by X-ray luminescent crystals. Non-luminescent crystals are characterized by:

1. Rather high, versus the luminescent crystal reference group, concentrations of P1 centers. For the Karpinskogo-1 pipe specimenssuch difference is more than 5 times; for the Arkhangelskaya pipe specimens the average concentrations of P1 centers for samplings of luminescent and non-luminescent crystals differ by more than two orders of magnitude.

Nevertheless, it is noteworthy absence of any "boundary" concentration of P1 centers to determine the possibility or impossibility of X-ray luminescence.

2. Sudden drop in the proportion of crystals containing P2 centers. The portion of P2 centers in the sampling of non-luminescent crystals slumps 24 times, from 3.91% (the reference group) to 0.16%, for two samplings of the Karpinskogo-1 pipe specimens. The proportion of P2 centers plunges from 22.84 to 0.3%, i.e. 76 times down, for two samplings of the Arkhangelskaya pipe specimens.

Moreover, note that only one crystal containing registered concentrations of P1 centers is present both in the non-luminescent crystal sampling from the Arkhangelskaya pipe, and in the sampling from the Karpinskogo-1 pipe; the reasons for their inclusion into a corresponding sampling may differ.

3. The ratio of average P1/P2 concentrations showing a degree of nitrogen disaggregability sharply increases for the non-luminescent samplings of crystals. For specimens from the Karpinskogo-1 pipe P1/P = 580 for non-luminescent crystals and 24 for the "reference" group crystals. The difference for two samplings from the Arkhangelskaya pipe is even more substantial: 320 for non-luminescent crystals and only 3 for luminescent ones.

4. In contrast to the "reference groups,"no W21 and N2 centers, typical to plastically deformed crystals with a high degree of nitrogen aggregation, were found in the samplings of non-luminescent crystals; it was expected as such centers are more specific to crystals with high concentrations of P2 center.

It may be noted that the principal distinction between non-luminescent and luminescent diamond crystals is about a substantial increase in the number of luminescent sampling crystals characterized by a medium degree of plastic deformation occurrence, and is linked to association of paramagnetic centers (W21, N2...) which P2 center "controls," i.e. for which it is indispensable. The origin of the empirically

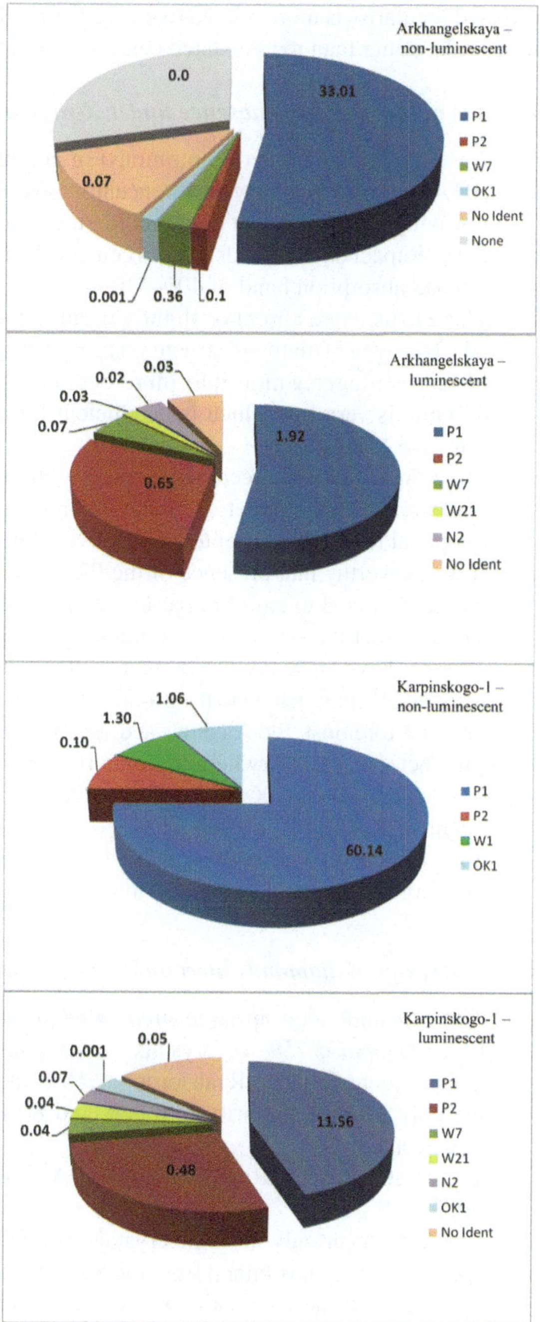

Fig. 4.38 Diagrams of the content of paramagnetic centers in luminescent and non-luminescent diamonds from the Arkhangelskaya and Karpinskogo-1 pipes. The area of the sectors in the diagrams represent the proportions of crystals in which the corresponding centers are found. The figures indicate the average concentrations of centers (ppm) for the studied samples. "No Ident" refers to crystals with unidentified centers. "None" refers to crystals in which the paramagnetic centers are not revealed

revealed regularity is more related to presence or absence of the P2 center samplings in crystals, rather than to associated centers related to plastic deformations.

Reasons of luminescence absence under X-rays are following

1. Absence (totally low concentrations) of complex aggregated nitrogen defects (B1, B2 types) and very low concentrations of N3 centers are a rationale for resolving the issue about diamond inertness to excitation by X-ray waves. The X-ray impact on diamonds excites centers close to N3 in a structure model, with a broad absorption band of 400–440 nm.
2. These centers are absent or slightly occur in the studied non-luminescent crystals. Nitrogen in diamond structure is generally present in single atoms; a degree of nitrogen aggregation into more complex defects, including N3 and 400–440 nm, is very low which is insufficient for glow to appear even in the most sensitive X-rays.
3. The primary factor directly associated with the ability of diamond crystals to emit luminescence under X-ray radiation is presence of paramagnetic P2 (N3 in optical microscopy) centers in its crystalline structure.
4. It is noteworthy that presence of the P2 (N3) centers is related to the ability of diamond crystal to emit luminescence under X-rays but is not its only reason.
5. The fact that the overwhelming majority of diamond specimens from various deposits have an X-ray luminescence slow component shows that the centers responsible for it belong to the group of the most widespread structural defects—first and foremost, the A center and, secondly, P2.
6. The fact that the overwhelming majority of diamond with high content of P1 (C) centers have no X-ray luminescence slow component is attributed to low quantity of A and P2 centers.
7. *Some crystals are characterized by luminescence "quenching"* at the expense of a large numbers of carbonate and water microinclusions.

Characteristics of diamonds inert under X-rays are following

1. *The diamonds with nitrogen predominantly as C-center high concentrations* (20–200 ppm.): 25% of crystals non-luminescent under X-rays contain C-defects identified by IR absorption. 10% of the crystals have C-defects as the only form of impurity nitrogen, and 80%—as dominating together with A-defects.
2. The crystals whose IR absorption have A and B2 bands together with C were not found.
3. C defects occur only in single crystals with faint fluorescence in IR absorption spectra of diamonds mined from the Karpinskogo-1 and Arkhangelskaya pipes. This confirms the assumption that complex defects close to N3 and B2 centers in configuration are responsible for luminescence under X-rays.
4. By their physical type "non-luminous" octahedron-dodecahedroid diamonds shall be Ia, IaAB types, and differ from luminescent by presence of subsidiary

absorption bands linked to microinclusions, which apparently affect X-ray luminescence intensity and accumulation.

5. All the systems have low luminescence intensity in the sampling of non-luminescent crystals from the Arkhangelskaya and Karpinskogo-1 pipes. N3 system occurs in the registered luminescence spectra; it was found in crystals with B2 band in the absorption spectra or in low-nitrogen or nitrogenless crystals.

The similarity of IR spectral types and presence of the weak N3 band in the photoluminescence spectrum suggests that these crystals have faint X-ray luminescence but as these are mainly splinters (plane specimens), they were just shielded from the RL source/detector by other minerals or got no chance to accumulate RL charge sufficient enough to be registered.

6. The EPR studies revealed several differences of diamond samplings (luminescent and non-luminescent under X-rays):
   - Higher P1 center concentration in crystals of non-luminescent samplings;
   - Nearly complete absence of P2 centers in crystals of non-luminescent samplings;
   - Very high values of the nitrogen disagreeability coefficient for crystals of non-luminescent samplings;
   - Absence of W21 and N2 centers in crystals of non-luminescent samplings.

Crucial points in gemological evaluation are following

1. Large crystals are rare among non-luminescent ones, diamonds −12+11 and −11+9 size grades are occasional.
2. All the diamonds from the collection of crystals inert under X-rays are characterized by low gemological characteristics: these are mostly chips and splinters of octahedron-dodecahedroid crystals which cannot be used for cutting into brilliants.
3. Compact dark-colored cubes and crystals in thick coat not used in jewelry prevail. These crystals are applied in abrasive tool manufacturing.
4. Tetrahexahedral and dodecahedral diamonds with microinclusions.

References

Afanasiev, V.P., Lobanov, S.S., Pokhilenko, N.P. et al.: Polygenesis of diamonds in the Siberian craton. Geol. Geophys. **3**, 335–353 (2011)

Beskrovanov, V.V.: Diamond Ontogeny, p. 264. Nauka, Novosibirsk (2000)

Dean, P.J.: Bound excitons and donor-acceptor pairs in natural and synthetic diamond. Phys. Rev. **139**(A588), 588–602 (1965)

Dean, P.J., Male, J.C.: Some properties of the visible luminescence excited in diamonds by irradiation in the fundamental absorption edge. J. Phys. Chem. Solids **25**, 1369–1383 (1964)

Dishler, B.: Handbook of Spectral Lines in Diamond, 467 p. Springer (2012)

Galimov, E.M.: Diamond isotopic composition variations and their relationship link to diamond origination conditions. Geochemistry **8**, 1091–1117 (1984)

Galimov, E.M., Zakharchenko, O.D., Maltsev, K.A., Makhin, A.I.: Isotopic composition of carbon in diamonds from kimberlite pipes of Arkhangelsk region. Geochemistry **1**, 74–76 (1994)

Hainschwang, T., Fritsch, E., Notari, F., Rondeau, B.: A new defect centre in type Ib diamond inducting one phonon infrared absorption: the Y centre. Diamond Relat. Mater. **21**, 120–126 (2012)

Khachatryan, G.K.: Classification of diamonds from kimberlites and lamproites according to distribution of the nitrogen centers in crystals. Ores Miner. **2**, 46–60 (2010)

Kriulina, G.Yu.: Constitutional characteristics of diamond fields of the Arkhangelsk and Yakutian diamondiferous provinces. Abstract from Ph.D. in geology and mineralogy thesis, p. 24. M.: MSU Publications (2012)

Kudryavtseva, G.P., Posukhova, T.V., Verzhak, V.V., Verichev, E.M., Garanin, V.K., Golovin, N.N., Zuev, V.M.: Atlas: morphogenesis of diamonds and their mineral-satellites from the kimberlites and other relative rocks from the Arkhangelsk diamondiferous province, 1st edn., 624 p. M.: Polyarny Krug Publications (2005)

Massi, L., Fritsch, E., Collins, A.T., Hainschwang, T., Notari, F.: The, "amber centres" and their relation to the brown colour in diamond. Diamond Relat Mater. **14**, 1623–1629 (2005)

Makeev, A.B., Kriulina, G.Yu,, Lyutoev, V.P., Ivannikov, P.V.: Peculiarities of Diamond Cuboids from the Arkhangelskaya Pipe, No. 3, pp. 2–6. Newsletter of the Komi Geology Institute Sc.C. of Urals Department of RAS (2011)

Mineeva, R.M., Speransky, A.V., Titkov, S.V., Zudin, N.G.: The ordered creation of paramagnetic defects at plastic deformation of natural diamonds. Phys. Chem. Minerals. **34**(2), 53–58 (2007)

Naletov, A.M., Klyuev, Yu.L., Nepsha, V.I.: Critical stress intensity factor—energy indicator of diamond toughness, pp. 108–115. 60 Years to VNIIALMAZ: Jubilee Compendium. M., Research and Development Institute of Natural, Synthetic Diamonds and Tools, JSC. VNIIALMAZ JSC (2007)

Natural Diamonds of Russia.: Kvaskov, V.B.M. (eds.), 303 p. Polyaron (1997)

Orlov, Yu.L.: Mineralogy of Diamonds, 263 p. M.: Nauka (1984)

Rassulov, V.A., Nikitin, M.V., Patsiansky, F.A., Goryachev, B.E., Kolenchenko, V.V.: Studying spectral and kinetic characteristics of diamonds and associated minerals from Arkhangelskaya pipe. Gorny J. **6**, 84–86 (2009)

Serov, R., Shelementiev, Yu.: Colored diamond from Anabar alluvial deposits, Western Yakutia. Classification on the base of spectroscopic features and colorimetrical data. In: 9th International Kimberlite Conference Extended Abstract (2008)

Taylor, W.R., Jaques, A.L., Ridd, M.: Nitrogen-defect aggregation characteristics of some Australasian diamonds: time-temperature constraints on the source regions of pipe and alluvial diamonds. Am. Mineral. **75**, 1290–1310 (1990)

Taylor, W.R., Milledge, H.J.: Nitrogen aggregation character, thermal history and stable isotope composition of some xenolith-derived diamonds from Roberts Victor and Finch. In: 6th International Kimberlite Conference Extended Abstract, pp. 620–622. Novosibirsk (1995)

Titkov, S.V., Gorshkov, A.I., Solodova, Yu.P., et al.: Mineral microinclusions in cubic diamonds from the Yakutian deposits based on analytical electron microscopy data. Report RAS **410**(2), 255–258 (2006)

Titkov, S.V., Shiryaev, A.A., Zudina, N.N., Zudin, N.G., Solodova, Yu.P.: Defects in cubic diamonds from the placers in the northeastern Siberian platform: results of IR microspectrometry. Geol. Geophys. **56**(1–2), 455–466 (2015)

Vasiliev, E.A., Kriulina G.Yu., Garanin V.K.: The nature and genetic interpretation of weak bands in IR absorption spectra of diamonds. In: Reports of the 10th International Conference "New Ideas in Geosciences", RGGRU, vol. 1, p. 148. M.: Extra-Print (2011)

Vins, V.G.: Optically active defects in diamond—regularities of origination and mutual transformation. Abstract of D.Sc. in Physics and Mathematics Thesis, 40 p. Altay State Technical University, Barnaul (2011)

Yelisseyev, A., Kanda, H. Optical centers related to 3d transition metals in diamond. New Diamond Front. Carbon Technol. **17**(3), 127–178 (2007)

Zaitsev, A.M.: Optical Properties of Diamond: A Data Handbook, 502 p. Springer, Berlin (2001)

Zedgenizov, D.A.: Composition and evolution of crystallization media of fibrous diamonds of lithospheric mantle of the Siberian Craton. In: Abstract of D.Sc. in Geology and Mineralogy Thesis, 46 p. Novosibirsk (2011)

Zinchuk, N.N., Koptil, V.I.: Typomorphism of Diamonds of the Siberian Platform, 603 p. M.: Nedra-Biznestsentr OOO (2003)

Chapter 5
Constitutional Characteristics and Patterns of Diamond Formation in the ADR

In this Chapter petrochemical types of ADR kimberlites are described and correlation with diamond grade is indicated. Diamonds morphological characteristics of diamonds from Yakutian Diamondiferous Province and Arkhangelsk Diamondiferous Region are described after comparative study. Crystallization and evolution of diamonds from ADR analyzed based on constitutional characteristics and general patterns of diamond formation.

5.1 Petrochemical Types of Kimberlites

Currently the scientists investigating the natural diamonds genesis try to determine thermodynamic conditions of their possible genesis, not only look for new evidences of their deep-seated origin. Experiments showed that the diamond phase can crystallize in practically any carbon-saturated environment with the respective thermodynamic parameters. This hypothesis is now confirmed by results of new research studies of natural diamonds and their inclusions, as well as by experimental investigations in the area of diamond synthesis (Zhimulev et al. 2002, 2015; Chepurov et al. 2005, 2009; Pal'yanov et al. 2005; Korsakov et al. 2015; Tomilenko et al. 2015).

First and foremost, we need to determine patterns of crystallization and evolution of the natural diamond matter depending on the rock type. It is common knowledge that the chemistry spectrum of kimberlite rock is quite wide, ranging from magnesian and iron–titanium to titanium-calcium rocks. Kimberlite and related rocks include three essential components: Mg#, Ti#/Fe#, and Ca#.

Thus, numerous bulk chemical analyses of kimberlite pipes were carried out for Yakutian Diamond Province (YDP) and results of cluster analysis revealed that all chemical analyses can be clearly divided into three groups (Vasilenko et al. 1997). These and further research studies (Bogatikov et al. 2007, 2010; Kostrovitsky 2009) demonstrated that chemistry was connected to the rocks petrology and mineralogy, including that of various oxide minerals (Garanin et al. 2009). Later researchers

V. Garanin et al., *Diamonds from the Arkhangelsk Province, NW Russia*, Springer Mineralogy, https://doi.org/10.1007/978-3-030-35717-7_5

attempted to figure out the link between diamond grade and matter composition of kimberlite rocks (Kriulina 2012; Kriulina et al. 2017).

Therefore, we will consider the new results related to identifying petrochemical types of rocks and their link to diamond-bearing capacity.

Diamond grade of kimberlites correlates with their chemistry. Variations of kimberlite composition, as well as their potential diamond grade, can be mainly put down to inhomogeneity of the mantle and their sources present in the region. The diamond integrity and, ultimately, the actual diamond-bearing capacity are affected by secondary factors causing redistribution of chemical components in kimberlites, i.e. all the processes of kimberlite melt fractionation as the latter ascends from the mantle depths.

Identifying kimberlites variations modern researchers (Bogatikov et al. 1999; Kononova et al. 2007, 2011; Kostrovitsky 2009; Tretyachenko 2008; Garanin et al. 2009; etc.) give priority to indicative role of TiO_2, as its low content reflects high pressure of kimberlite formation (Vasilenko et al. 2005). The general petrological classification has been offered (Bogatikov et al. 2007, 2010), which is based on dividing kimberlites into three petrogeochemical types notionally dubbed low titaniferous ($TiO_2 < 1.0$ wt%) (LTT), moderate titaniferous ($1.0 < TiO_2 < 2.5$ wt%) (MTT) and high titaniferous ($TiO_2 > 2.5$ wt%, HTT). We added a brief mineralogical specification to description of each type. It gives a clear idea of primary distinctions of the types in terms of their petrology, geochemistry and mineralogy.

The first group of bodies with the largest diamond potential has the highest Mg#, with the low titanium content (usually < 1 wt% TiO_2) in the rock (high magnesian, up to 35 wt% MgO, low titaniferous). Peridotites prevail among xenoliths, ilmenite differences of mantle rocks are practically non-existent, magnesian and magnesian-ferriferous eclogites are present as well, pyrope-chromite paragenesis is strongly represented, heavy minerals outcrop is low when chromite level is increased. Rocks are characterized by low heavy minerals outcrop. Percentage of pyrope and chromite of diamond paragenesis is very high in the heavy fraction, especially the proportion of the latter. The total content of microcrystalline oxides in kimberlite rocks is low, especially in highly diamondiferous rocks (<1 vol.%). Ti-chromites (up to 95%) with high chrome content (50–55 wt% Cr_2O_3) and low titan content, less than 2–3 wt% TiO_2 prevail in highly diamondiferous rocks, ferriferous and manganiferous ilmenite are found in minor quantities (from 0–10 to 40–60%).

Diamond crystals from low-titaniferous kimberlites (LTT, less than 1 wt% TiO_2) from the M. V. Lomonosov deposit (ADR) and the Botuobinskaya, Nyurbinskaya and Internatsionalnaya pipes (YDP) are characterized by a wide range of nitrogen concentration ($50 < Ntot < 3000$ at. ppm) with low aggregation (NB $\leq 30\%$ on average), low platelets absorption ratios were registered ($2 < P < 5$ cm^{-1}) when maximum position of the absorption band is shifted to the short-wave region of the IR spectrum (1364–1370 cm^{-1}), which proves the diamond crystals are formed at relatively low temperatures (Vasiliev 2007) and also a short period of high-temperature post-crystallization annealing in which defects can be transformed. Deposits are characterized by a high concentration of hydrogen impurity (3107 cm^{-1}), carbonates are present as well (1340 cm^{-1}).Hydrogen impurity in crystals is attributed to the growth mechanism. In diamond crystals formed under the tangential growth mechanism, hydrogen centers are present in negligent quantities, which is typical of Yakutia diamond deposits (Kriulina et al. 2011). Diamonds with fibrous and zonal-sectorial internal structure are typical of the M. V. Lomonosov deposit and are characterized by the highest hydrogen concentrations (Kriulina et al. 2011). These structural features are revealed in fancy variations of yellow and gray tints, zonal-sectorial structure with normal and tangential growth mechanisms (Kriulina 2012).

The second group of kimberlite bodies, the most abundant one, with moderate titanium content (1–3 wt% TiO_2) and moderate Mg# (up to 25–27 wt% MgO) (magnesian, moderate titaniferous); ilmenite ultrabasites are widely spread among xenoliths, pyrope-almandine-picroilmenite paragenesis is strongly represented, heavy minerals grade is high, ilmenite being the most abundant. The total content of microcrystalline oxides in the kimberlite matrix is elevated (up to 5–20 vol.%). The closest ratio of microcrystalline buildups of these minerals (spinel 35–55%, ilmenite 35–50%) is typical of the matrix for highly diamondiferous kimberlites with low perovskite (< 10–15%) and rutile (< 15%) content. In the matrix of less diamondiferous kimberlites, the perovskite content is elevated (from 20–30 to 80–85%), occurrence of ilmenite phases is much lower (ilmenite from 5–10 to 30%), spinel from 10–15 to 50–70%). For the binding mass of high and medium diamondiferous kimberlites, high-chromium picrochromites are typomorphic. They contain 43.1–52.6 wt% Cr_2O_3; 3.7–7 wt% TiO_2; 7.7–13.6 wt% MgO; 1.5–10.1 wt% Al_2O_3; 0–1.3 wt% MnO; 4–17 wt% Fe_2O_3.

Diamonds from the kimberlites of moderate titaniferous type (MTT, 1–1.5 wt% TiO_2) (the Udachnaya, Komsomolskaya, Yubileynaya, Mir pipes, YDP and V. Grib pipe, ADR) mostly originated in the layer-by-layer growth (Beskrovanov 2000). Crystals grow layer by layer at higher temperatures. They are characterized by low nitrogen concentrations (Ntot < 500 at. ppm.), colorless or slightly yellow tinted octahedral crystals. Diamonds were exposed to extended post-crystallization annealing: nitrogen is mostly present in the aggregated B-form (50 < NB < 95%). The model age of MTT kimberlites' enrichment with diamondiferous rocks under the pipes of Daldyn-Alakit region makes 600–700 million years (Kononova et al. 2011).

The high platelets absorption ratio (0–44, on average 8–10 cm^{-1}) and its maximum being located in the IR spectrum ~ 1364 cm^{-1} in diamonds of Daldyn-Alakit region is indicative of diamond growth at highest temperatures. The study of chromites chemical composition confirmed this conclusion (increased content of Cr_2O_3 up to 66 wt%) in diamond inclusions (as reported by N. V. Sobolev). Diamonds are characterized by low concentration of hydrogen impurities (3107 cm^{-1}) and presence of the hydroxyl group (broad peak ~ 3440 cm^{-1}).

Diamonds are mostly represented by octahedrons. This group of bodies includes the renowned Mir, Udachnaya, V. Grib and other deposits. Their diamond potential is also quite considerable, though slightly lower compared to diamond bodies of the first group. They are adjacent to bodies of kimberlites of ADR's Kepina field with low diamond potential and high outcrop of picroilmenite, as well as similar bodies from YDP, for example, Dalnaya, Morkoka, etc.

The third group of bodies is located in the northern part of YDP (Nizhne-Oleneksky and Kuonapsky regions). Their petrochemical composition considerably deviates from typical YDP kimberlites. These bodies are characterized by high titanium content (> 2 wt% TiO_2) and heavy rare earths (HREE Er-Lu = 1.7–6.6 ppm). Picroilmenite is abundant. They are also characterized by high content of microcrystalline oxide minerals (up to 20 vol.% in Montichellitovaya dyke), which are represented by spinels and perovskite only, in approximately equal quantities. The characteristic feature of the ensemble of microcrystalline spinels from the binding mass of the mentioned rocks is significant predominance of ulvospinels and titanomagnetites, the range of compositions being quite wide. Chrome-spinels are either absent whatsoever, or are present in minor quantity. Among chrome-spinels, there are practically no high-chromium variations with low titanium content (over 40–42 wt% Cr_2O_3, < 6 wt% TiO_2), typical of the diamondiferous kimberlites matrix. Diamond potential is extremely low, most pipes are non-diamondiferous and in some pipes there are ore columns with low diamond content, diamonds being strongly corroded (Garanin et al. 2009; Kostrovitsky 2009). Such a group of bodies has not been registered in ADR yet.

In each of these groups of bodies, as the longer the kimberlite column ascent, the more actively diamonds dissolve and rounded crystals develop, diamond potential drops and chromite-ulvospinel-magnetite solid solutions start crystallizing in the binder mass of rocks, the range of compositions being quite wide. On a large scale, perovskite crystallizes in the matrix of kimberlite rocks with low diamond potential.

Earlier, we, along with a number of other researchers, compared characteristics of diamonds and kimberlite rocks containing these diamonds found in Arkhangelsk diamondiferous region and Yakutian diamondiferous province. The comparative analysis showed correlation between typomorphic diamond groups and a certain petrochemical type of kimberlite rocks. Diamonds from the same field and of same petrochemical type of kimberlite rocks displayed most similar parameters (Evans 1992; Taylor and Milledge 1995; Beskrovanov, 2000; Vasiliev 2007; Kopchikov 2009; Kriulina 2012). Overall, the studies revealed consistent decrease in age of kimberlites enrichment as titanium oxide content grows and diamond potential of these kimberlite rocks declines. Moreover, content of diamonds with a high degree of nitrogen aggregation in B-form and platelets concentration goes up. In other words, there is a certain link between properties and quality of diamonds and those of kimberlite rocks containing such diamonds. While evaluating reserves, it is crucial to count quality of rough diamonds, but not just quantity (millions of karats) and content (carats per ton of ore). In that way, value of the raw materials produced can be determined more precisely.

5.2 Morphogenetic Groups of Diamonds from the Deposits in Arkhangelsk Diamondiferous Region and Yakutian Diamondiferous Province

Vast factual basis (Beskrovanov and Spetsius 1991; Beskrovanov 2000; Palazhchenko 2008; Khachatryan 2003, 2010, 2017; Bulanova et al. 2010) obtained as a result of studying distributed nitrogen impurities in diamonds with inclusions of various paragenesis, also from the mantle xenoliths, enabled to identify crystals developed in mantle substrate of a certain composition. Concentration of nitrogen impurities in diamonds of the ultrabasic type usually does not exceed 500 at. ppm with mode 100–200 at. ppm (Khachatryan et al. 2008). Diamonds of eclogitic paragenesis are characterized by "more blurred" bimodal nitrogen distribution (from 0 to 1500 at. ppm), with maximums of 0–100 at. ppm and 300–600 at. ppm. In this paper, we analyze the content of defect and impurity centers in diamonds from LTT and MTT kimberlites (Bogatikov et al. 2010; Kriulina et al. 2011; Kriulina 2012).

Seven morphogenetic groups of diamonds have been identified for kimberlite diamond deposits in Russia. Each group differs for physical, chemical and PT-conditions and formation duration. (Kriulina 2012; Khachatryan 2010, 2017) The groups were identified empirically, and this division was confirmed by cluster analysis, in view of the diamond mineralogical peculiarities, under the petrochemical types of kimberlites. This is illustrated by the diagram showing concentration ratio of total nitrogen and proportion of nitrogen aggregated in B-form (Fig. 5.1). The patterns

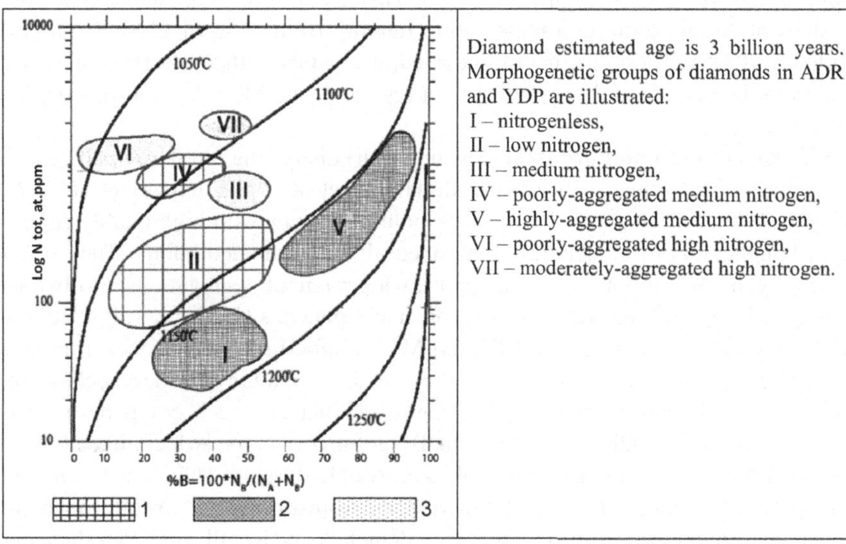

Fig. 5.1 Nitrogen concentrations for crystals of IaAB type diamonds from YDP and ADR based on the data from Taylor and Milledge (1995). Diamond groups identified: 1—specific for LTT and MTT, 2—typical of MTT, 3—typical of LTT

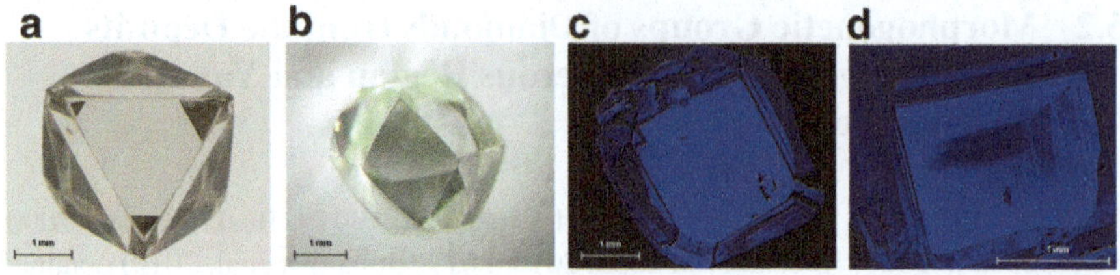

Fig. 5.2 Group I crystals from the Mir (**a**, **c**) and Grib (**b**, **d**) pipes. Morphology of crystals: **a** octahedron, **b** dodecahedroid; internal structure: **c** homogeneous, **d** straight-line octahedral zoning

revealed determine diamond potential of kimberlite bodies and help indirectly assess quality of rough diamonds.

In LTT kimberlites, nitrogenless diamonds are rare, their quantity does not exceed 3% both in the ADR pipes and in Yakutia. In MTT kimberlites, the quantity of nitrogenless crystals is slightly higher (up to 8%).

Crystals of groups II, IV are found in kimberlites of two types. For the Udachnaya, Komsomolskaya and Grib pipes, group II diamonds is principal (group IV is secondary), while for the Arkhangelskaya, Snegurochka, Botuobinskaya, Internatsionalnaya and Mir pipes, group IV is principal, while group II is secondary.

Diamond group I, which we conventionally dubbed "nitrogenless", includes crystals with total nitrogen content of maximum 100 at. ppm ($20 \leq \%N_B \leq 55$, $P \leq 4\ cm^{-1}$ and $H \leq 1\ cm^{-1}$). Such diamonds are most typical of plane-face diamonds of ultramafic paragenesis of the Udachnaya and Yubileynaya pipes, they are also found in the Mir and Grib pipe (Fig. 5.2). This group of crystals underwent post-crystallization annealing at a temperature ranging from 1150 to 1200 °C (Taylor, Milledge, 1995). These are mainly octahedral crystals, either colorless or with a slight tint. Diamonds of this group are not typical of the M. V. Lomonosov deposit and other LTT pipes.

Crystals of diamond group II (low nitrogen) displaying ultrabasic paragenesis are typical of all kimberlites worldwide (Ilupin et al. 1982; Klyuev et al. 1972; Sobolev 1978; Khachatryan 2010), they include crystals with high quality (gemological) indicators of octahedral, dodecahedral and combined habits. They reveal low nitrogen content (Ntot < 500 at. ppm), wide range of aggregations into B-form ($\%N_B \sim 13$–58), average values of hydrogen and platelets $H < 3$ см$^{-1}$, $P < 5\ cm^{-1}$. Such diamonds are abundant in LTT and MTT kimberlites. Their concentration in the Arkhangelskaya pipe is quite high; they are also found in the Snegurochka pipe (where the hydrogen concentration is slightly higher $H = 2.5$, compared to 0.4–0.7 cm^{-1} for other studied pipes) and the Botuobinskaya pipe (which, like all other pipes in Yakutia, contains no or minimal quantity of hydrogen). The group is principal among the MTT kimberlite pipes (the V. Grib, Komsomolskaya, Yubileynaya and Udachnaya pipes). According to references, (Zinchuk and Koptil 2003; Palazhchenko et al. 2006; Palazhchenko 2008; Khachatryan 2010), diamond crystals with inclusions of ultrabasic paragenesis dominate, but there are also crystals with inclusions of eclogitic paragenesis. We should note the optimal combination of defects due to which crystals from the Udachnaya, Yubileynaya and Komsomolskaya pipes possess enhanced toughness properties.

Group III (moderate nitrogen diamonds) is not numerous and is identified only among colorless and gray-tinted crystals of dodecahedral and combined habits from the Arkhangelskaya and Snegurochka pipes. According to Kopchikov (2009), similar crystals are found in the Koltsovskaya and Pervomayskaya pipes of the Zolotitsa field. This group is not present among diamond crystals from the YDP deposits.

Group IV crystals (poorly-aggregated medium nitrogen) are characterized by high nitrogen content ($400 < Ntot < 1500$ at. ppm, $10 < \%N_B < 40$ (23–25% on average) and low content of hydrogen (H ~ 0.5 cm^{-1}) and platelets (P ~ 3.5 cm^{-1}). This group of diamonds dominates in the pipes made by LTT kimberlites (Arkhangelskaya, Internatsionalnaya, Botuobinskaya) and is found in the MTT pipes (the V. Grib, Mir and partially Komsomolskaya pipes). Crystals are characterized by high degree of tempering, preservation of octahedral habits, often with the polycentric growth pattern and absence of etching traces (Fig. 5.3). Based on the data obtained (Fig. 5.1), we can assume that annealing of diamonds in highly diamondiferous bodies was brief, with the range of temperatures being quite limited, from 1050 to 1100 °C (Taylor, Milledge, 1995). According to Garanin et al. (2009), magmatic diamondiferous zones of the Botuobinskaya, Internatsyonalnaya and Mir pipes were most deep-seated and ascent of kimberlite magma was most dynamic. These conditions probably contributed to preserving octahedral habits and small degree of defects aggregation. Tangential mechanisms of growth, layer-by-layer-zonal internal structure and minimum hydrogen content are indicative of diamonds formation in non-aggressive conditions.

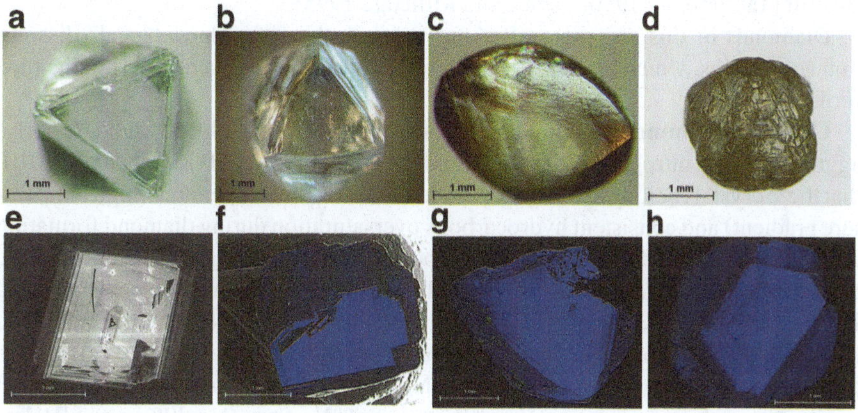

Fig. 5.3 Group IV crystals. Morphology of crystals (**a**–**d**) and their internal structure (**e**–**h**): **a**, **d** octahedron with straight-line octahedral zoning (V. Grib pipe); **b**, **f** combined-shape crystal with octahedral faces and dodecahedral surfaces with straight-line stepped zoning (Arkhangelskaya pipe); **c**, **g** dodecahedroid with traces of plastic deformation and complicated internal structure (Karpinskogo-1 pipe); **d**, **h** combined variety IV crystal, tangential growth mechanism being replaced by normal one

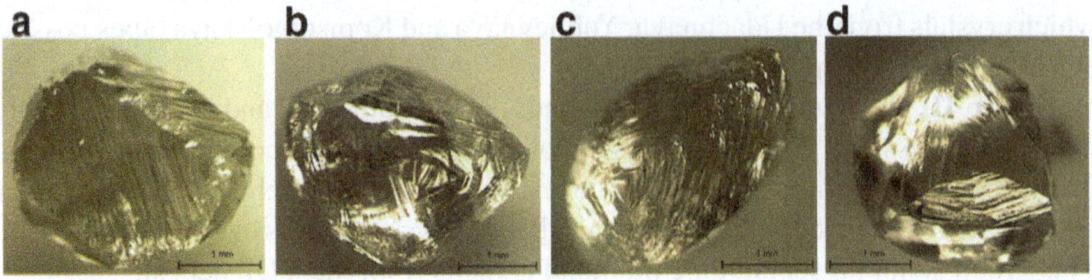

Fig. 5.4 Group V crystals from Komsomolskaya (**a**), Yubileynaya (**b**, **c**), and Udachnaya (**d**) pipes

Diamonds with plastic deformation traces refer to the lower boundaries of this group. Variety IV diamonds form a link between medium nitrogen and high nitrogen crystals in upper area of group IV on the chart (Fig. 5.1). Transparent and translucent octahedrons and combined forms of crystals dominate in the Botuobinskaya pipe, cubes and non-transparent dodecahedroids prevail in the Arkhangelskaya pipe.

Group V of diamond crystals (highly-aggregated medium nitrogen) is typical only for multiphase pipes with complicated evolution comprised by MTT kimberlite: the Udachnaya, Yubileynaya and Komsomolskaya pipes. These pipes contain diamonds with various nitrogen content in the form of A- and B-centers (200 < Ntot < 1500 at. ppm, 60 < $\%N_B$ < 90 (65% on average), with high hydrogen content for YDP ($H_{mean\ value} = 2.5\ cm^{-1}$) and platelets (P ~ 18 cm^{-1}).

Predominantly homogeneous internal structure of diamonds from the Udachnaya and Komsomolskaya pipes (Fig. 5.4) indicates their growth under quite tranquil (equilibrium) conditions, with extended (1150–1200 °C) post-crystallization annealing (Bogush et al. 2008) causing transformation of the considerable part of nitrogen into B-form (Taylor et al. 1990; Taylor and Milledge 1995).

Diamonds of this group are often curve-faced, they often have “local” dissolution accessories, which implies gradual ascent of kimberlite magma during the pipe formation.

Group VI diamonds dominate in all pipes of the Zolotitsa field (poorly-aggregated high nitrogen) (Fig. 5.5). Empirical findings proved (Khachatryan 2010) that these crystals belong to eclogitic paragenesis. Drop in temperature (high temperature gradient) and consistent hydrocarbons oversaturation during diamond formation in the pipes of the M. V. Lomonosov deposit contributed to development of cubic crystals with normal and combined growth mechanisms. In the Arkhangelskaya and Karpinskogo-1 pipes, crystals with these characteristics have cubic or tetrahexahedral habits. Growth rate of these crystals is maximum and impurity capture is also high: high concentrations of nitrogen defects (N_{tot} 800–1500 at. ppm) and hydrogen (H > 1 cm^{-1}), with no or low content of B-centers ($\%N_B$ from 0 to 30%) and platelets (P < 5 cm^{-1}). It is crucial to emphasize that crystals of this group are quite high nitrogen and are not typical for other kimberlite pipes elsewhere in the world. Post-crystallization annealing of crystals from this group was quite brief and proceeded at low temperatures ~ 1050 °C (Taylor, Milledge 1995).

Specific crystals (Fig. 5.6) from the Arkhangelskaya and Snegurochka pipes of dodecahedral (rarer octahedral) habits and the gray color form **group VII**

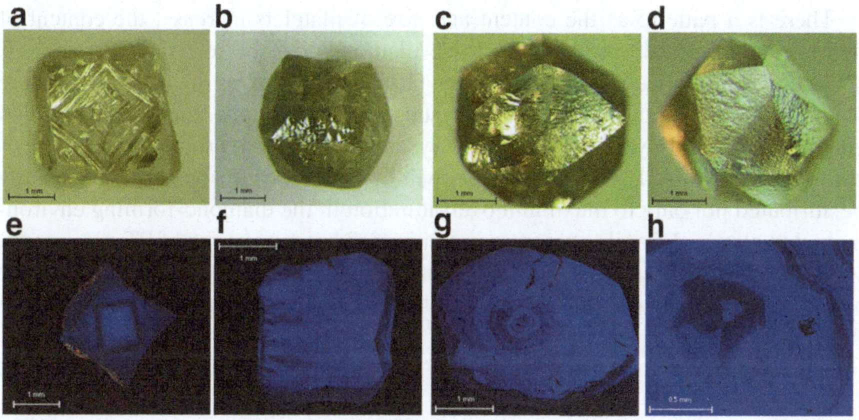

Fig. 5.5 Group VI crystals. Morphology of crystals (**a**–**d**) and their internal structure (**e**–**h**): **a**, **e** cube with saddle-like faces with an octahedral seed, tangential growth mechanism being replaced by normal one (Arkhangelskaya pipe); **b** combined-shape crystal with cubic and tetrahexahedral faces (Arkhangelskaya pipe); **c** dodecahedroid with a drop-shaped sculpture (Karpinskogo-1 pipe), **d** tetrahexahedroid (Karpinskogo-1 pipe); **f**–**h** complicated zonal structure with multiply alternating tangential and normal growth mechanisms

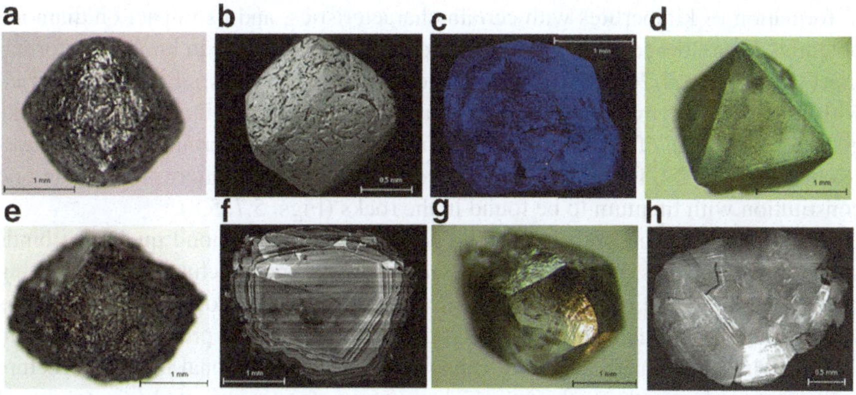

Fig. 5.6 Group VII crystals. **a**–**c** dodecahedroid (**a**) with the surface covered by etch channels and small etch patterns (**b**), mixed and normal growth mechanisms form zonal-sectorial structure (**c**); **d** octahedron with the "graphitized" surface; **e**, **f** combined crystal with octahedral and pseudo rhombic dodecahedral faces, with "graphitized" surface, layer-by-layer zoning in the central zone and the layer-by-layer structure being replaced by fibrous one (?) at the diamond periphery; **g**, **h** dodecahedroid with defective near-surface zone and complicated zonal-sectorial structure

(moderately-aggregated high nitrogen); they are unmatched among diamonds from other pipes. Numerous inclusions, impurities and defects are attributed to the structure of these crystals: the highest content of nitrogen, hydrogen and platelets (Ntot > 1000 at. ppm to 3000 at. ppm, $\%N_B$—25–65, P—12–32 cm^{-1}, $0.6 < H < 22\ cm^{-1}$).

There is a pattern: as the content and size of platelets increase, the content of hydrogen impurity drops. Visually these diamonds can be referred to variety V, according to classification of Yu. L. Orlov (1984). However, as the internal structure was studied, among macroscopically crystals of the same type, there were revealed zonal-sectorial and layer-by-layer-zonal crystals with various carbon isotopic composition ($\delta^{13}C$ from − 10 to − 25‰). Growth of crystals of this group can probably be attributed not only to maximum oversaturation in the diamond-forming environment, but also to dynamic processes accompanied by sharp surge of PT-parameters, which, on the one hand, contributed to partial graphitization of the crystal surface and, on the other, facilitated emergence of planar defects, microclusters and severe fracturing.

5.3 Constitutional Characteristics and General Patterns of Diamond Formation from Low-Titaniferous and Moderate Titaniferous Kimberlites

The research papers of the scientific group supervised by Kononova and Bogatikov (Bogatikov et al. 2007, 2010; Kononova et al. 2011) emphasize the role of titanium in formation of kimberlites with certain characteristics, and its impact on diamond potential. First, low content of titanium indicates high pressure in kimberlite formation (Vasilenko et al. 2005); second, as the temperature increases, silicate melts enrich with titanium and zirconium (Rotman et al. 2005), third, titanium content in rocks goes up, as the age of kimberlites enrichment with diamondiferous rocks decreases (Kononova et al. 2011). Consequently, we can expect certain patterns in the diamond constitution with titanium to be found in the rocks (Figs. 5.7, 5.8).

Titanium has crucial properties that positively affect diamond quality: it binds nitrogen and prevents it from penetrating into the diamond structure, thus decreasing oversaturation of the environment with carbon. Diamonds from MTT kimberlites were formed under calm high temperature conditions in the presence of titanium in the environment, and have homogeneous, layer-by-layer-zonal internal structure with tangential growth mechanisms. These parameters contributed to integrity of diamond octahedral habits despite longer post-crystallization annealing at elevated temperatures, which resulted in enhanced toughness properties of crystals from the Udachnaya, Yubileynaya and Komsomolskaya pipes.

Color. Diamonds from LTT kimberlites are characterized by a maximum spectrum of colors and maximum occurrence of colored gems among diamonds of octahedral-dodecahedral series. Yellow, gray, smoky-brown tinted crystals prevail. It must be stressed that fancy-colored crystals can be found as well: pink-violet, blue, orange (auburn) and yellow. The proportion of colorless gems is minimum (below 10%); together with slightly tinted diamonds they account for 20–25%, while crystals with a yellow tint of various intensity dominate (30–54%).

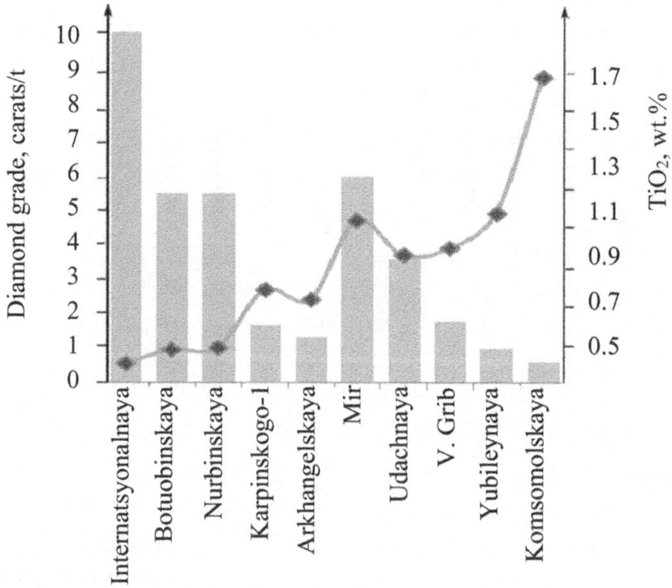

Fig. 5.7 The ratio of diamond grade in the pipes (according to ALROSA Annual Report 2016) and TiO_2 concentration (Garanin et al. 2009)

In MTT kimberlites, among diamonds of octahedral and dodecahedral habits, colorless and slightly tinted crystals dominate (33–60%). The fraction of yellow-tinted crystals with a visible tint is negligible (9–19%). There is a significant number of colored smoky-brown gems (from 14% in the V. Grib pipe to 45% in the Udachnaya pipe) with plastic deformation as consistent with the dislocations gliding mechanism. Among LTT kimberlites, there are rare pink-violet crystals deformed under the mechanism of mechanic fine-scale twinning related apparently to high pressure during kimberlite formation (Kononova et al. 2011; Mineeva et al. 2009).

Shape. Variety I diamonds, dominating in all pipes, are potential raw materials for jewelry; variety II and IV crystals fit for gem-cutting are rarer. Consequently, quantity of variety I diamonds should be considered to evaluate quality of diamonds in the pipe. In LTT kimberlites pipes, variety I diamonds account for 77–88% (except for Internatsionalnaya pipe, 98%). Typomorphic peculiarity of diamond crystals from the Zolotitsa field is considerable concentration of diamonds of cubic habit II, III, presence of IV and "kimberlite" V + VII varieties of Yu. V. Orlov classification (1984). In diamonds of the M. V. Lomonosov field from large size and weight groups, the content of variety III, IV, and V diamonds increases, while proportion of variety II diamonds remains approximately the same. Typomorphic for pipes of the Nakyn field is occurrence of up to 10% of variety IV diamonds. As the sizes increase, so does occurrence frequency of variety IV diamonds and variety VIII polycrystalline aggregates with slightly shrinking proportion of variety I diamonds. In the Internatsionalnaya pipe, the content of variety I diamonds (97–98%) and variety VIII

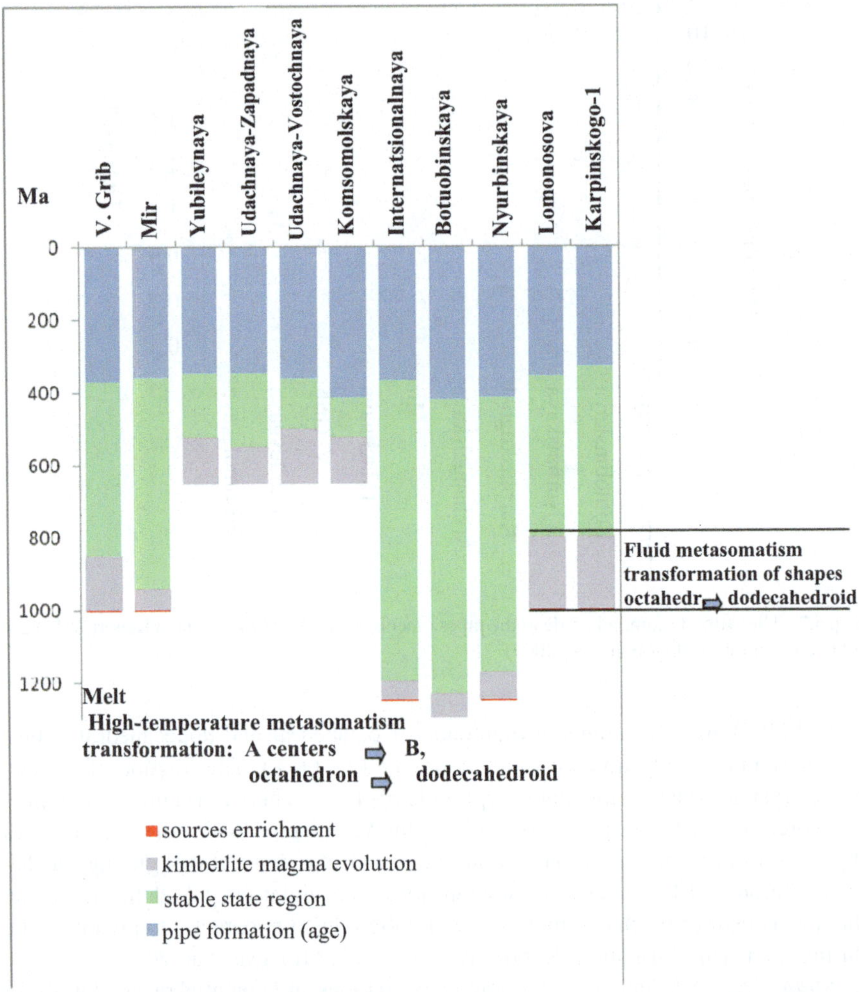

Fig. 5.8 Schematic diagram of kimberlite melt evolution time under the pipes, after the data (Garanin et al. 2009; Kononova et al. 2011)

(1.3–2.2%) diamonds, as well as single variety III crystals, is virtually unchanged in all studied size and weight groups.

In all reviewed MTT kimberlite pipes, variety I diamonds, according to classification of Yu. L. Orlov, dominate (up to 95%) (Zinchuk and Koptil 2003; Palazhchenko et al. 2006), the only exception being the Yubileynaya pipe. Its typomorphic peculiarity lies in considerable proportion of variety IV diamonds (up to 15% in large classes). In this pipe, polyzonal crystals with polycrystalline cubic segments are found. Their inner peculiarities and appearance match those of similar diamonds from the Arkhangelskaya pipe. Variety II diamonds are not typical of MTT kimberlites.

In the investigated kimberlite pipes, as productivity drops, mineralogical types of diamonds become more diverse, consequently, the proportion of raw materials for gem cutting purposes shrinks.

Moderate content of titanium in diamond-forming environment has a positive impact on quality of diamonds as it reduces their ability to be enriched with nitrogen and alleviates the environment oversaturation with carbon. Diamonds from moderately titaniferous kimberlites (diamond potential ranging from high to poor) are mostly colorless, characterized by octahedral habit, almost complete absence of hydrogen and low level of nitrogen (Ntot < 500 at. ppm.) with a large proportion in B-form (25–80%), which meets the conditions of prolonged high temperature annealing.

Diamonds from low-titaniferous kimberlites (diamond potential ranging from extremely high to poor), on the contrary, are relatively low quality, which can be attributed to exposure to lower temperatures during formation and brief annealing.

The largest estimated age of kimberlites enrichment with peridotites and eclogites correlates with small duration of diamond post-crystallization annealing when defects can be transformed and the degree of nitrogen defects aggregation is low. Pipes with maximum enrichment age are characterized by domination of diamonds of certain morphogenetic groups, depending on the type of kimberlite rocks: nitrogenless and low nitrogen for MTT types and medium nitrogen and high nitrogen for LTT bodies (Kriulina 2012). Dormant mantle diamond potential of kimberlite bodies is determined by physical and chemical criteria described in papers by V. S. Sobolev, N. V. Sobolev, A. A. Marakushev et al. (Sobolev 1971, 1974, 1983; Sobolev et al. 1972; Marakushev and Bobrov 2005).

We can predict that in pipes with lower titanium oxide content in kimberlite and enrichment age of over 1 billion years, there is a higher number of yellow-tinted diamonds and there are more chances of finding pink-violet and purple (M2 defect) diamonds. If enrichment age goes down to 0.6 billion years, there is a higher number of colorless (with defects in complex A-, B-, P-forms) and smoky-brown (N2, W7 defects) gems.

In case of long ascent of diamond matter to the surface, conditions for its oxidation are created, that is, transformation of forms into dodecahedral ones with loss of weight until they dissolve completely, as well as recrystallization: growth of diamonds of cubic habit and coated diamonds (varieties II and IV).

Let us consider the diagram (Fig. 5.8) of kimberlite melt temporary evolution under pipes of the Zimny Bereg area.

Early formation process is typical of the kimberlite bodies from the V. Grib and M. V. Lomonosov deposits. The model age of diamondiferous rocks entrainment by kimberlite mass is about 1 billion years (Kononova et al. 2011), that is diamonds spent relatively little time in the mantle after formation (crystallization). Our diamond studies confirm this idea: the crystals are characterized by high temperature annealing of short duration, according to the IR spectrometry findings. As a result, we can see that diamonds of syngenetic color (nearly colorless and pink) dominate, while proportion of epigenetically-tinted brown diamonds is quite small. During initial intrusion, the overlying mantle rocks exert pressure over kimberlite and diamonds as such, but temperatures are lower than those of the mantle, which probably is the reason why some crystals have acquired a pink tint.

Then comes gradual advancement of protokimberlite substrate with entrained diamondiferous ultramafic and eclogitic rocks. This stage can have different duration, subject to presence of aulacogens and toughness of overlying rocks. We register duration of this stage under development degree of secondary minerals in the kimberlite microcrystalline oxide mass and presence of reaction edging formations on spinelides, garnets and other indicator minerals (Garanin et al. 2009). Thus, rocks of the Grib pipe have signs of high temperature metasomatic transformation. In the rocks of the pipes from the M. V. Lomonosov deposit, we can see systemic transformation of minerals from quite high-chrome chromites to ulvospinel, that is, quite a long trend from high temperature to low temperature spinelide varieties. Long gradual cool down with enhanced oxygen fugacity and increased role of fluids results in diamond dissolution and the transformation of forms. In the M. V. Lomonosov deposit we register the maximum number of dodecahedroids compared to all studied deposits. As a result, despite initially favorable mantle conditions, the actual (real) diamond potential of the pipes plummets.

Thus, actual productivity of kimberlites (that is possible drop in mantle diamond potential) depends on the duration of additive impact of high temperature mantle and lower temperature carbonatization inside the pipe. These processes reveal in the kimberlite rocks matrix by diversity, spinelide trends outstretch, reaction minerals development and in the diamonds, by the degree of nitrogen aggregation and curve-facet forms abundance.

5.4 Crystallization and Evolution of Diamonds from ADR Bodies

The shape of crystals depends on external crystallization conditions as well, not just on internal features (crystal structure, nature of defects, etc.). Therefore, morphology of minerals can sometimes be indicative of their development conditions. Based on well-established theories on crystal morphology of natural minerals, the diamond structure includes three main flat grids of carbons parallel to the faces of the cube {100}, rhombic dodecahedron {110} and octahedron {111}. For the diamond, theoretically equilibrium is face {111} and non-equilibrium are faces {110} and {100}. Morphological stability of faces {111} and {100} mainly depends on the degree of the environment oversaturation with carbon. In case of low oversaturation, faces {110} are stable, while as oversaturation increases, faces {100} become more important. Equilibrium faces have maximum capacity for layer-by-layer growth (tangential), while non-equilibrium faces are characterized by normal (fibrous) growth mechanism. Octahedrons growth region adjoins the diamond-graphite equilibrium curve and they crystallize at relatively high temperatures. Cubic diamond crystals generate at relatively low temperatures. They are characterized by normal growth and an

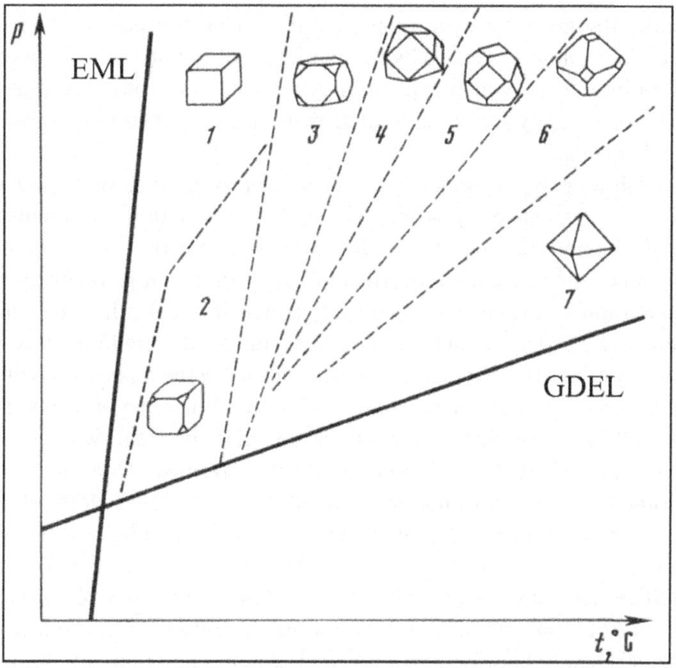

Fig. 5.9 Evolution of the diamond growth habit forms depending on thermodynamic parameters according to (Beskrovanov 2000), EML—eutectic melt line, GDEL—graphite-diamond equilibrium line

increased number of defects compared to octahedrons. Evolution of growth habits of diamonds is believed to take place when thermodynamic parameters drop in the direction of: octahedron {111} → cuboctahedron {111 + 100} → cube {100} (Fig. 5.9).

The surface of rhombic dodecahedron {110} can form both during growth (Ansheles 1955, 1957; Litvin et al. 2004) and during dissolution (oxidation) of diamond crystals which were initially of octahedral and cubic habits (Kukharenko 1955; Chepurov et al. 1985; Khokhryakov and Pal'yanov 1990, 1999; Khokhryakov 2000). Kukharenko (1955) established that in conditions of fast crystallization strongly oversaturated systems give growth to octahedrons with rough platelet face sculpture, octahedron aggregates, octahedral skeleton forms. Normal octahedrons develop at smaller oversaturation. In poorly undersaturated systems, crystals slowly dissolve to form octahedroids. Dodecahedroids represent a final product of long dissolution under conditions of almost equilibrium undersaturated systems.

Ansheles (1955, 1957) came up with theoretical justification for emergence of rhombic dodecahedrons as a result of growth: the growth rate of the most probable faces of the diamond is the same as the rate of their proliferation. Emerging octahedron faces will be flat and will overlap as they keep growing and forming crystals under the layer-by-layer tangential growth mechanism. Further complication of face morphology is caused by accumulation of impurities in front of growing and proliferating layers, which leads to insufficient growth of octahedral faces and development of layer-by-layer sectorial growth of faces under normal or combined growth mechanisms.

Thus, combining typomorphic peculiarities of ADR diamonds and results of genetic analysis we have proposed the evolution diagram for diamonds from Arkhangelsk diamondiferous region shown at (Fig. 5.10). The diagram includes three main stages of the diamond formation: crystallization (growth) (stage I) and partial dissolution in the mantle conditions (stage II), crystallization, dissolution, deformation and mechanical impact on crystals in the moving kimberlite melt. The diagram shows diversity of crystallization conditions at the mantle growth stage and their replacement with possible formation of various simple habit forms (Figs. 5.10, 5.11). Dissolution in the mantle conditions includes the stage with alternation of initial crystal forms. The stage of kimberlite melt movement determines diversity of the environment ascent, deformations overlap, as well as alternation of thermodynamic parameters during movement. The final stage at the diagram shows diversity of morphologic types of diamonds at actual ADR deposits (Fig. 5.10).

As noted above, with low oversaturation of diamond-forming environment with carbon, faces {111} are stable and demonstrate maximum ability to layer-by-layer (tangential) growth. Therefore, in high temperature in-depth ultrabasic sources, plane-face octahedrons formed with rhombohedral faces development (Fig. 5.11) (let us dub these conditions favorable). Such crystals refer to the first population and were found among larger crystals at the V. Grib and M. V. Lomonosov deposits.

Similar morphological features of crystals and nitrogen and hydrogen defects indicate similar stable thermodynamic and chemical conditions of octahedron formation for all ADR pipes and bodies (Fig. 5.11). These are most favorable conditions for forming gem quality diamonds. Various deposits in the province were exposed to favorable (equilibrium) conditions for different periods of time, such periods shrinking as follows: the V. Grib → M. V. Lomonosov deposit pipes, Snegurochka → pipes and bodies of Zolotitsa, Kepina, Verkhotina and Izhmozero fields. This probably explains different proportion of large and small crystals of this diamond population (the most productive and high quality one) among the indicated bodies and it is one of the reasons why ADR pipes have different diamond potential (Kopchikov 2009).

Crystallization is not an uninterruptible process, as confirmed by "coated diamond" crystals and inclusions like "diamond-in-diamond". Origination of "coated diamond" crystals and "diamond-in-diamond" was caused by earlier diamond generation, which, after crystallization was interrupted and growth resumed, overgrows with a thin diamond coat or is entirely included into the new crystal (Bulanova et al. 1986; Barsanov et al. 1988; Kudryavtseva et al. 2005), which proves that crystallization was a complicated and multistage process.

Hydrogen centers do not just characterize the growth mechanism, but also indicate crystallization rate (Blinova 1987)—the higher content of this impurity is, the faster crystals grow. It is obvious that higher crystallization rate leads to smaller number

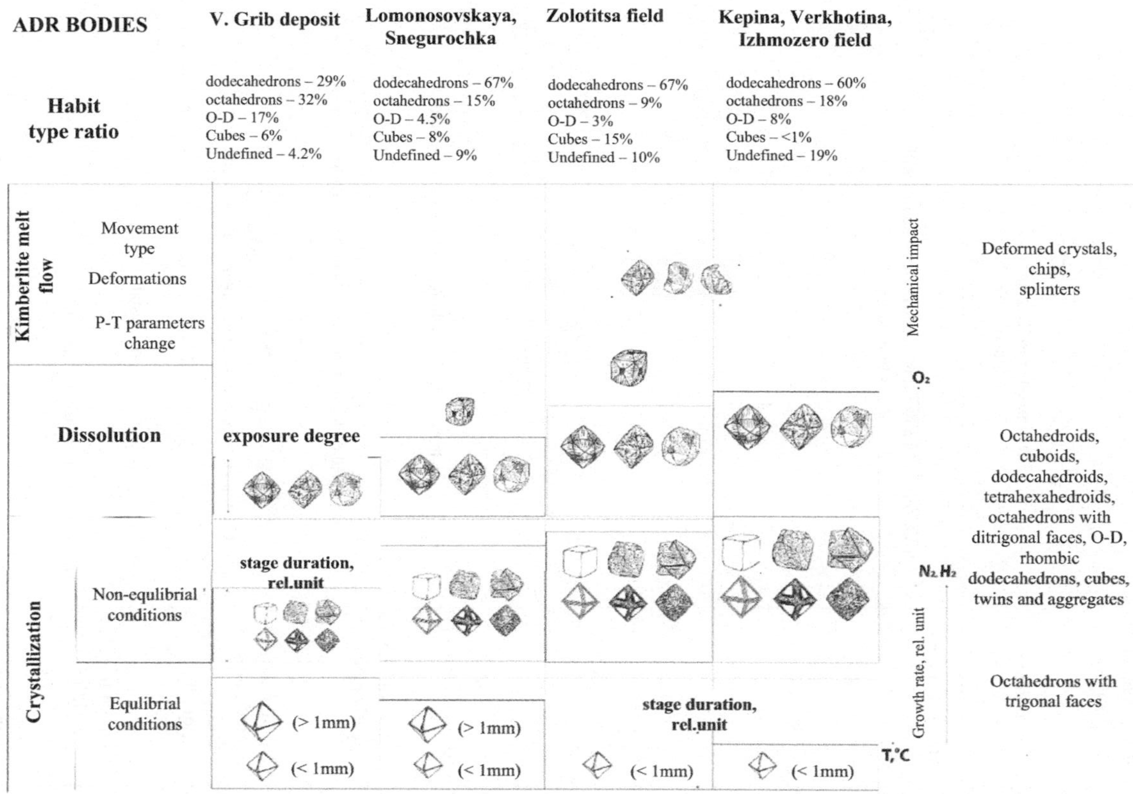

Fig. 5.10 The diagram of evolution of diagrams in the Arkhangelsk diamondiferous province (currently ADR), (Kopchikov 2009)

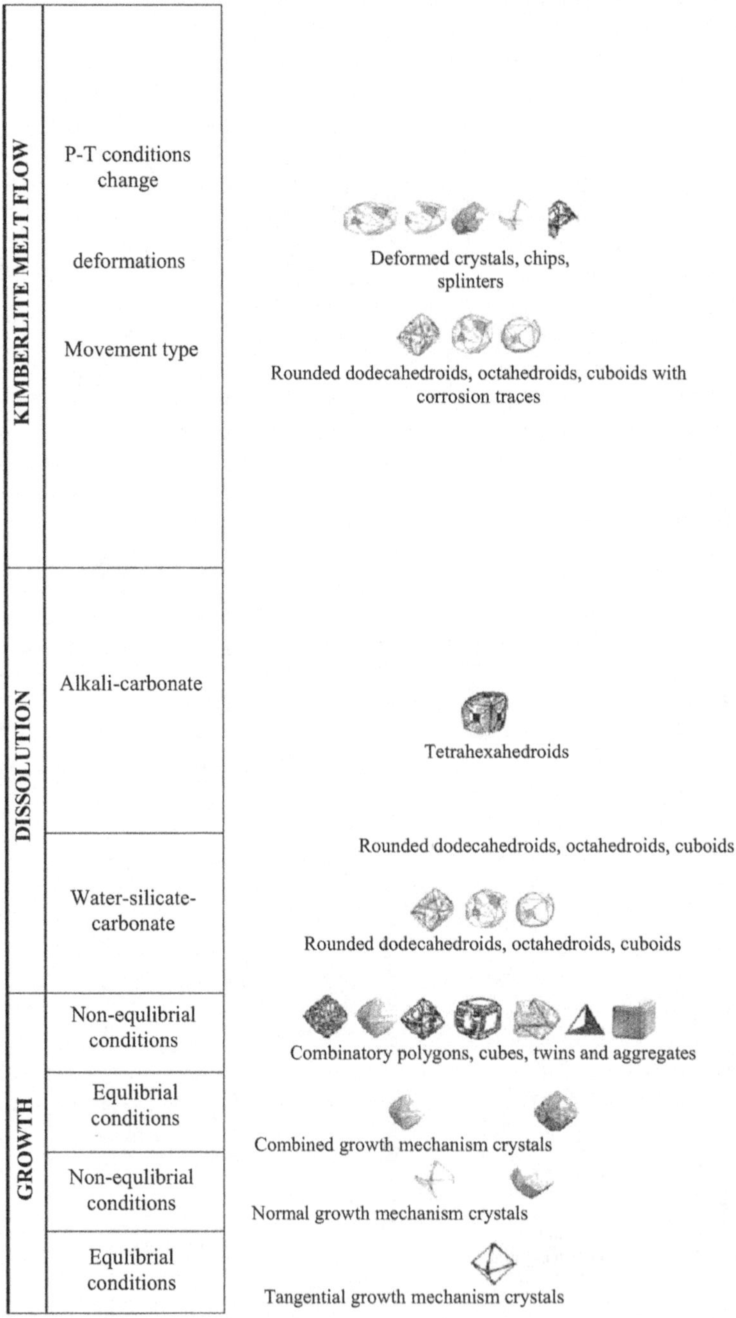

Fig. 5.11 Evolution diagram of simple habit forms of diamonds from ADR (Kopchikov 2009)

of large diamonds with tangential layer-by-layer structure and larger number of those with normal, combined structure. Most crystals of the Grib deposit contain a minor quantity of hydrogen structural impurity and are therefore characterized by a relatively slow growth rate. On the other hand, diamonds from the pipes of the Izhmozero and Verkhotina fields contain hydrogen centers in quite high quantities compared to the V. Grib pipe, that is the growth rate increases as follows: the V. Grib → Lomonosov, Snegurochka → pipes and bodies of Zolotitsa, Kepina, Verkhotina and Izhmozero fields. This is another reason for various ratios of habit types in ADR bodies, with various degrees of diamond potential (Fig. 5.10, 5.11).

Thus, peculiarities of diamond crystallization in ADR include wide variations of nitrogen concentrations ($N_{tot} \approx 0$–2000 at. ppm) and elevated hydrogen content (up to 12 relative units) within one pipe, dominance of crystals with combinatory and normal growth mechanism, as well as sharp distinction in properties of diamonds from the bodies located closely to each other. Most diamonds from the pipes of Yakutian diamondiferous province are characterized by prevailing layer-by-layer tangential growth mechanism, dominance of octahedral crystals and considerably lower levels and fewer variations of nitrogen and hydrogen (Khachatryan 2003; Argunov 2005).

Along with the revealed peculiarities of diamond crystals growth in the mantle, dissolution and corrosion were equally important for their formation in the ADR bodies, which is not typical of most kimberlite pipes worldwide.

Findings of experimental research (Chepurov et al. 1985; Pal'yanov et al. 1985, 2001,2005; Khokhryakov 2000) show that at the initial stage of diamond dissolution, negatively oriented trigons and ditrigonal dissolution layers originate on faces {111}, while tetragonal pits form on faces {100}. Crystals develop rounded surfaces with sheaf-shaped or hackly striations, drop-shaped hillocks and face seam. During dissolution, rounded surfaces expand and finally fully eliminate initial crystal faces. Initial faces of octahedral crystals disappear completely, while the crystal still keeps its octahedral habit and can be characterized as an octahedroid. Initial faces of natural cubes acquire a characteristic form of a tetrahexahedroid. Dissolution of laminar rhombic dodecahedrons exhibits in gradual rounding of pseudo-faces {110}. On all types of crystals there are ditrigonal layers around axes [111] and face seams. During further dissolution, crystals round even more and lose their initial habit completely, acquire a round shape typical of "Ural" type diamonds, rounded dodecahedroids.

In view of this, the number of curve-faced and rounded diamond crystals will primarily depend on how long they were exposed to dissolution environment, and on chemistry of this environment (oxidation–reduction balance). As indicated above, crystals from the V. Grib pipe are characterized by local development of dissolution processes, with systemic and general development of such processes being typical of the pipes from other fields. Dissolution duration and solution aggressiveness, as well as oxygen fugacity, major factor of the diamonds safety (Rudenko et al. 1979), seem to be increasing as follows: the V. Grib → Zolotitsa field pipes, → bodies of Kepina, Verkhotina and Izhmozero fields (Fig. 5.11), which is exhibited by various ratio (dominance) of typical dissolution and corrosion structures. High content of tetrahexahedroids (7–17%) (Zolotitsa field pipes) can imply duration of the diamond dissolution process (Kudryavtseva et al. 2005). Tetrahexahedroids form in the conditions of extended and relatively calm diamond dissolution at final stages of formation in the carbonate melt (Kudryavtseva et al. 2005).

Thus, various diamond potential of rocks, ratio of habit types, as well as safety of crystals directly correlates with intensive dissolution of diamonds in ADR, which resulted in dissolution of small crystals and intensive destruction of large diamonds.

Peculiarities of kimberlite melt flow are also important. As established (Palazhchenko 2008), the V. Grib kimberlite pipe was formed while melt moved quite fast, which accounts for diamonds remaining undamaged.

Diatremes of the M. V. Lomonosov deposit and probably weakly diamondiferous pipes of the Zolotitsa field were formed in more contrasting conditions (occurrence in stages, slower ascent rate, alternation of P&T parameters) of kimberlite melt flow, which is confirmed by a considerable number of rounded (dissolved) crystals, and distorted crystals compared to the V. Grib pipe.

A more complicated and multiple stage process of formation, presumably, with overlay of deformations (Figs. 5.10, 5.11) and alternating abrupt change of thermodynamic parameters are typical of pipes with low diamond potential that mostly represent distorted crystal chips and splinters, which, along with complicated aggressive dissolution, predetermined their low productivity and rough diamonds quality.

Thus, we can determine the main stages of the diamonds generation from ADR bodies and their evolution:

1. Initial formation of earlier diamond crystals at depth (about 200 km) where the lower mantle borders the asthenosphere. Octahedrons growth under the tangential mechanism with rhombohedral faces development—crystals of the first population which at this research stage can be preliminarily considered the first generation.
2. Minor dissolution of early crystals as the composition and physical and chemical conditions changed during growth and crystallization of further diamond generations which built up on earlier crystals with a considerable time gap.
3. Formation of diamonds of dunite-harzburgite and eclogitic associations at the depth of some 150–200 km at P $\approx$ 45–55 kbar and T $\approx$ 1150 °C. As a result of continued growth (additional formation) of deep-earth diamond crystals, structures like "diamond-in-diamond" and "coated diamond" emerged, growth mechanisms changed and reductive conditions were replaced by oxidizing ones.
4. Additional growth and evolution (dissolution) of diamonds in mantle conditions with dissolution mainly taking place in oxidizing environment.
5. Various nature of the kimberlite melt ascent resulted in various ratios of dissolved and corroded diamond crystals. Abrupt change of PT-conditions and deformation contributed to mechanical impacts over diamonds and led to emergence of deformed crystals, chips and splinters. Varying speed of the kimberlite melt ascent resulted in different degrees of safety of diamond crystals in ADR pipes and bodies.
6. Gradual increase in pressure and temperature in the kimberlite melt during the pause before explosion could contribute to crystallization of octahedral crystals

without signs of reabsorption and some macrocrystals during a short period of time from the kimberlite melt in conditions of higher P and T.

7. Explosion, ascent of the kimberlite melt, formation of bodies with local and systemic weak dissolution of diamond macro- and microcrystals at the last stage of formation of ADR pipes.

The combination of the above processes, especially aggressive dissolution, can be indicative of quite specific conditions for the Arkhangelsk province diamond evolution, which resulted in these diamonds being different from those formed in other provinces worldwide.

The processes of ADR diamond evolution reveal that the ratio of habit types of diamond crystals, distribution of structural nitrogen and hydrogen defects in them, quality properties and productivity of ADR bodies are determined by diverse conditions of crystallization, dissolution and nature of the kimberlite melt flow.

References

Ansheles, O.M.: Determining habits of crystals based on their atomic structure. Rep. AS USSR. **101**(6), 1109–1112 (1955)

Ansheles O.M., Towards a theory of crystal growth. Scientific notes of the Leningrad State University: No 215. Geol. Sciences Ser. **8**(84), 179 (1957)

Argunov, K.P.: Diamonds of Yakutia–Novosibirsk, SO RAN Publication, Geo Branch, 2005, 402 p

Barsanov, G.P., Garanin, V.K., Kuznetsova, V.P.: Diamond-in-diamond inclusions from kimberlite pipes of Yakutia. Proceedings AS USSR. Geol. Series. **3**, 70–75 (1988)

Beskrovanov, V.V.: Diamond ontogeny, p. 264. Nauka, Novosibirsk (2000)

Beskrovanov, V.V., Spetsius, Z.V.: Morphology and physical properties of diamonds from mantle xenoliths. Mineral. J. **13**(5), 31–41 (1991)

Blinova, G.K.: Structural impurities as growth mechanism indicators in natural diamond crystals. Reports of AS USSR. **294**(4), 868–871 (1987)

Bogatikov, O.A., Kononova, V.A., Nosova, A.A., Kondrashov, I.A.: Kimberlites and lamproites of the East European Craton—petrology, geochemistry. Petrology **15**(4), 339–360 (2007)

Bogatikov, O.A., Kovalenko, V.I., Sharkov, E.V.: Magmatism, tectonics, geodynamics of Earth: time and space relationship. Nauka, 606 p (2010)

Bogatikov, O.A., Garanin, V.K., Kononova, V.A., et al.: Arkhangelsk diamondiferous province, 522 p. MSU Publishing (1999)

Bogush, I.N., Pomazansky, B.S, Koval'chuk, O.E., Vasiliev, E.A.: Structural features of natural diamonds with varying face morphology from some deposits of Yakutia. In: Problems of predicting and prospecting diamond deposits in closed territories. Yakutsk, pp. 151–157 (2008)

Bulanova, G.P., Michael, J.W., Smith, C.B., Kohn, S.C., Armstrong, L.S., Blundy, J., Gobbo, L.: Mineral inclusions in sublithospheric diamonds from Collier 4 kimberlite pipe, Juina, Brazil: subducted protoliths, carbonated melts and primary kimberlite magmatism. Contrib. Mineral. Petrol. 22 (2010). (Springer, Berlin)

Bulanova, G.P., Varshavsky, A.V., Leskova, N.V., Nikishova, L.V.: Central inclusions as indicators of growth conditions of natural diamonds. Comp. Phys. Prop. Mineral. Natl. Diamond. Yakutsk, pp. 29–45

Chepurov, A.I., Khokhryakov, A.F., Sonin, V.M., et al.: On dissolution forms of diamond crystals in silicate melts at high pressure. Reports of AS of USSR. **285**(1), 212–216 (1985)

Chepurov, A.I., Sonin, V.M., Fedorov, I.I., Chepurov, A.A., Zhimulev, E.I.: Generation of microinclusions in synthetic crystals of diamond under exposure of high P and T parameters. Ores Metals **4**, 49–53 (2005)
Chepurov, A.I., Zhimulev, E.I., Yelisseev, A.P., Sonin, V.M., Fedorov, I.I.: On genesis of low nitrogen diamonds. Geochemistry **5**, 551–555 (2009)
Evans, T.: Aggregation of nitrogen in diamond. In: The properties of natural and synthetic diamond, pp. 259–290. Academic (1992)
Garanin, V.K., Bovkun, A.V., Garanin, K.V., Rotman, A.Y., Serov, I.V.: Microcrystalline oxides from kimberlites and related rocks of Russiav, 498 p. MSU Publication (2009)
Ilupin, I.P., Yefimova, E.S., Sobolev, N.V., Usova, L.V., Savrasov, D.I., Kharkiv, A.D.: Inclusion in diamond from diamondiferous dunite. Reports from AS of USSR, vol. 264. No. 2, pp. 454–456 (1982)
Khachatryan, G.K.: Improved method of evaluating nitrogen concentrations in diamond and its practical application. In: Geological aspects of mineral resources base at ALROSA: state-of-the-art, prospects, solutions. Mirny, pp. 319–322 (2003)
Khachatryan, G.K.: Classification of diamonds from kimberlites and lamproites according to distribution of the nitrogen centers in crystals. Ores Minerals **2**, 46–60 (2010)
Khachatryan, G.K.: Nitrogen and hydrogen in world diamonds as indicators of their genesis and tool for prospecting of primary diamond deposits. Abstract of DSc in geology and mineralogy thesis. TsNIGRI, Moscow, 48 p (2017)
Khachatryan, G.K., Palazhchenko, O.V., Garanin V.K., Ivannikov, P.V., Verichev, E.M.: Genesis of "non-equlibrated" diamonds crystals from Karpinskogo-1 pipe according to cathode luminescence and IR spectroscopy data. Newsletter of Moscow University, series 4. Geology, No. 2. pp. 38–45 (2008)
Khokhryakov, A.F.: Experimental study of the formation of rounded diamond crystals. Newsletter of OGGGGN RAS. No. 5. Issue 15. vol. 1, pp. 80–88 (2000)
Khokhryakov, A.F., Baev, D.A., Pal'yanov, Y.N.: Forms of growth and dissolution of diamond crystals in the $CaCO_3$-C system. In: Works of the 4th international conference "Crystals: growth, actual structure, application". Town of Aleksandrov, vol. 1, pp. 337–341 (1999)
Khokhryakov, A.F., Pal'yanov, Y.N.: Morphology of diamond crystals dissolved in the aqueous silicate melts. Min. J. **12**(1):14–23 (1990)
Klyuev, Y.A., Nepsha, V.I., Dudenkov, Y.A., et al.: Absorption spectra of diamonds of various types. Reports from AS of USSR, 1972. M. 203. No. 5, pp. 1054–1057 (1972)
Kononova, V.A., Golubeva, Y.Y., Bogatikov, O.A., Kargin, A.V.: Diamond content of Zimneberezhnoe Field kimberlites (Arkhangelsk region). Geol. Ore Deposits **6**, 483–505 (2007)
Kononova, V.A., Bogatikov, O.A., Kondrashov, I.A.: Kimberlites and lamproites: criteria for similarity and differences. Petrology **19**(1), 35–55 (2011)
Kopchikov, M.B.: Typomorphic peculiarities of diamond from Arkhangelsk diamondiferous province. Abstract of Ph.D. in geology and mineralogy thesis. MSU, 17 p (2009)
Korsakov, A.V., Zhimulev, E.I., Mikhailenko, D.S., Demin, S.P., Kozmenko, O.A.: Graphite pseudomorphs after diamonds: an experimental study of graphite morphology and the role of H_2O in the graphitisation process. Lithos **236–237**, 16–26 (2015)
Kostrovitsky, S.I.: Mineralogy and geochemistry of kimberlites of West Yakutia. Abstract of DSc in geology and mineralogy thesis. Institute of geochemistry RAS, Irkutsk, 44 p (2009)
Kriulina, G.Y.: Constitutional characteristics of diamond fields of the Arkhangelsk and Yakutian diamondiferous provinces. Abstract from Ph.D. in geology and mineralogy thesis, p. 24. MSU Publications (2012)
Kriulina, G.Y., Garanin, V.K., Rotman, A.Y., Koval'chuk, O.E.: Peculiarities of diamonds from the commercial deposits of Russia. Newsletter of Moscow University, Series Geology, No. 3, pp. 23–34 (2011)
Kriulina, G.Y., Garanin, V.K., Vasiliev, E.A., Zedgenizov, D.A., Bobrov, A.V., Vasiliev, R.V., Vyatkin, S.V.: Microinclusions in diamonds from deposits of different genetic kimberlite types.

Abstract of 11th International Kimberlite Conference. Abstract No 11IKC-004614. Botswana (2017)

Kudryavtseva, G.P., Posukhova, T.V., Verzhak, V.V., Verichev, E.M., Garanin, V.K., Golovin, N.N., Zuev, V.M.: Atlas: Morphogenesis of Diamonds and Their Mineral-Satellites from the Kimberlites and Other Relative Rocks from the Arkhangelsk Diamondiferous Province, 1st edn, 624 p. Polyarny Krug Publication (2005)

Kukharenko, A.A.: Diamonds of Urals. Gosgeoltekhizdat. 514 p (1955)

Litvin, Y.A., Butvina, V.G., Bobrov, V.G., Zharikov, V.A.: The first synthesis of diamond in sulfide-carbon systems: the role of sulfides in diamond genesis. Reports of the AS, vol. 382. No. 1, pp. 106–109 (2002)

Marakushev, A.A., Bobrov, A.V.: Genetic types of diamondiferous rocks. Comp.: Geology of diamonds—the present and the future, pp. 528–541. VSU Publication, Voronezh (2005)

Mineeva, R.M., Titkov, S.V., Speransky, A.V.: Structural defects in natural plastically deformed diamonds: evidence from EPR spectroscopy. Geol. Ore Deposits **51**(3), 261–271 (2009)

Orlov, Y.L.: Mineralogy of diamonds. Nauka, 263 p (1984)

Pal'yanov, Y.N., Chepurov, A.I., Khokhryakov, A.F.: Growth and morphology of antiskeleton crystals of synthetic diamonds. Miner. J. **7**(5), 50–61 (1985)

Pal'yanov, Y.N., Shatsky, V.S., Sokol, A.G., et al.: Crystallization of Metamorphic Diamond: An Experimental Modeling. Reports of AS, vol. 380. No. 5, pp. 671–675 (2001)

Pal'yanov, Y.N., Sokol, A.G., Sobolev, N.V.: Experimental modeling of mantle diamond-forming processes. Geol. Geophys. **46**(12), 1290–1303 (2005)

Palazhchenko, O.V.: Diamonds from the deposits of Arkhangelsk diamondiferous province. Abstract of Ph.D. in geology and mineralogy thesis. MSU, 24 p (2008)

Palazhchenko, O.V., Verichev, E.M., Garanin, V.K., Kudryavtseva, G.P.: Morphological and spectroscopic peculiarities of diamonds from V. Grib deposit from Arkhangelsk diamondiferous province. Article 1. Morphology of diamond crystals. In: Proceedings of IHLs. Series Geology and exploration. No. 2, pp. 14–22 (2006)

Rotman, A.Y., Zinchuk, N.N., Ashepkov, I.V., Egorov, K.N.: Kimberlitic magmatism and diamond grade questions. In: Diamonds geology—present and future. VSU, Voronezh, pp. 856–892 (2005)

Rudenko, A.P., Kulakova, I.I., Shturman, V.L.: Oxidation of natural diamond. New data on minerals of the USSR. Issue. 28. M., Science, 1979, pp. 105–125

Sobolev, N.V.: Deep-seated inclusions in kimberlites and the problem of the upper mantle composition, 264 p. Nauka, Novosibirsk (1974)

Sobolev, N.V.: Diamond paragenesis and the problem of abyssal mineral formation. In: Proceedings of the RMS, vol. 112. Issue No. 4, pp. 389–397 (1983)

Sobolev N.V. Nitrogen centers and growth of natural diamond crystals. Problems of Earth crust and upper mantle petrology, pp. 245–255. Nauka, Novosibirsk (1978)

Sobolev, N.V.: On mineralogical criteria of a diamond potential of kimberlites. Geol. Geophys. **3**, 70–80 (1971)

Sobolev, V.S., Dobretsov, N.L., Sobolev, N.V.: Classification of deep-seated xenoliths and types of upper mantle. Geol. Geophys. **12**, 37–42 (1972)

Taylor W.R., Milledge H.J.: Nitrogen aggregation character, thermal history and stable isotope composition of some xenolith-derived diamonds from Roberts Victor and Finch. Extended Abstract of the 6-th International Kimberlite Conf. Novosibirsk, pp. 620–622 (1995)

Taylor, W.R., Jaques, A.L., Ridd, M.: Nitrogen-defect aggregation characteristics of some Australasian diamonds: time-temperature constraints on the source regions of pipe and alluvial diamonds. Am. Mineral. **75**, 1290–1310 (1990)

Tomilenko A.A., Chepurov A.I., Sonin V.M., Bul'bak T.A., Zhimulev E.I., Chepurov A.A., Timina T. Yu., Pokhilenko N.P. The synthesis of methane and heavier hydrocarbons in the system graphite-iron-serpentine at 2 and 4 GPa and 1200 °C. High Temperatures-High Pressures, vol. 44, pp. 451–465 (2015)

Tretyachenko V.V. Mineragenic zoning of kimberlite area of South Eastern White Sea region. PhD in geology and mineralogy thesis, 28 p (2008)

Vasilenko, V.V., Zinchuk, N.N., Kuznetsova, L.G.: Petrochemical models of diamond deposits of Yakutia, 574 p. Nauka, Novosibirsk (1997)

Vasilenko V.B., Zinchuk N.N., Kuznetsova, L.G., et al.: Rare elements and other trace and volatile elements in light of structural features of kimberlites and their diamond-bearing capacity (exemplified by Aykhal pipe). In: Geology of diamonds—the present and the future, pp. 773–785. Voronezh State University, Voronezh (2005)

Vasiliev, E.A.: Planar optically active centers of diamonds as indicators of diamond formation conditions. Abstract of Ph.D. in geology and mineralogy thesis. Plekhanov State Mining Institute of Saint Petersburg. SPb., 21 p (2007)

Zhimulev, E.I., Sonin, V.M., Chepurov, A.I.: Formation of diamond crystals with faces protruding due to etching. In: Proceedings of the RMS. No. 1, pp. 111–113 (2002)

Zhimulev, E.I., Chepurov, A.I., Sonin, V.M., Pokhilenko, N.P.: Migration of molten iron through an olivine matrix in the presence of carbon at high P-T parameters (experimental data). Reports of AS. vol. 463, Issue No. 1, pp. 72–74 (2015)

Zinchuk, N.N., Koptil, V.I.: Typomorphism of diamonds of the Siberian Platform. Nedra-Biznestsentr OOO, 603 p (2003)

Chapter 6
Method of Morphogenetic Studies of Diamond and Associated Indicator Minerals, Its Application to Solve the Problems of Typing Search Halos and Evaluation of the Degree of Diamondiferous Kimberlite Rocks Archangelsk Diamondiferous Region

Crystal morphology is one of the most important diagnostic features of minerals, since the shape they take is determined primarily by the features of the crystal structure (Honigman 1961). However, the shape of crystals is no less influenced by the conditions of their growth (Buckley 1954; Anshles 1957; Barton et al. 1959; Sunagawa 1986, 1994), so the shape of the crystals may change depending on the conditions of formation (Kan 1967; Kuroda et al. 1977). Even (Fersman 1955) wrote: "the Crystal inevitably bears traces of previous moments of its existence, and by its shape, by the sculpture of its faces, by the small details and features of its surface, we can read its past." Excellent examples of such changes are described for quartz (Askhabov 1979), beryl (Feklichev 1970), zircon (Chervinskaya 1982), phlogopite (Glikin 1982) and other minerals. All of them show that the shape of mineral crystals reacts sensitively to changing environmental conditions (Voitsekhovsky and Mokievsky 1971).

6.1 Theoretical Foundations and Development of Morphogenetic Studies of Diamond and Indicator Minerals

The property of mineral crystals and their aggregates to preserve in specific morphological features information about the conditions of their origin (Sheftal 1947, 1977) and further transformations is the basis of crystallomorphological analysis (Geguzin 1975; Glazov 1981), the principles of which are considered in the works of (Grigoriev 1961) and (Zhabin 1979), (Evzikova 1984) and others. Special attention in such investigations should be paid to the microtopography of mineral grains (Strickland 1971).This allows us to take into account such diverse phenomena as the appearance of simple forms in the process of growth and dissolution of crystals on the way to the formation of a "final form" (Shubnikov 1975), that is, macromorphology, and short-term kinetic processes of dissolution, regeneration, and true

V. Garanin et al., *Diamonds from the Arkhangelsk Province, NW Russia*, Springer Mineralogy, https://doi.org/10.1007/978-3-030-35717-7_6

etching (Tiller 1959), involving only near-surface zones, that is, the actual micromorphology (Feklichev 1970). A detailed investigation of the microtopography of mineral grains, in combination with other research methods (Hirsch 1968), allows, despite the convergence of a number of morphological features, to identify and study typomorphic features inherent in various genetic types of minerals (Zagalskaya, Litvinskaya 1976; Dorokhova, Kaplunnik 1986). Being an energy-sensitive characteristic (Greg and Sing 1970; Nishida 1973), the microtopography of mineral grains reflects the behavior of minerals in various natural processes, so it should play an important role in the complex study of minerals and their formation processes (Murowchick 1987).

Accumulation of a large quantity of factual material and the use of advanced electron microscopic and spectroscopic methods (Soos and Dodony 1990; Burns 1993) showed that learning dependencies: a form of selection—composition—structure—properties, it is possible to recover, to reconstruct the conditions of appearance and disappearance of minerals and identify their typomorphic properties (Chernov 1980; Chesnokov 1974; Chukhrov 1983). This dependence of the mineral forms on the conditions of their genesis (Shafranovsky 1961, 1974) is the basis of morphogenetic analysis, the essence of which was perfectly formulated by A.E. Fersman (1955): "A crystal is not just a geometric body; … it is organically linked to an infinite number of factors and phenomena surrounding it during the period of crystallization, and all these factors and conditions impose their imprint on its surface."

According to D.P. Grigoriev (1975), the concept of "genesis of minerals" includes:

1. comprehension of the origin, growth, and change of individual mineral crystals;
2. physical and chemical mechanism of formation of mineral communities (associations of minerals);
3. geological processes that lead to the formation of mineral deposits and rocks.

Based on this definition, it is clear that a large amount of material for genetic reconstructions is provided by studying the relationship of minerals with each other (Cullen et al. 1973), which allows us to identify and study the stages and phases of mineral formation in deposits of various genetic types (Greer 1978; Deicha 1984; Goshka et al. 1994). Analysis of textures and structures of ores and rocks (Khvorova, Dmitrik 1972; Brodskaya et al. 1986) is an inseparable part of any mineralogical research. For example, it is the shape and size of the inclusions that allow petrographers to distinguish igneous porphyry rocks (Gimeno and Pijolrin 1994) from metamorphic porphyrites. All genetic reconstructions of the stages and phases of mineral formation in hydrothermal ore deposits (Kreiter 1948; Bonev and Radulova 1994; Sergeeva 1986; Jafarov, 1970; Tarbaev and Ovchinnikova 1990; Bogush 1991) and in pegmatites (Gbelsky 1980; Popov 1983; Melchorre et al. 1994) are based essentially on morphometric analysis-the study of the relationship of minerals with each other.

The conditions of joint origin and further joint growth of minerals and their aggregates were studied by (Lemmlein 1945), (Grigoriev and Zhabin 1975) and formed the basis of a special direction in mineralogy-ontogeny of minerals. In the generalizing works of these and other authors (Mokievsky 1983; Pavlishin et al. 1988; Mosing 1980) the theoretical foundations and main trends of changes in the habitus and appearance of crystals, both individual and complex mineral aggregates, in the formation of paragenetic associations that crystallize together in a single process under the same or similar P–T conditions were considered in detail (Belov et al. 1982; Chukhrov et al. 1983). Paragenetic analysis is widely used for reconstruction of mineral formation conditions (Putnis and McConnell 1983; Berry et al. 1987). Its application makes it possible to determine the P–T parameters of crystallization based on the study of the features of the composition of coexisting phases, which was made possible by using the principles of thermodynamics in mineralogy and petrography (Perchuk 2000; Bulakh 1978).

Thus, the study of mineral formation conditions for paragenetic associations, widely used in various fields of mineralogical knowledge, includes two main components: morphometric (ontogenic) analysis, which allows to identify and study the sequence of crystallization of phases (Yushkin 1976, 1977), and chemical-genetic analysis which allows to identify and study the features of the composition, properties and P–T conditions of their formation (Talanov et al., 1975; Garrels and Christ 1968).

A completely different case is the joint presence in the geological body of minerals that have crystallized in different processes and combined into a single mineral complex as a result of post-crystallization changes. A typical example of such combinations is the association of minerals in placers. The joint finding of minerals in this case is also natural and is due to the specific conditions of these post-crystallization processes. Such associations (Bartoshinsky 1984) proposed to call “parasteric". Since the concentrate-mineralogical method of searching for mineral deposits is one of the most important components of geological exploration, this was the impetus for the development of morphometric analysis of minerals in such parasteric associations formed in loose sediments (Gravenor 1981). Work was carried out to study the forms of gold in placers (Popenko 1993). Other minerals were also studied, including diamond and its satellites minerals.

Large-scale searches of diamond deposits conducted in the USSR made it possible to collect and generalize a huge amount of factual material on changing the shape of diamond crystals and its satellites during the formation of scattering halos and placers. At the first stages, the search was conducted on the basis of large-volume testing of loose terrace deposits and all the diamonds extracted from these samples were studied. This approach made it possible, even at the initial stages of the search for diamond deposits in Yakutia, to generalize the obtained material, draw a conclusion about the genetic unity of the primary sources of diamond on the Siberian platform and localize its possible location in the Vilyuy river basin (Bobrievich et al. 1959). This forecast was made possible, in particular, thanks to a competent morphometric analysis of extracted diamonds based on a single scheme of their morphological description (Gnevushev and Shemanina 1970).

In the subsequent period, search operations were conducted using the method of “pyrope mapping”, developed by (Sarasadskikh and Popugaeva 1955), and the attention of researchers switched to studying the morphology of grains of minerals-satellites of diamond, mainly garnets and chromespinelides (Kvasnitsa et al. 1988).

Schemes of morphological description of diamond satellite minerals (pyrope, picroilmenite, and chromspinelides) were developed (Afanasiev et al. 1980, 1984, 2001; Kharkiv et al. 1995).The stages and phases of changes in their morphology were also considered, both during the weathering of kimberlites and during their transfer and deposition in continental and marine conditions. The regularities of changing the color scheme of garnets and granulometry depending on the distance of transportation were studied, and the main forms of hypergenic corrosion and mechanical wear of grains were described.

In parallel, morphometric studies of the diamond continued. In the classic works of (Orlov 1963; Kukharenko 1961; Bartoshinsky 1984), classification schemes were developed and various morphological types of single crystals and polycrystalline diamond splices were identified. These studies allowed us to identify individual regions in Yakutia that differ in the prevailing habitus types of diamond (Bartoshinsky 1961, 1984; Zinchuk and Koptil 2003). Diamonds of non-kimberlite genesis were identified and studied: metamorphogenic (Lavrova 1991; Xu Shi et al. 1992) and impact diamonds (Masaitis 1972), which differ from mantle kimberlite diamonds in composition (Walter et al. 1990), physical properties (Martovitsky et al. 1987) and, most importantly, in size and shape of separation (Nadezhdina et al. 1990, 1993). An ontogenic approach was again used to explain the observed differences in diamond populations in different fields. Studies conducted in the late 80's of the last century (Garanin et al. 1991; Beskrovanov 1992; Taylor et al. 1995) using color cathodoluminescence methods revealed the complex internal structure of most mantle diamonds and the discrete nature of the diamond formation process. In order to take into account the internal heterogeneity of diamond crystals, it was proposed to distinguish in them quasi-homogeneous homogeneous areas with physical properties, called an ontogenic type of diamond (Beskrovanov 2003). Certain types of diamond can form the central, intermediate, or peripheral part of an inhomogeneous crystal (Table 6.1), reflecting the changing conditions of its growth (Garanin et al. 1991). The study of the internal heterogeneity of diamond crystals allows us to identify the stages and phases of their ontogenetic history (origin, growth, change), and the study of the total characteristics allows us to identify the generations and populations of diamonds in individual geological bodies.

The practical solution of the problems of typing and localization of scattering halos gradually led researchers to realize the need to study not only the shape of diamond crystals, but also the grains of satellite minerals in kimberlite tubes. This generalizing work on the example of Yakutia was carried out by (Afanasyev et al. 2001). The main stages and phases of morphogenesis of diamond satellite minerals in kimberlites were identified (Fig. 6.1).

Unfortunately, this scheme does not take into account the changes that occur at these stages with diamond crystals. Also, the forms of change in clinopyroxenes, which are also important genetic satellites of diamond in kimberlites, are not described. In addition, there are no stages in the scheme that precede the entry of xenocrystals of satellite minerals into the kimberlite melt. It is assumed that all of them had an idiomorphic form before.

Table 6.1 Types of diamond substance according to V. V. Beskrovanov's classification 2003

Property	The area of the crystal		
	Central	Intermediate	Peripheral
Morphology	Rounded, octahedron, cube, pyramids $< 100 >_{\alpha}$	Rough-layered octahedron, sharp-edged octahedron, pyramids $< 111 >_{\beta}$	Sharp-edged octahedron
Perfection of the crystal structure	Imperfect, fibrous. Dislocations	Relatively perfect	Perfect
Birefringence	Very high	The alternation of zones with high and low	Low or absent
The absorption of UV radiation	$\lambda < 225$ nm + N9, $\lambda < 260$–300 nm	$\lambda < 280$ nm + N3. Alternating high and low optical density zones	$\lambda < 300$ nm + A
The absorption of IR radiation	K_{B1}, $K_A = K_{B2} = 0$, $K_A \approx K_{B1}, >> K_{B2}$ 3107 system	$K_A \approx K_{B1}, \approx K_{B2}$	$K_A > K_{B1}, >> K_{B2}$
Photoluminescence	Green, yellow-green, yellow, orange	Blue	Not excited
Selective etching	Strong, continuous etching	Strong etching along the borders of growth zones and pyramids	A very faint etching

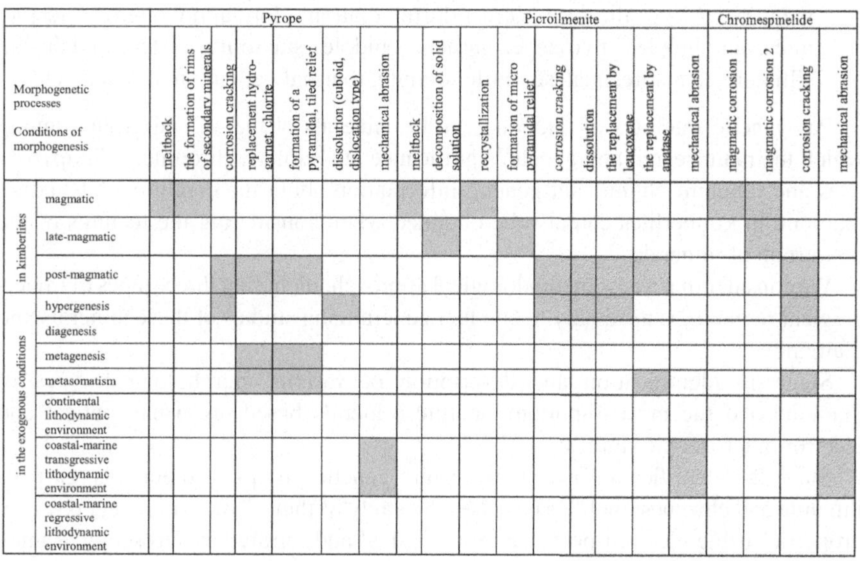

Fig. 6.1 Summary morphogenetic diagram of indicator minerals of kimberlites (according Afanasyev et al. 2001)

It should also be noted that studies of diamond and its satellite minerals were conducted in parallel and independently, and as a result, there was a paradoxical situation in which a significant gap was formed in the scientific literature, which consists in the absence of generalizing works that consider the morphogenesis of diamond and its satellite minerals in a single evolutionary process. We tried to take these shortcomings into account in our research, considering the stages and phases of morphogenesis of diamond and its satellites together from their origin in the depths of the mantle to processing in hypergenic conditions. Our methodological developments are based on the understanding that kimberlites are a complex hybrid rock that combines various mineral associations: high-pressure ones formed in mantle conditions, and low-pressure ones formed in the earth's crust. Based on this, kimberlite minerals are divided into several genetic groups (Kovalsky et al. 1981).

1. Primary kimberlite minerals that crystallize during the fractionation of a kimberlite melt are represented by rock-forming (olivine, phlogopite, calcite) and accessory (Cr-Ti-containing spinelids, perovskite, rutile, etc.) kimberlite minerals.
2. Minerals are products of disintegration of deep mantle rocks of ultrabasic and eclogite composition: olivine, phlogopite, Cr-spinelids, Mg-ilmenite, garnets, clino- and orthopyroxenes.
3. Postmagmatic minerals that develop in the primary or form a fractured-pore mineralization of the hydrothermal or hypergenic stage: saponite, sepiolite, serpentine, talc, mica, leucoxene, carbonates, hematite, chalcedony, barite, gypsum, iron hydroxides.
4. Xenogenic minerals captured by kimberlite magma from metamorphic and sedimentary rocks of the earth's crust during diatrema formation: quartz, feldspar, mica, amphiboles, pyroxene, garnets, epidote, stavrolite, disten, andalusite, sillimanite, apatite, anatase, rutile, titanite, tourmaline, zircon, ilmenite, etc.

Paragenetic satellites of diamond can be minerals of the second group, among which the main search and genetic significance are garnets, chromites, clinopyroxenes and ilmenite. Significant genetic information about the conditions of finding diamond in kimberlites can also be obtained when considering the features of the first group of minerals.

We consider that a new methodological approach, including three stages of mineralogical research, is necessary to conduct generalizing studies of these minerals and diamond.

Stage 1—identification and description of various morphological types of diamond and the most important satellite minerals based on visual, optical and electron-microscopic studies;

Stage 2—identification and division into genetic groups of diamonds, garnets, ilmenite and chromespinelides based on the study of their composition and physical properties using electron-probe analysis, optical and infrared spectroscopy, Raman spectroscopy, electronic paramagnetic resonance and thermomagnetic analysis;

Stage 3—generalization of the obtained materials, identification and description of the stages and phases of morphogenesis of diamond and its paragenetic satellites in the bedrock and loose sediments.

At each stage of such complex research, original methodological approaches are used, which are based on the identification of typomorphic features of diamonds and satellite minerals, the main of which will be discussed below.

6.2 The Methodology for the Study of Morphogenetic Peculiarities of Diamond Crystals

At the first stage, in accordance with the proposed method, the identification and description of various morphological types of diamond is carried out on the basis of visual, optical and electron-microscopic studies.

The method of mineralogical research of diamonds is based on the description of typomorphic features of diamonds. It includes methods and algorithms for conducting relevant mineralogical and instrumental studies. The method is based on a comprehensive study of the mineral and geochemical properties of diamond from different deposits in Russia and the mineral and petrochemical characteristics of the kimberlite rocks that form these deposits.

The classification of diamonds and their division into 7 morphogenetic groups are used. On the basis of mineralogical characteristics and defect-impurity composition of the diamond, the dependence of the distribution of diamond groups in kimberlites of low-titanic and medium-titanic types is established. These groups are illustrated in (Fig. 6.2).

A comprehensive study of the mineralogical features of diamond includes:

Mineralogical study (optical microscopy method):

- determination of crystallomorphology, the nature of formation, and the degree of preservation of crystals,
- description of topomineralogy of the crystal surface,
- color estimation and determination of the color nature;
- determination of the type of distribution of inclusions and their mineral composition;

Study of the defect-impurity composition of diamond (spectroscopic methods):

- photoluminescence (visual assessment of the color and intensity of the glow),
- photoluminescent spectroscopy,
- infrared spectroscopy;
- EPR spectroscopy.

This method establishes the possibility of predicting the presence of diamonds with specific characteristics in a particular deposit, taking into account the different

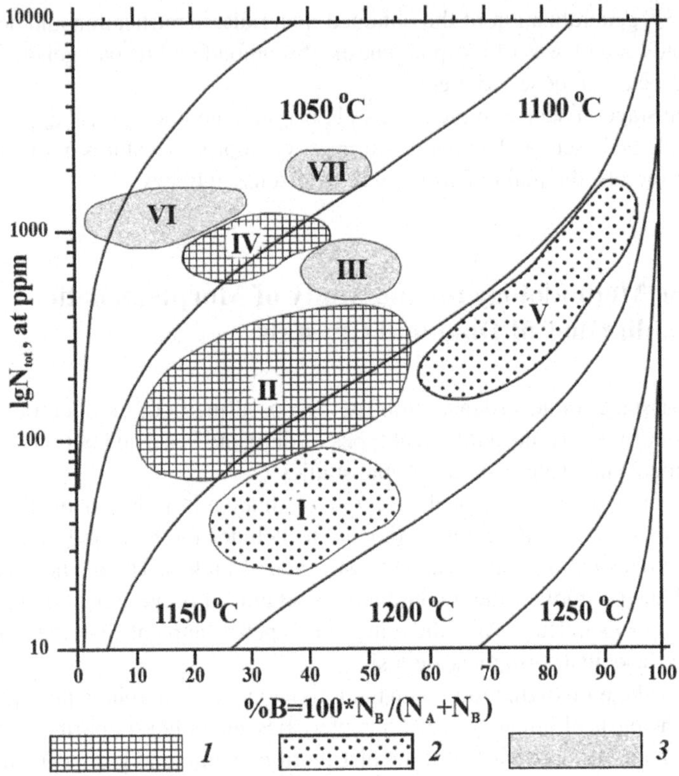

Fig. 6.2 Diagram of the ratio of total nitrogen concentration ($N_{tot} = N_A + N_B$) and the degree of nitrogen aggregation in B1-form (%I) for diamond crystals of the IaAB type from kimberlite bodies of YDP and ADR (Kriulina 2012). Diamond groups that are typical for the main petrochemical types of kimberlites are identified: 1-typical for kimberlite bodies of low-titanium and medium-titanium types of kimberlites, 2-typical for kimberlite bodies of medium-titanium type, 3-typical for kimberlite bodies of low-titanium type. Morphogenetic groups of diamond from ADR and YDP are applied (Kriulina et al. 2013): I—"no nitrogen", II—low nitrogen, III—moderately nitrogen, IV—medium nitrogen, low aggregated, V—medium nitrogen, high aggregated, VI—enriched with nitrogen, low aggregated, VII—enriched with nitrogen, moderately aggregated. The diagram is made using data (Taylor et al. 1995), the estimated age of the diamond is 3 billion years

petrochemical type of kimberlite rocks in the deposits of Yakutia and the Arkhangelsk region.

A comprehensive study of the mineralogical features of a diamond is performed in order to identify the main parameters that allow this crystal to be assigned to one of the seven morphogenetic groups.

Morphological types of diamond crystals

The external shape of diamonds is diverse and has long been the subject of study and discussion. In Russian language literature, the most detailed morphological descriptions of diamond are given in the works of (Fersman 1955, Kukharenko 1955, Gnevushev 1961, Orlov Yu. L., 1963 and Bartoshinsky 1983), and abroad—in the works of (Sunagawa 1986 and Harris et al. 1975).

The first classical morphogenetic classification of natural diamonds was proposed by A. E. Fersman in 1911. He divided them into 2 large groups: 1-crystals of pure growth; 2-crystals whose formation ended in the dissolution stage (Fersman 1955). In the genetic classification proposed by (Yu. l. Orlov 1984), mineralogical varieties of diamond are divided according to the difference in planar growth forms and other typomorphic features that arose during the crystallization of crystals. Each variety is characterized by a visual degree of transparency, absorption characteristics in the IR, visible and UV parts of the spectrum, the nature of the glow in UV rays, and the presence of typical optically active centers.

According to the mineralogical classification of Yu. l. Orlov, potentially jewelry raw materials are diamonds of the I variety, dominating in all tubes, less often there are crystals of the II and IV varieties, suitable for cutting. Therefore, to assess the quality of diamonds in the tube, it is necessary to take into account the number of diamonds of the I variety.

Variety I. Transparent crystals in the form of an octahedron or dodecahedroid, colorless or yellowish. The octahedral habitus of crystals can be severely distorted due to the appearance of combinational uneven surfaces instead of sharp edges. The highest concentrations of A and B centers in the absorption spectra in the IR region are established in the crystals of the I variety. The N3 center is typical. By physical classification—diamonds of type Ia: IaA, IaB, IaAB, IIa, IIb.

Variety II. Transparent cubic crystals with an intense amber-yellow or green color. When observed in a microscope, no zonal structure is detected. All of them contain nitrogen impurity in significant concentrations (type Ib diamonds). Dominated by A and C centers. The crystals show yellow and yellow-orange luminescence.

Variety III. Translucent, colorless, or to varying degrees gray or almost black opaque stones. Possess cubic or occasionally combinational shape (octahedron + the rhombic dodecahedron + cube). Irregular intergrowths, as well as twin intergrowths, are characteristic. Internal structure: in the center of them there is a colorless transparent zone, and in the outer part there are microscopic inclusions. These inclusions cause the gray color of the crystals. Due to the presence of many defects in the outer zone, a large number of small etching patterns develop on the crystal faces when dissolved. The IR absorption spectra are dominated by centers A, sometimes small concentrations of B1 are recorded, in the absence of C and B2. According to the physical classification—type IaA diamonds.

Variety IV. Diamonds in shells ("coated diamond"), in which the outer zone is very different from the inner core. The outer zone is usually cloudy, milky white, grayish, or to varying degrees colored yellow or green. The inner core is usually in the form of a transparent crystal. They usually have the shape of an octahedron or cube. According to the content of impurity centers of nitrogen, the inner parts of diamonds in the shells have the same characteristics as the crystals of the I variety.

Variety V. Dark or completely black due to graphite impurities crystals with a transparent and colorless central part. Usually have the form of an octahedron.

Variety VI. Hidden crystal spherical formations of the "ballas" type.

Variety VII–IX. Polycrystalline intergrowths of crystals, "bort".

Yu. I. Orlov's classification is a complex one that allows taking into account both the morphology of the diamond and the special properties of the crystals that are characteristic of the special conditions of their formation, and it is good for solving certain genetic mineralogical problems. The disadvantage of this classification is that all single crystals of diamond of mantle genesis are combined into one 1-st variety, which does not allow typing of indigenous sources and search halos on the basis of this classification.

A very detailed, logically constructed, crystallomorphological classification of diamonds from kimberlites was proposed by (Bartoshinsky 1983). It identifies 12 groups of crystals (53 types), each of which differs in the predominance of individual simple forms, characterized by different surface curvature {110} and features of the structure of the faces. This classification covers all the morphological diversity of diamonds (Fig. 6.3).

Each group has several types that are transitive from an octahedron to a dodecahedron. The main attention is paid to the individual's habitus, the shape of the growth layers, the way they are layered, and the curvature of the surfaces {110}. It should be recognized that the classification proposed by Z. V. Bartoshinsky is better suited for solving search and exploration problems than other classifications. In addition, it is based on common classification features, which allows it to be used for typing search halos, that is, for linking them to already known kimberlite bodies. However, this classification takes into account mainly the external shape of crystals, and less attention is paid to such important search, genetic and gemological features as granulometry, color, luminescence and other physical properties.

From a mineralogical point of view, the gemological system for evaluating diamond raw materials "SITY"is also of particular interest. The main classification features in this system are: dimension; shape and degree of distortion of the shape; the nature of the face surface; defect (quality) and color of diamonds. This classification is designed to describe large crystals. It should be noted, however,that it allows the most objective separation of groups of diamond crystals, and such sorting, as a rule, gives the most reproducible results, that is, it can be considered the most objective. Thus, this classification can become the basis for mineralogical classification, if it is extended to smaller classes of rough diamonds, unifying the classification features laid down in it.

Since both large and small diamonds that carry important information are often found in indigenous deposits and placers, it is obvious that when describing the morphology of diamonds, it is advisable to use several classifications at once, unifying them. The most objective results of the morphological features of minerals can be obtained by optical and raster electron microscopy with digital imaging. At the same time, often when identifying typomorphic features of diamond crystals, it

Fig. 6.3 Morphological types of diamond crystals according to the classification of Z. V. Bartoshinsky (1983)

is enough to visually describe them using a single simple universal scheme, which allows you to unify this procedure and avoid subjectivism when describing grains by different researchers.

As a result of our research, we have identified the main typomorphic features that allow us to distinguish crystals from different deposits. Since the morphology of crystals is a reflection of the internal structure and conditions of formation of diamonds, the first important property of crystals is their habit.

A planar sharp-edged octahedron is formed with a uniform layer-by-layer growth of the diamond, this is rather an ideal case. Strictly crystallographic, the vast majority of natural diamond crystals contain only octahedral faces, which are almost always observed in combination with sculptured surfaces (often rounded), which are located in place of the edges and vertices of the octahedron. These surfaces in the first approximation correspond to the faces of a rhombododecahedron and a cube, but they are not faces in the conventional sense, so when labeling such surfaces from our point of view, you should add the prefix "pseudo", for example, a pseudo-cube, a pseudo-rhombododecahedron.

Among the minor simple forms, it is necessary to distinguish between active and passive forms of growth, as well as forms of dissolution. They are identified when studying the anatomy of large diamond crystals of different habit, when a violation of the ideal zoning on (Shafranovsky 1961) of the central and intermediate parts of the crystals is detected. The imperfect nature of the faces of many minor diamond forms (surface matte, partial curvature, etching marks) indicates their possible occurrence as a result of dissolution. Faces with smooth and shiny surfaces can be actually growth forms and have growth pyramids. Most of these forms occur due to the degeneration of structurally important octahedral faces and are so-called deceleration faces. The degeneracy is indicated by incomplete shapes and weak morphological significance in the structural and area plans. The reason for this phenomenon may be a decrease in the concentration of the useful component in the melt.

Rounded crystals of rhombododecahedral habit are the result of dissolution processes. Their characteristic feature is the trigon-trioctahedral habitus at the initial stages of dissolution, which is later replaced by a dodecahedral one. Crystals whose appearance is formed during dissolution have rounded face surfaces and require the suffix "-oid" in the name, meaning "rounded". For example, dodecahedroid, tetrahexahedroid.

For individuals of the octahedron—rhombododecahedron series, the main feature of the faces is the distinct laminar structure of the ditrigonal shape. Semicircular crystals also make up an extensive group of individuals, diverse in structure of faces and properties. The degree of their laminar structure varies widely—from coarse-layered individuals with polycentric growing faces to polyhedra, whose laminar structure is detected only under a microscope.

For rounded individuals, a latent laminar structure is characteristic, appearing only near the exits of the triple axes of symmetry in the form of the thinnest sheaf-like hatching; the central part of the faces {110} appears mirror- smooth at normal magnifications and, as a rule, does not show signs of laminarity. These are individuals of rhombododecahedral habitus, on which flat faces {111} are rarely found as insignificant sites that do not affect the habit of crystals. Even more rarely, the outputs of the (Pavlishin et al. 1988) axis are blunted by small rough surfaces {100} or groups of square depressions.

The crystallomorphology of diamond inherits the results of growth processes and the elements superimposed on them that indicate dissolution. In this book, it is customary to specify the shape of the crystal, implying the predominance of growth or dissolution processes in the formation of the diamond's appearance (Fig. 6.4).

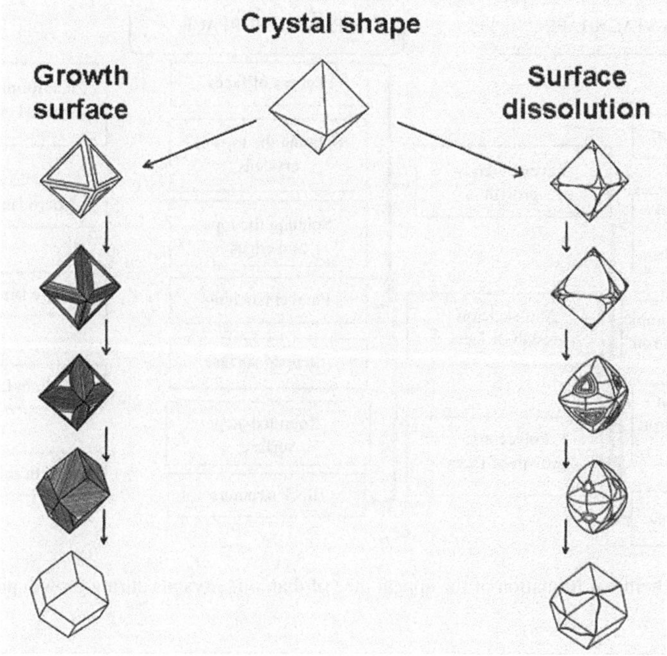

Fig. 6.4 A diagram illustrating the morphogenetic approach in the classification of the diamond

The absolute majority of crystals with the growth form has an octahedron with a layer-by-layer growth mechanism—these are diamonds of the I variety according to the classification of (Y. L. Orlov 1984).

Crystallomorphological description of the diamond

Growth process

As a result of growth processes on the crystals, an anti-skeleton mechanism of growth of faces, polycentric growth of faces can appear, which leads in the initial manifestation to splitting of the crystal vertices up to a sharp step (Figs. 6.5, 6.6, 6.7). In this case, the combination surfaces of the rhombododecahedron are formed and as a result, the crystal takes the form of a pseudo-rhombododecahedron. In this case, the thickness of the end stages that form the surfaces of the rhombododecahedron can be from micro-laminar to significant layering.

During growth processes, an octahedron (a crystal with a layer-by-layer growth mechanism) can be transformed into a crystal of combinational shape:

- with the faces of the octahedron and rhombic dodecahedron surfaces with anti skeleton growth,
- when the octahedron faces are completely replaced, a pseudo-rhombododecahedron is formed,

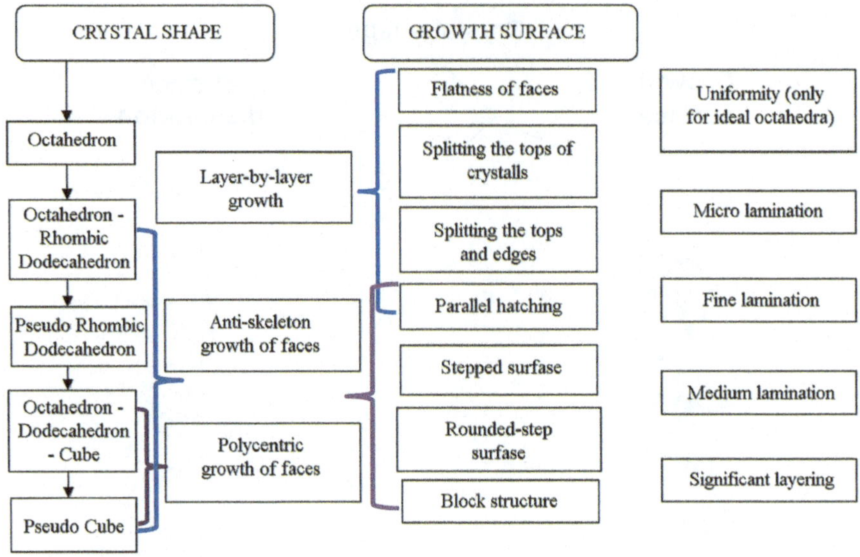

Fig. 6.5 Scheme of formation of the appearance of diamond crystals during growth processes

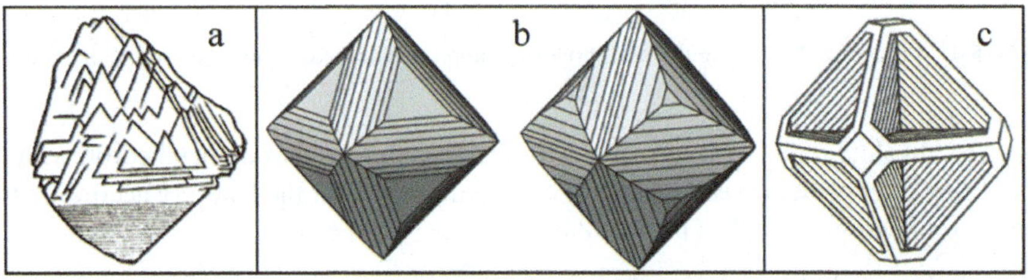

Fig. 6.6 Morphology of crystals with a layer-by-layer growth mechanism: **a**—polycentric character of face growth; **b**—anti-skeleton character of face growth; **c**—skeleton character of face growth (rare)

- with simultaneous anti-skeleton and polycentric growth of the faces—a crystal is formed with octahedron faces and rhombododecahedron and cube surfaces,
- with simultaneous anti-skeleton and polycentric growth of the crystal, as well as with complete replacement of the octahedron faces, a pseudo cube is formed.

Dissolution process

During the further evolution of the diamond, oxidative smoothing dissolution may occur, which leads to rounding of the crystal. The initial form is the octahedron (or octahedron—rhombododecahedron), and the ending—dodecahedroid. Form of transition is matching the shape of the crystal with octahedron faces and surfaces of dodecahedroid formed when dissolved. When dissolved, hatches and sculptures

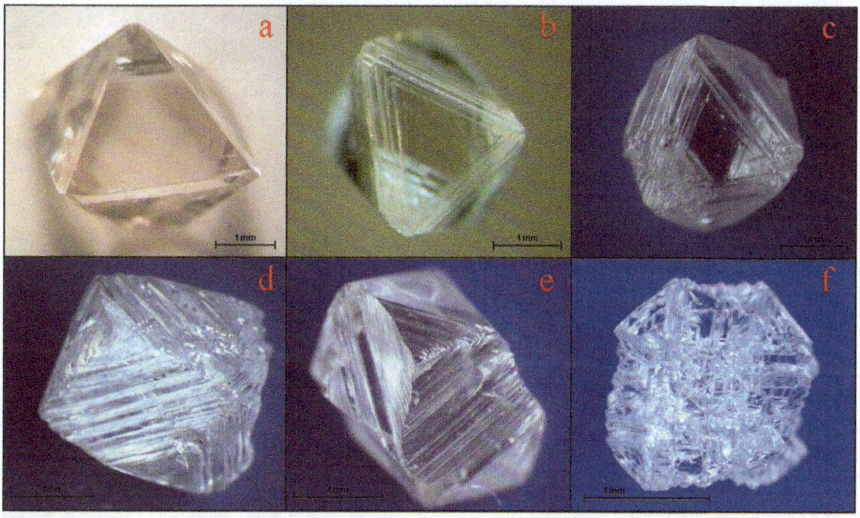

Fig. 6.7 A number of diamond crystals in the expression of anti-skeleton growth: **a**—sharp-edged octahedron; **b**—octahedron with parallel hatching; **c**—the crystal faces of the octahedron and pseudo rhombic dodecahedron surfaces; **d**—crystal with a uniform development of the octahedron faces and pseudo rhombic dodecahedron surfaces; **e**—pseudo rhombic dodecahedron formed by an anti-skeleton growth mechanism; **f**—crystal with a combination of octahedron, pseudo rombic dodecahedron, and pseudo-cube faces formed by an anti-skeleton growth mechanism combined with the polycentric character of the face growth

appear on the surface. The final form of dissolution is a "a curved dodecahedroid of the Ural type" (Figs. 6.8, 6.9, 6.10, 6.11, 6.12, 6.13, 6.14, 6.15, 6.16, 6.17, 6.18, 6.19, 6.20, 6.21 and 6.22).

Elements of the crystal surface

The description of the elements of the crystal surface is used to clarify the crystallomorphological characteristics of crystals, which allow us to conclude about their growth and dissolution.

When describing the crystal surface, the following characteristics are determined:

- Step form of surface. The degree of development of laminar individual layers during growth.
- Traces of dissolution. The dissolution hatches are formed depending on the laminarity of the crystal and its initial structure.
- Traces of plastic deformation processes: sliding lines, shagreen surface, plastic deformation bands.
- Traces of mechanical wear (only visible on crystals from placer deposits).

According to the nature of the surface, depending on the degree of development of sculptural forms on the faces, diamond crystals are divided into 3 main groups:

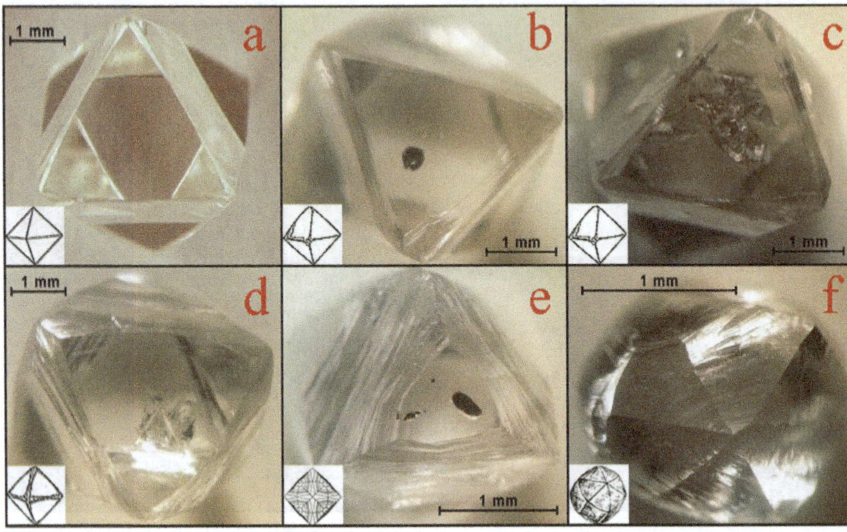

Fig. 6.8 Diamond crystals with various degrees of dissolution processes: **a**—octahedron with flat faces and sharp edges; **b**—octahedron with rounded vertices; **c**—octahedron with rounded vertices and edges with splinter hatching; **d**—octahedron with sheaf- shaped hatching; **e**—combinational shape crystal with octahedron faces and dodecahedroid surfaces; **f**—dodecahedroid

- smooth-sided crystals;
- crystals on the surface of which the sculptural forms are expressed weakly, slightly;
- sharply sculpted crystals, on the surface of which various sculptural details are clearly displayed.

The first group includes crystals with a smooth glossy surface—all smooth-sided sharp-edged octahedra and curved forms with a thin-laminar surface structure, including very thin-layered sheaf-shaped and concentric striations (Figs. 6.8, 6.23, 6.24, 6.25, 6.26, 6.27, 6.28, 6.29, 6.30, 6.31).

The second group includes crystals that combine smooth and sculptured forms-octahedral crystals with trigonal and ditrigonal growth sculptures (groups II and III according to the classification of (Bartoshinsky 1983) and curved dodecahedroids with clearly manifested splinter, teardrop or mosaic-block sculptures (Figs. 6.8, 6.15, 6.16, 6.17, 6.18, 6.19, 6.23, 6.24, 6.25, 6.26, 6.27, 6.28, 6.29, 6.30, 6.31, 6.32, 6.33, 6.34, 6.35, 6.36, 6.37, 6.38, 6.39, 6.40, 6.41, 6.42, 6.43, 6.44, 6.45, 6.46, 6.47).

The third group includes crystals that do not have smooth faces, with numerous and diverse sculptures on the surface, characterized by sharp stepped surfase. These are diamonds that usually correspond to the categories Cleavage, Irregulars and Makeables according to the gemological classification (Figs. 6.12, 6.13, 6.14, 6.25, 6.26, 6.27, 6.30). 8

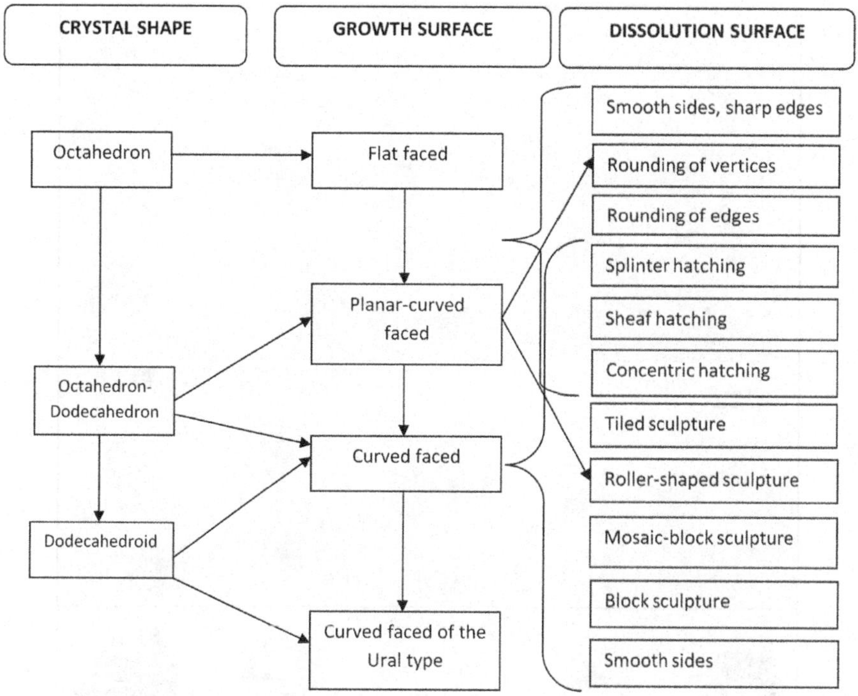

Fig. 6.9 Diagram of changing the shape of a diamond crystal during growth and dissolution processes

Figures 6.23, 6.24, 6.25, 6.26, 6.27, 6.28, 6.29, 6.30, 6.31, 6.32, 6.33, 6.34, 6.35, 6.36, 6.37, 6.38, 6.39, 6.40, 6.41, 6.42, 6.43, 6.44, 6.45, 6.46, 6.47, 6.48, 6.49, 6.50, 6.51, 6.52 show the main types of face morphology.

Degree of preservation.

Diamond crystals are divided into three groups according to safety classes: whole and damaged crystals; broken crystals and fragments; fragments.

The first group—"whole and damaged crystals" - includes diamonds that have completely preserved their primary shape (Figs. 6.53, 6.54 and 6.55), as well as crystals with small natural and man-made defects that make it easy to restore their primary appearance. The second group includes diamonds with only partially preserved primary surface (Fig. 6.56), and it is not possible to reconstruct their initial habitus from the preserved fragments of the primary cut. The third group includes diamonds that have no primary surface at all, diamonds that are entirely formed only by chipped surfaces (Fig. 6.57).

Fig. 6.10 Crystals with varying degrees of lamination from a thin- layered octahedron to a pseudocube with significant layering

Color

The color of a diamond is a property that is perceived visually, and this perception is influenced by several parameters: the actual color shade (tone), its intensity, the distribution of the volume of the crystal, and the transparency of the crystal (Fig. 6.58.)

In the genetic interpretation of color, it is necessary to distinguish the color formed in the crystal during growth (syngenetic) and superimposed during the evolution of the diamond after its growth (epigenetic). In this regard, the mineralogical description of the color provides for the separation of the color (color tone) of the crystal into syngenetic and epigenetic components.

According to the color shade, diamonds with the following colors are traditionally distinguished: Syngenetic colors:

- Colorless—perfectly transparent colorless diamonds or with a slight tint, very faintly visible against the background of white paper.
- Yellow—crystals that have all shades of yellow, including honey-yellow, lemon-yellow.

Fig. 6.11 Crystal matching shape with the octahedron faces and surfaces of the pseudo rhombic dodecahedron. Formed with anti-skeleton growth

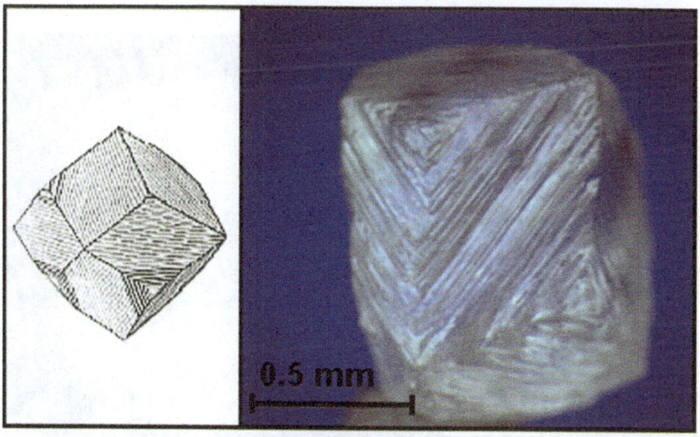

Fig. 6.12 Pseudo rhombic dodecahedron. The crystal is formed by the mechanism of anti-skeleton growth, it surface is composed of the ends of trigonal octahedral plates. Crystal without a facet seam

- Gray and black are diamonds with numerous minute dark-colored inclusions, usually translucent or opaque crystals of technical quality, with a zonal distribution of color, identified by Y. L. Orlov in the III, V, VII varieties.
- Blue—rare diamonds, the color of which is presumably associated with the presence of an impurity of boron in the structure of the crystal.

Epigenetic colors:

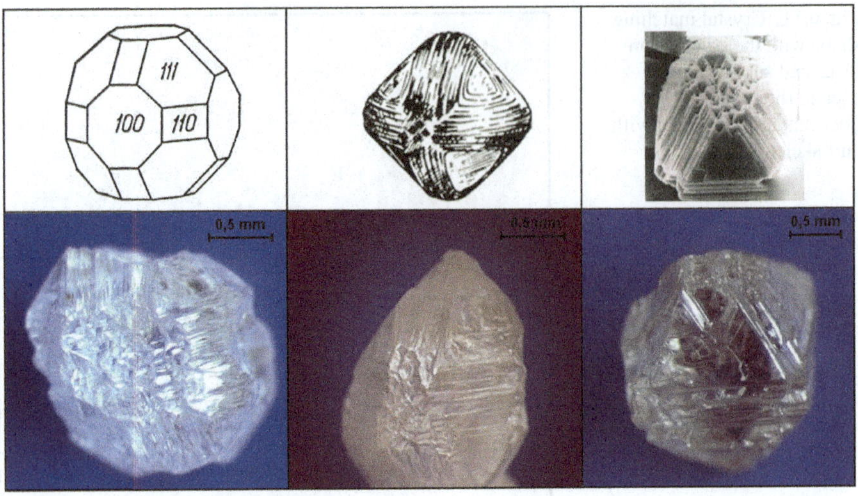

Fig. 6.13 Combination shape crystals with octahedron faces and pseudo- rhombododecahedron and pseudo-cube surfaces. Formed with simultaneous anti-skeleton and polycentric character of development of the octahedron faces

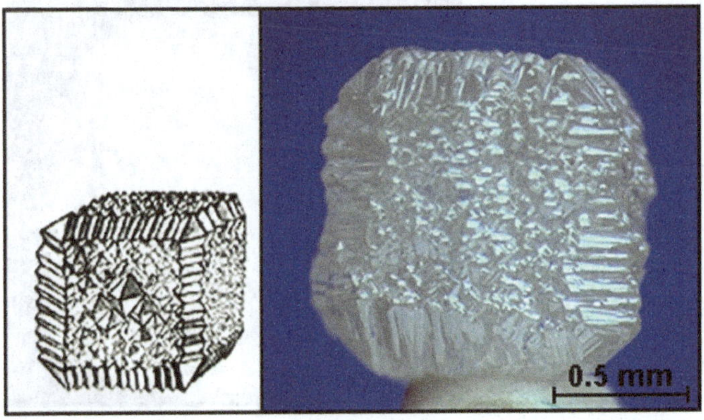

Fig. 6.14 Pseudo-cube. Formed by the simultaneous anti-skeleton and polycentric character of development of the octahedron faces

- Brown, smoky-brown, purple-brown—a group of transparent and translucent diamonds with traces of plastic deformation processes.
- Pink, pink-brown-crystals with sharp bands of plastic deformation.
- Green (yellow-green, bluish-green)—diamonds with characteristic spots of pigmentation on the surface, resulting from natural radiation exposure.

Different deposits differ markedly in the content of colorless and yellow-colored (syngenetic type of color), and brown (epigenetic type of color) crystals. When

Fig. 6.15 Combination crystals with octahedron faces and dodecahedroid surfaces. Formed by the smoothing dissolution of octahedra

applying brown color to colorless and gray, the visually perceived color of the crystal is smoky (gray)-brown, and yellow—tobacco-brown.

The intensity of color is usually estimated with the following gradation:

- colorless,
- with a slight shade of color,
- visible color shade,
- color (obvious color)
- intense color.

The color distribution over the volume of a diamond crystal is obviously divided into two types: uniform and non-uniform. The heterogeneous distribution is manifested, among other things, in several typical variants:

- diamonds in "shells" ("shirts"), differing in color and transparency;
- spots pigmentation;
- zonal color;

Fig. 6.16 A dodecahedroid is a rounded 12-sided crystal with a face seam, the surface formed as a result of polishing dissolution. The final form of dissolution of diamond crystals

Fig. 6.17 Cube of the II variety according to Yu. I. Orlov's classification. Crystals are formed under the normal mechanism of growth of faces

- uneven color associated with plastic deformation.

Transparency is an important factor that affects the perception of others. In the mineralogical description, transparency is evaluated by the following gradation: highly transparent, transparent, translucent, light-transmitting, opaque.

Studies of mineral and gas–liquid inclusions in diamonds are also of primary importance for the purpose of reconstructing the conditions of diamond genesis (Takaoka and Ozima 1978). The study of the chemical composition of solid inclusions

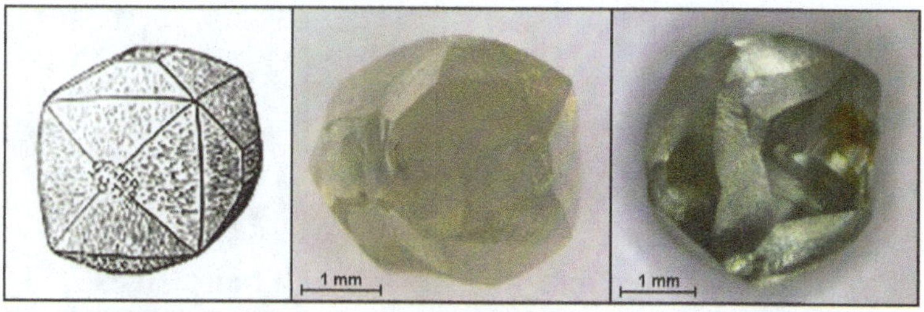

Fig. 6.18 A combination shape crystal with cube and tetrahexahedron faces of the II variety according to Yu. l. Orlov's classification

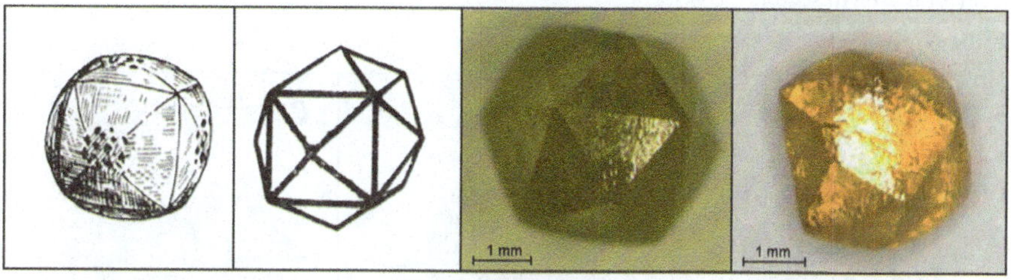

Fig. 6.19 Tetrahexahedron of the II variety according to Yu. l. Orlov's classification

Fig. 6.20 Cube of the III variety according to Yu. l. Orlov's classification. Crystals are formed by a zonal-sectorial growth mechanism

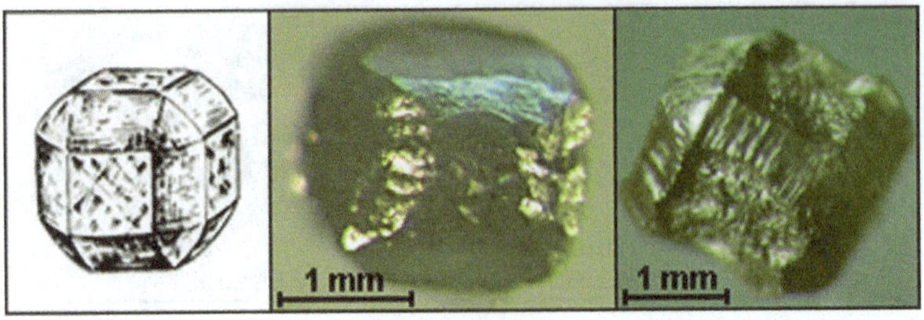

Fig. 6.21 Cube with combination surfaces of a dodecahedroid of the III variety according to Yu. l. Orlov's classification. Formed with a zonal-sectorial growth mechanism, it may have fragments of octahedral growth sectors

Fig. 6.22 Polycrystalline aggregates of VII-IX varieties according to the classification of Yu. l. Orlov (1984); **a**—clearly grained; **b**—fine-grained; **c**—micro-grained

(Sobolev et al. 1997; Meyer 1987; Bulanova et al. 1993) showed that kimberlite diamonds contain minerals belonging to two paragenetic associations: ultrabasic and eclogite (Schuize et al. 1996). Inclusions in diamonds from associations of the first type are mainly represented by olivine, pyrope, diopside, chromite, less often enstatite, picroilmenite, and zircon. Inclusions in diamonds from associations of the second type are represented by pyrope-almandine garnet and omphacite, as well as rutile, ilmenite, coesite, kyanite (disten), corundum, as well as magnetite and sanidine. The prevalence of various types of inclusions in diverse deposits is different (Kharkiv et al. 1992,1995) (Fig. 6.59).

Additional information about the genesis of diamonds can also be obtained by studying their isotopic composition. Studies (Galimov 1984, Galimov et al. 1989) have shown that diamonds are divided into 2 groups according to this indicator. Ultrabasic paragenesis diamonds have a narrow range of variation of $\delta^{13}C$ (relative to the reference distribution) from $-19^0/_{00}$ to $+2^0/_{00}$, corresponding to the mantle

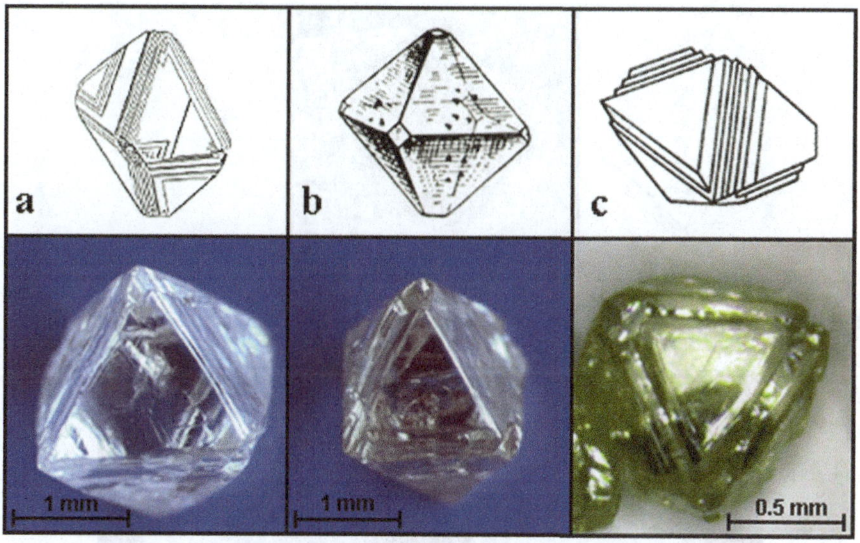

Fig. 6.23 **a**—octahedron with split edges; **b**—octahedron with negative vertices; **c**—octahedron with grooves on edges—typical of crystals of octahedral or transitional habit, along the edges, grooves with parallel or sheaf-like striation on borts are developed.

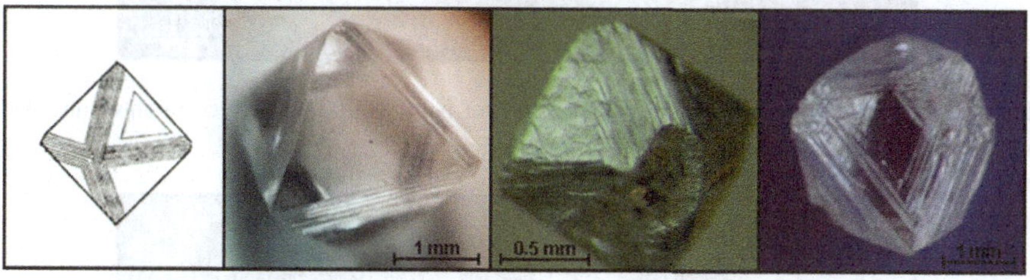

Fig. 6.24 Parallel striation crystals of octahedral habit, composed of trigonal layers with rectilineal contours, the area of each layer being smaller than that of the underlying layer

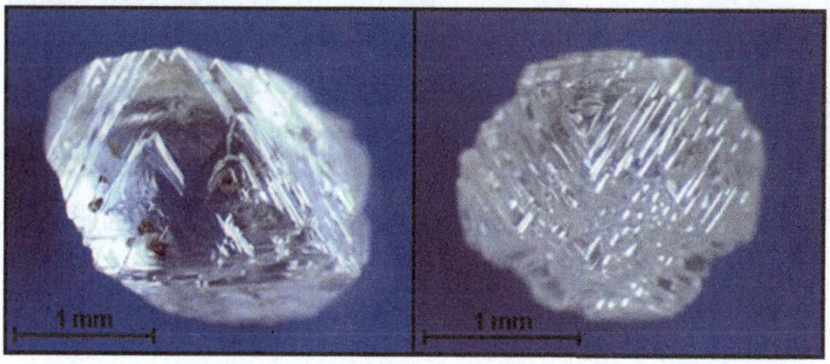

Fig. 6.25 Polycentric crystals of octahedral or transitional habit, on the faces of which packs of trigonal layers are developed. These packs are chaotically shifted relative to each other. Because of polycentrism, the octahedron corners are chipped, and become serrate

Fig. 6.26 Anti-skeleton character of facet growth. It is noted for uneven development of faces

Fig. 6.27 Rounded-stepped crystals. Thick packs of poorly ditrigonal layers on an octahedron, whose end, forming steps, are rounded

Fig. 6.28 Vertex crystal. A crystal with clearly defined vertices at the exit points of the L3 axes, formed during antiskeletal growth and subsequent dissolution

Fig. 6.29 A curved dodecahedroid of the Ural type. Smooth-sided crystal of the final degree of dissolution, there is no relief on the surface of the crystal

Fig. 6.30 Stepped crystals: **a**, **b**—micro layering. Thin layers on the faces that are separated by some distance from the edges of the octahedron; **c**—finely stepped crystal; **d**—medium stepped crystal; **e**, **f**—sharply stepped, lamellar crystals

source. Eclogite genesis diamonds are characterized by a wider range of variation $\delta^{13}C$ from $-34^0/_{00}$ to $+ 1^0/_{00}$.

Also, information describing the processes of growth and evolution of diamonds can be obtained by studying the fluid inclusions in the diamond. Evaporating components were analyzed in diamonds of different habitus. Today, various fluid inclusions have been found in diamonds of Canada, Zaire, South Africa, Botswana, Brazil, and Yakutia. (Fig. 6.60) gives an idea of these inclusions and their composition.

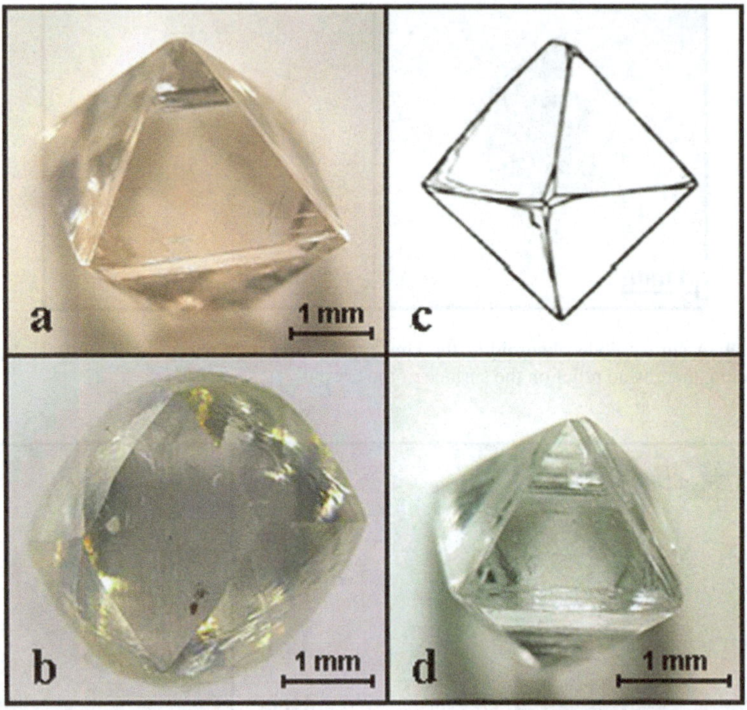

Fig. 6.31 Dissolution surfaces; **a**—smooth-sided octahedral crystal; **b**—smooth-sided dodecahedral crystal; **c**, **d**—octahedron with rounded vertices

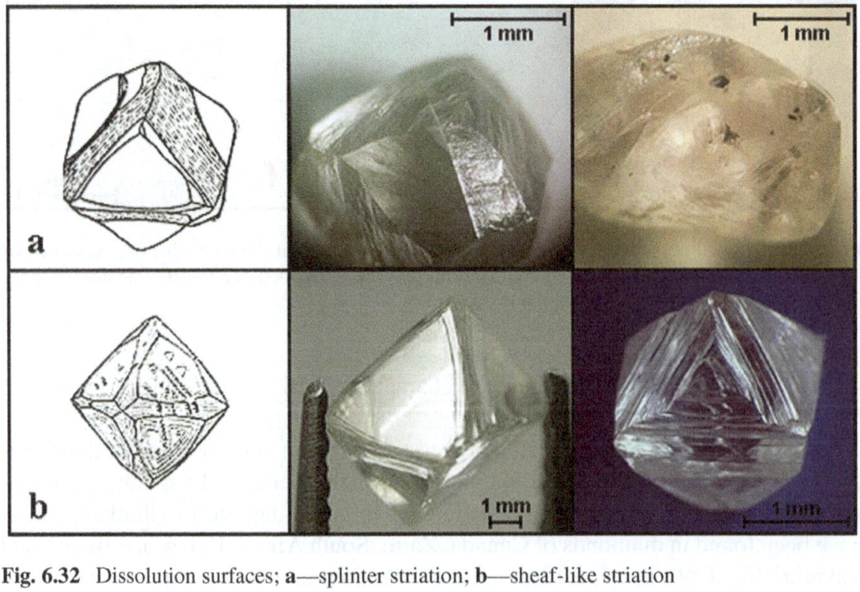

Fig. 6.32 Dissolution surfaces; **a**—splinter striation; **b**—sheaf-like striation

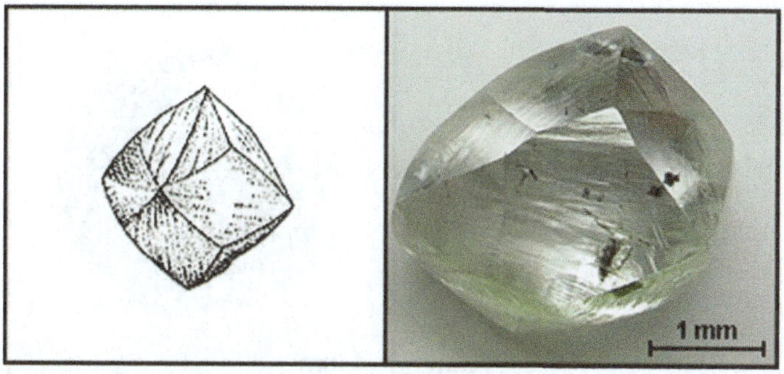

Fig. 6.33 Dissolution surfaces; splinter—sheaf-like striation

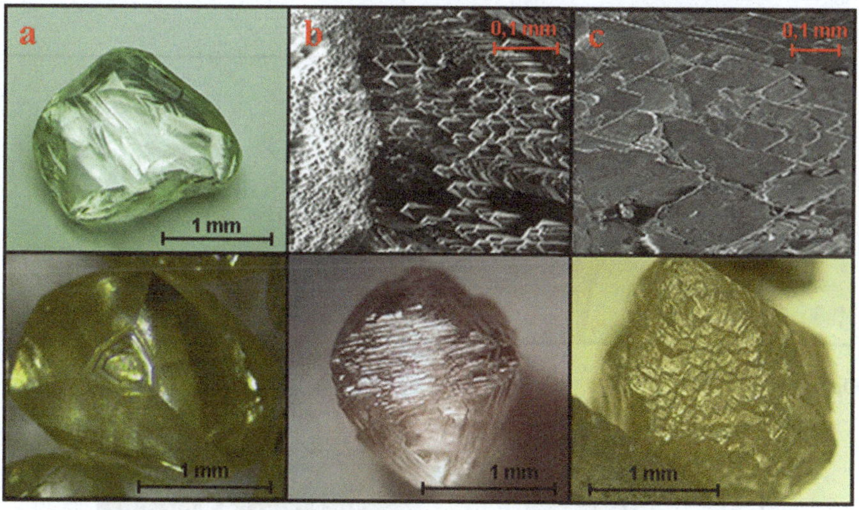

Fig. 6.34 Dissolution surfaces; **a**—concentric striation; **b**—splinter- columnar striation; **c**—tiled striation

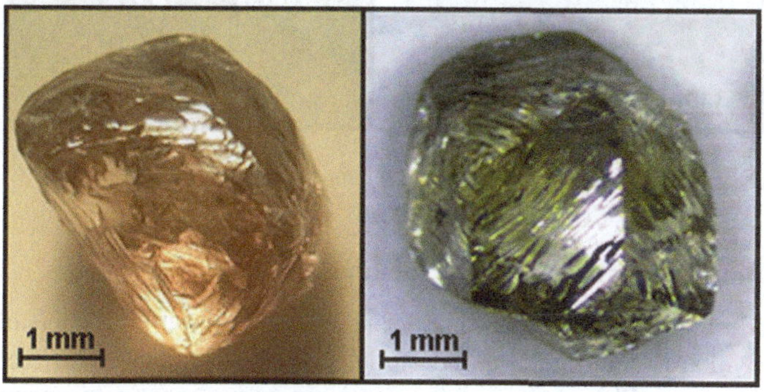

Fig. 6.35 Dissolution surfaces; teardrop-tiled sculpture

Fig. 6.36 Dissolution surfaces; teardrop-block sculpture

Fig. 6.37 Dissolution surfaces; rolls-form sculpture

Fig. 6.38 Dissolution surfaces; mosaic-block sculpture—concave, similar to chips

Study of diamond by spectroscopic methods

The second important group of properties is the defect-impurity composition of the diamond, which is determined based on the study of samples by Infrared (FTIR), Photoluminescent (PL) spectroscopy, electronic paramagnetic resonance (EPR); also, the method of spectroscopy in the visible range can be used additionally.

Fig. 6.39 Dissolution surfaces; block sculpture

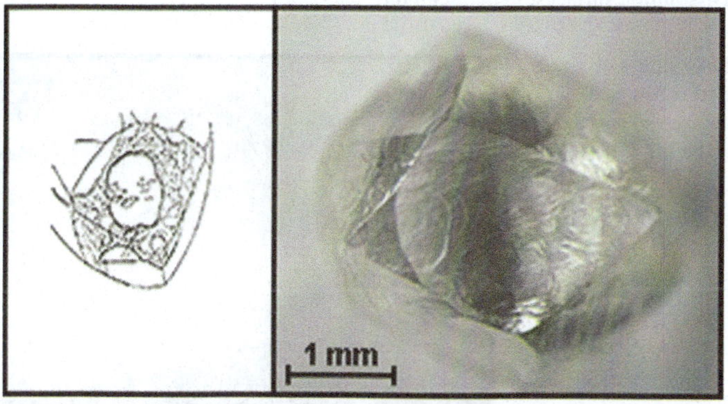

Fig. 6.40 Dissolution surfaces; disc sculpture—common mainly to rounded crystals, has a shape of convex discs and ovals, or irregular shapes with rounded contours

Fig. 6.41 Dissolution surfaces; icicle sculpture

Fig. 6.42 Dissolution surfaces; lattice sculpture

Fig. 6.43 Dissolution surfaces; teardrop-shaped humps on the curved surface of the dodecahedroid

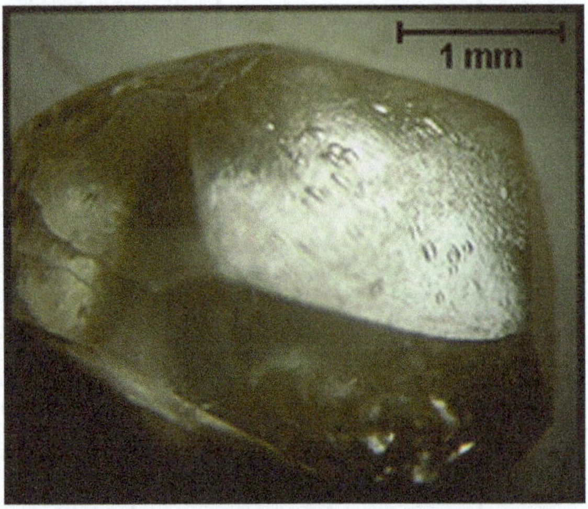

Fig. 6.44 Fibrous growth (striation along the normal to the face) on crystals of cubic habitus

Fig. 6.45 The topography of the surface of cubic crystals: **a**—a lattice surface formed by a pattern of small tetragonal pyramids on the edges of a cube; **b**, **c**—teardrop-shaped, hummocky sculpture on the edges of the cube; **d**—hummocky relief of edges of the tetrahexahedron; **e**—wavy surface of the cube

At the second stage of research, in accordance with the proposed method, diamonds are divided into genetic groups based on the study of their composition and physical properties by various spectroscopic methods.

The selected crystallomorphological types of diamond differ not only in their external shape. They also differ in their physical properties, primarily in color, the degree of visual transparency, the nature of absorption in the IR, visible and UV parts of the spectrum, and the nature of the glow in UV rays. The revealed differences are related to the peculiarities of their composition and structure and are due to the presence of intrinsic and impurity defects, which are formed depending on the growth conditions of the diamond.

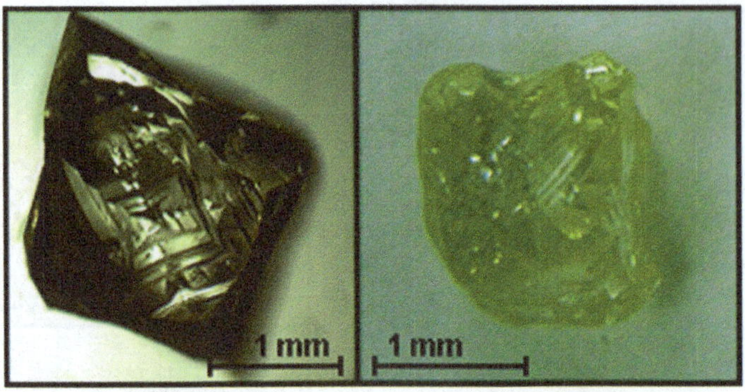

Fig. 6.46 Zonal-sectorial growth. Saddle faces on cubes of the II variety according to the classification of Yu. l. Orlov (1984) are the result of antiskeleton growth, they are formed by large stepped tetragonal pyramids

Fig. 6.47 Cellular sculpture on faces of cubes of the III according to the classification of Yu. l. Orlov (1984)

Fig. 6.48 Cellular sculpture on faces of crystals of the IV variety according to the classification of Yu. l. Orlov (1984)

Fig. 6.49 Diamonds in shells of variety IV according to the classification of Yu. l. Orlov (1984); **a**, **b**—the shell is very thin; **c**—the shell is thin, and the core of the crystal is visible under it; **d**—the shell is of medium density, the core of the crystal is not visible under it; **e**, **f**—the shell is dense, thick; the crystal is opaque, rounded or of cubic habit

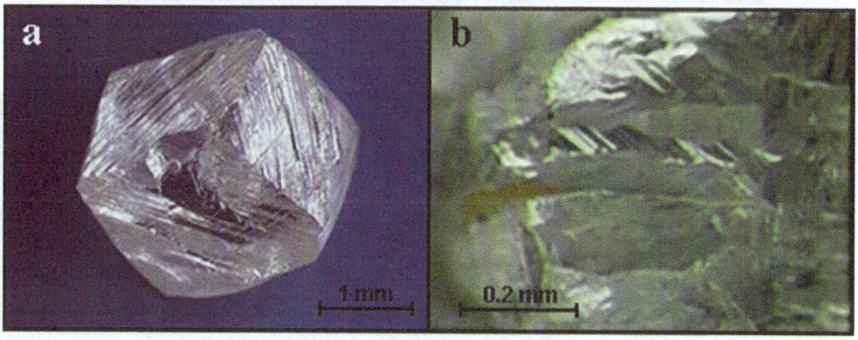

Fig. 6.50 Lines of plastic deformation; **a**—lines perpendicular to the L4 axes; **b**—sharp bands of plastic deformation; the relief ledges of pseudo faces formed due to sliding steps

It has been shown experimentally that under thermobaric conditions at temperatures above 1600° C and under pressure, C—centers transform into A—defects, followed by their aggregation into B_1 and B_2 defects (Vince 2008).

Information about the transformation kinetics obtained in experiments allows us to estimate the growth conditions of diamonds—depending on the accepted assumptions—or the temperature of their crystallization and the time spent in the mantle

Fig. 6.51 Lines of plastic deformation; sliding lines

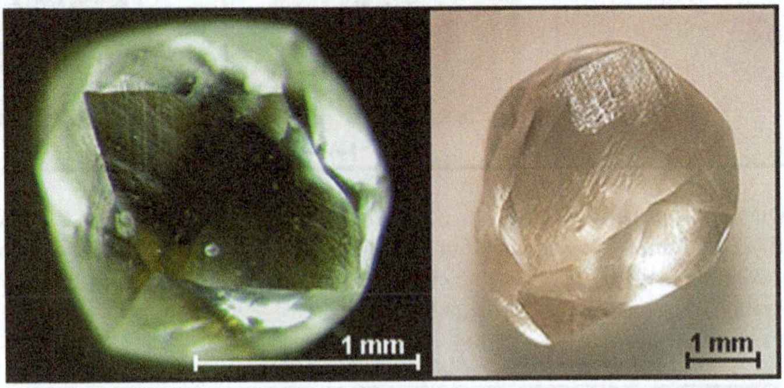

Fig. 6.52 Plastic deformation; crystals with shagreen

Fig. 6.53 Degree of preservation; a whole crystal

Fig. 6.54 Degree of preservation; a crystal with a slight chip; the chip is up to 1/4 of the original crystal and does not make it difficult to determine the original shape of the diamond

Fig. 6.55 Degree of preservation; crystal with a chip; the chip is from 1/3 to 2/3 of the volume of the original crystal

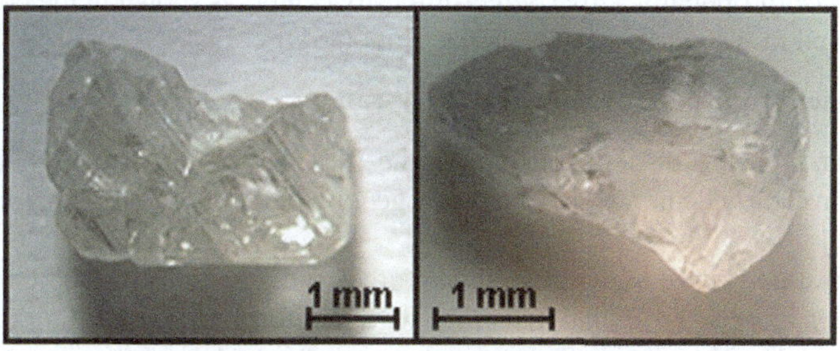

Fig. 6.56 Degree of preservation; broken crystals. The chip is more than 2/3 of the original crystal, the definition of the original shape is very difficult or impossible

Fig. 6.57 Degree of preservation; splinter; the chip is more than 2/3 of the volume of the crystal, there are no sections of the primary cut, and it is impossible to determine the original shape

Fig. 6.58 Diamonds of various color shades

at a constant temperature. In addition to impurity defects, there are widely developed defects in diamonds that are associated with dislocations or are located nearby. The presence of optically active defect-impurity centers is shown in the color of the diamond and in the color of fluorescence (Fig. 6.61.)

Fluorescence and spectroscopy in the UV range

Fluorescence occurs in diamond due to the presence of impurities, their uneven distribution clearly demonstrates the heterogeneity of diamonds, their possible zonal structure.

UV radiation is used to diagnose the nature of a diamond. Natural diamond fluoresces when excited by radiation with a wavelength of 365 nm, when excited by short-wave UV (254 nm), it fluoresces very rarely. At UV excitation of 254 nm, synthetic diamonds grown by NRNT and CVD methods fluoresce.

In standard gemological practice, fluorescence is evaluated under a gemological UV lamp with wavelengths of 254 and 365 nm.

The intensity of the glow is estimated on a 5-point scale: absent, weak, medium, intense, very intense.

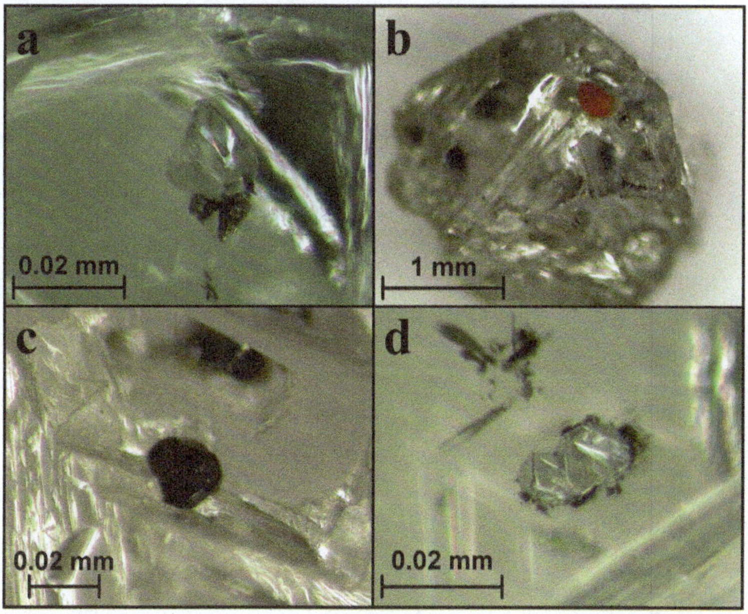

Fig. 6.59 Inclusions in diamonds: **a**—olivine; **b**—spessartine garnet; **c**—chromite; **d**—diamond in diamond

Diamonds can have different colors of fluorescence: blue, sky blue, pink, yellow, orange, green, yellow-green. The color of the glow is associated with the presence of certain defect-impurity centers in diamond crystals (Table 6.2).

Photoluminescence spectra are recorded on a spectrofluorimeter when excited at a wavelength of 350 nm at room temperature. The analysis is performed for the main photoluminescence centers: H3, H4, N3, N4, S1, S2 and S3, as well as additional absorption peaks at wavelengths 326, 376, 379, 380, 384, 390, 480, 500–650, 500–700, 550–600, 550–650, 550–700, 608, 620, 830.

It should be noted that intense fluorescence contributes to the visual component of the color. Crystals with a slight yellow tinge and strong blue fluorescence look more colorless in natural daylight. Orange fluorescence contributes to the staining of yellow stones of type II. These parameters are taken into account when evaluating the quality of raw diamonds and when marking the cut.

Infrared spectroscopy

The IR spectroscopy method allows us to determine the presence of structural nitrogen, nitrogen-vacancy centers (defects), non-structural hydrogen impurities, hydroxyl groups, carbonates in the diamond, and the presence of additional unidentified absorption peaks that are typomorphic for some deposits. The typical IR spectrum of a diamond crystal is shown in Fig. 6.62.

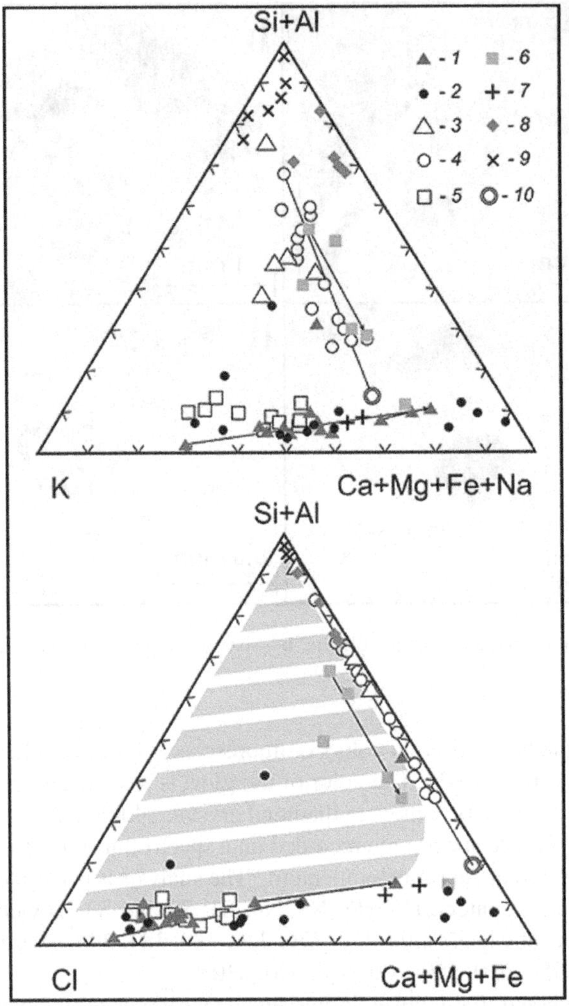

Fig. 6.60 Variations in the composition of molten and fluid inclusions in diamonds from various regions of the world: 1, 2—Canada: 1—Diavik (Klein-Ben David et al. 2004,2006); 2—Panda (Tomlinson et al. 2006); 3—Zaire (Navon et al. 1988); 4—Botswana (Navon et al. 1988; Schrauder and Navon 1994); 5—South Africa (Koffifontein tube) (Izraeli et al. 2001, 2004); 6—Brazil (Shiryaev et al. 2005); 7–9—Yakutia: 7—Jubileynaja tubes (Logvinova et al. 2008; Klein-Ben David et al. 2006); 8—Aikhal tubes (Bulanova et al. 1993); 9—Mir tubes (Bulanova et al. 1988; Novgorodov et al. 1990). The composition of the carbonatite boundary term of inclusions in diamond (Schrauder and Navon 1994), used in experiments (10), is shown. Connodes connect the extreme members of the inclusion compositions, and the inclusion of different zones of diamond crystals (arrows show the evolution of the compositions of inclusions from center to edge of crystals). The assumed area of incompleteness in the chloride-silicate-carbonate system is shaded, according to (Safonov et al. 2007)

Fig. 6.61 Color and different colors of fluorescence of diamond crystals (0.35–0.40 carats) from the Arkhangelsk tube of the Lomonosov deposit

Table 6.2 The main centers of fluorescence and glow color in natural diamonds

Defect	Center model	UV wavelength, nm	Visual perception	
			The color of the fluorescence	Visible color
C	N	270	Yellow	Yellow, orange-yellow
637	NV^-	637	Red	Pink
C +	NV^0	575	Orange	Yellow
N3	N_3—V_C	415,2	Blue, sky blue	Straw yellow
440	C–N–O–C	440,3	Sky blue	Straw yellow
H3	$2N_C$—V_C	503,2	Green	Green
H4	N_3–V–V–N	495,8	Green	Different
S1	C–N–C–O–C	Zero-phonon lines: 503,4; 510; 523; 530; 536; 548; 562; 580	Yellow	Different
S2	C_2N–$Ni_{V\text{-}V}$–N_2C	Three bands with zero-phonon lines 470; 477; 489	Yellow-green	Different
S3	$(C_2N$–$Ni_{V\text{-}V}$–$NC_2)^-$	497	Yellow-green	Different

Note Models and wavelength locations of centers are given by (Dishler 2013; Titkov 2018)

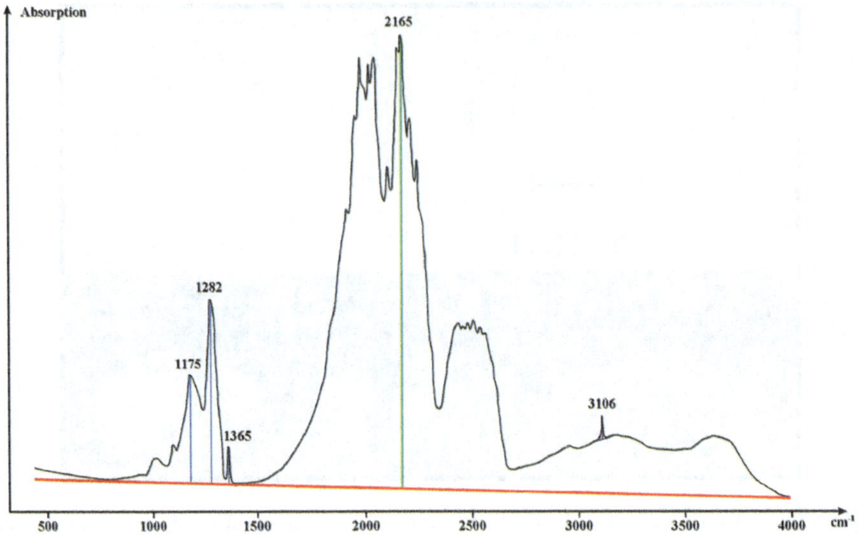

Fig. 6.62 IR spectrum of natural diamond type Iib. The base line drawn at two points with minimal absorption is highlighted in red. Blue indicates the relative intensity of the defect lines: A-1282 cm^{-1}, B-1175 cm^{-1}, B2-1365 cm^{-1}. Green indicates the relative intensity of the line 2165 cm^{-1} for calculating the effective width of the sample. For the CH line (3106 cm^{-1}), its own baseline is drawn, along which its intensity is measured

In the works (Chrenko et al. 1967; Klyuev et al. 1971, 1979; McMillan and Hofmeister 1988; Natural diamonds…, 1997; Khachatryan 2003, 2010; Vasilyev 2013), the principal possibility of evaluating the physical and chemical conditions of crystallization and post-crystallization temperature regime for diamond by the total nitrogen content and the ratio of concentrations of the main nitrogen defects differing in the degree and forms of nitrogen aggregation is established. These characteristics of the diamond, which are typomorphic, can be obtained using IR spectroscopy, and used in basic research and exploration.

Calculation of the concentration of nitrogen defects in the form of C, A, B, P, H centers

C-centers are formed by single nitrogen atoms that isomorphically replace carbon atoms in the diamond structure (the main absorption band is 1135 cm^{-1}). They are registered in rare cases, mainly in crystals of cubic habitus.

A-centers are formed by two adjacent nitrogen atoms replacing carbon (the main absorption band is 1280 cm^{-1}) (Natural diamonds 1997).

B-centers are formed by nitrogen atoms tetrahedral grouped around the vacancy (the main absorption band is 1175 cm^{-1}) (Natural diamonds 1997).

Pleitelets (P, B2) (line in the area of 1365 cm^{-1}) are plate defects of interstitial nature, several atoms thick (Sobolev 1968). For B2, the absorption coefficient is determined (B2, cm^{-1}) and the exact position of the maximum absorption band (RzB2, cm^{-1}).

H-centers-defects involving hydrogen (CH, maximum about 3107 cm^{-1}) (Khachatryan, 2003).

To determine the absorption coefficients, the effective thickness of the samples is estimated. In this case, the lattice absorption in the range of 1500–3500 cm^{-1} is used as an internal standard for the study of diamonds of small size or irregular shape in order to calculate their thickness and account for reflection. As parameters of the internal standard, the absorption coefficients at the frequencies of 1973 cm^{-1} and 2500 cm^{-1} were selected, respectively $k_{1973} = 12.5\ cm^{-1}$, $k_{2500} = 4.9\ cm^{-1}$ (Klyuev 1971).

To determine nitrogen concentrations in A - and B - forms, the analytical dependencies given in the works (Boyd et al. 1994, 1995) are used. Nitrogen concentrations are calculated using the method proposed in the paper (Khachatryan 2003).

The content of plate defects (pleitelites) and hydrogen is estimated by the absorption coefficients of lines about 1365 and 3107 cm^{-1}, respectively. Approximate diamond formation temperatures using nitrogen defect concentrations are determined in accordance with (Taylor and Milledge 1995).

When calculating the nitrogen concentration using the method (Natural diamonds… 1997; Khachatryan 2003), it is assumed that most of the nitrogen atoms in natural diamond crystals are in the aggregated form of A—and B1-defects. In the IR spectra of type Ia crystals, C-band defects are usually not detected, and therefore the proportion of single nitrogen atoms is not taken into account in the calculations. The part of nitrogen that may be associated with planar B2 defects is also excluded from the calculations, due to the remaining uncertainty of their structure. According to the method used, the concentration is calculated in the following order:

1. Drawing the baseline at the points with the maximum transmission (Fig. 6.62).
2. Measurement of the intensity of the absorption bands of the lattice (2000–2200 cm^{-1} and 2500 cm^{-1}), and the main bands of defects: A (1282 cm^{-1}), B1 (1175 cm^{-1}).
3. The intensity of the band absorption (lattice) in the region 2500 cm^{-1} corresponds to an absorption coefficient of 4.6 cm^{-1} (Natural diamond… 1997), and in the region 2000–2200 cm^{-1}—absorption coefficient of 12.8 cm^{-1}; based on this, write a proportion to calculate the absorption coefficient for the bands A-(1282 cm^{-1}) and B1-defects (1175 cm^{-1}).
4. when calculating the absorption coefficients μ of overlapping main bands of A - and B1-defects according to their intensity, the empirical equations are used

$$\mu_A = 1{,}2\mu_{1175} — 0{,}49\mu_{1282},$$
$$\mu_B = 1{,}2\mu_{1282} — 0{,}51\mu_{1175}.$$

5. when calculating the nitrogen content of N_A and N_{B1} using the equations

$$N_A = K_A * \mu_A \text{ ppm},$$
$$N_{B1} = K_{B1} * \mu_{B1} \text{ ppm}.$$

in addition to the values of μA and μB1, the coefficients $K_A = 16.05$ and $K_{B1} = 21.55$ are used (Khachatryan 2003). To convert ppm values to the number of atoms per cm^3, the ratio.

$1*10^{19}$ at/cm^3 = 57 pm is used.

Then the fractions of aggregated nitrogen (%B) are calculated, the total nitrogen concentrations in A—and B—form, and their ratio is established.

The total nitrogen content is calculated using the formula $N_{tot} = N_A + N_{B1}$.

The proportion of nitrogen in the 1-form is determined by the formula B1,% = B1*100/N_{tot}.

During analysis, data can be plotted on a Taylor Diagram (Fig. 6.2).

The relative content of impurities and microinclusions of hydrogen, hydroxyl groups, and carbonates is determined. The ratio of the location of the maximum absorption band and the absorption coefficient is set.

Diamonds from deposits of each petrochemical type, despite wide variations in the gross nitrogen content, are characterized by close values of the degree of nitrogen aggregation in B-form and pleitelets. The obtained data allowed us to establish empirically that the ratio of the absorption coefficient B2 and the maximum position of this band is individual for diamond crystals from tubes of various deposits (Kriulina 2012). Diamonds from the low-titanic kimberlites ($TiO_2 < 1.0$ wt.%) and from medium-titanic kimberlites ($1.0 < TiO_2 < 2.5$ wt%) are separated into two fields in the diagram (Fig. 6.63).

For the reliability of the analysis, the average characteristics obtained by IR spectroscopic studies only for crystals of the I variety according to the classification of Yu.L. Orlov (1984) are taken into account, since they dominate in all tubes and are diverse in mineralogical, physical and structural characteristics and, therefore, most fully characterize the deposits.

Diamonds with the highest platelets absorption coefficient, proportional to the content of this defect in the crystal, and the location of the maximum absorption band in the long-wave region grew under the most high-temperature conditions. These are diamonds from Udachnaya, Komsomolskaya, and Grib pipes. Crystals from the tubes Internatsionalnaya, Mir, Botuobinskaya, Archangelskaya, and Karpinskogo-1 grew at lower temperatures.

Based on a comprehensive analysis of the mineralogy and defect-impurity composition of the diamond, the comparison and determination of the crystal's belonging to one of the 7 morphogenetic groups is carried out (Fig. 6.2).

This brief review shows that diamond is a polygenic mineral (Nakamuta and Aoki 2000), with a variety of physical characteristics and complex internal structure, so a rational method for studying its crystals can and should be varied depending on the goals and objectives of research.

The main parameters that determine the habit, sculptural features and physical properties of diamond are temperature, pressure, the degree of supersaturation of the melt, as well as the conditions for the formation of kimberlite pipes (the rate of rise, chemistry of the mineral-forming medium, redox potential). The variety of factors that affect the morphology and properties of diamond determines the quality of raw diamond. At the same time, it is possible to draw a definite conclusion about

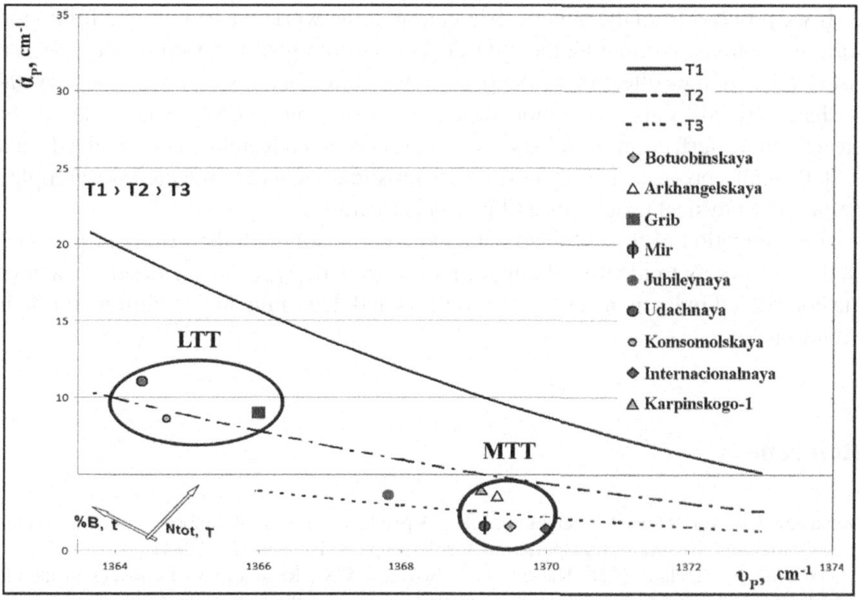

Fig. 6.63 Figurative points of averaged characteristics obtained during IR spectroscopic studies for diamonds of the I variety according to the Yu. L. Orlov classification of ADR and YDP tubes in the coordinates absorption coefficient ($ά_P$)—the position of the maximum of the absorption band P ($υ_P$)

the diamond crystallization medium. Mineralogical studies of syngenetic inclusions in natural diamonds have shown that the parent environment in which diamonds were formed is heterogeneous. Its phase composition is determined by silicates, aluminosilicates, oxides, sulfides of ultrabasic and basic parageneses, carbonates, phosphates, chlorides, carbides, metals, water, carbon dioxide, methane, strongly compressed fragments of carbonatite melts. Currently available large arrays of mineral-analytical and experimental-mineralogical information allow us to argue for the carbonate–silicate ("mantle-carbonatite") concept of growth medium for the dominant mass of natural diamonds in kimberlite deposits (Litvin et al. 2009).

For the purposes of typifying primary sources, localizing search halos, and evaluating the quality of raw diamonds, a visual and optical description of whole crystals according to the scheme described in this paper is usually sufficient. Special spectroscopic studies are required to solve genetic and some applied problems. They can be performed using non-destructive methods (x-ray tomography, scanning electron microscopy, EPR, Raman spectroscopy, magnetometry). After cutting, polishing and crushing crystals, research can bc carried out at a more local level using ultralocal methods of studying matter (IR-Fourier spectroscopy, visible and ultraviolet spectroscopy, cathodoluminescence, electron-probe microanalysis, etc.). The set of methods is determined by the specific nature of the tasks.

If we proceed from the genetic orientation of the work, we can use the following various methods. A mineralogical description is compiled for each of the crystals. All samples of the collection are studied under a binocular microscope. Each crystal is characterized by its size, color, degree of visual transparency, shape, degree of preservation, surface morphology, and presence of inclusions. The received data is statistically processed. The most characteristic crystals are selected for complex research of physical properties and internal structure.

Generalization of the obtained data and comparison with the results of previous studies allows us to identify the main morphogenetic types of diamond in each of the considered indigenous and placer objects and determine the conditions for their evolution.

References

Afanasiev, V.P., Varlamov, D.A., Garanin, V.K.: Dependence of wear of kimberlite minerals on the conditions and distance of transportation. Geol. Geophys. **10**, 119–125 (1984)

Afanasiev, V.P., Zinchuk, N.N., Harkiv, A.D., Sokolov, V.N.: Regularities of changes in mantle minerals in the weathering crust of kimberlite rocks. In the collection Mineralogy of the hypergenesis zone.- Moscow, pp. 45–50 (1980)

Afanasiev, V.P., Zinchuk, N.N., Pohilenko, N.P.: Morphology and morphogenesis of indicator minerals of kimberlites. Novosibirsk, Geo branch of the Siberian branch of the Russian Academy of Sciences, Publishing house "Manuscript", p. 276 (2001)

Anshles, O.M.: Some questions of connection of crystal forms with their structure: Crystal growth, pp. 67–73. Publishing house of the USSR Academy of Sciences, Moscow (1957)

Askhabov, A.M.: Regeneration of crystals. Leningrad, Nauka publishing house, (1979)

Barton, V., Cabrera, N., Frank, F.: Crystal growth and equilibrium structure of their surfaces. In sat. Elementary processes of crystal growth. Moscow, Foreign lit. Publishing house, pp. 11–109 (1959)

Bartoshinsky, Z.V.: Comparative characteristics of diamonds from various diamond-bearing regions of Western Yakutia. Geol. Geophys. **6**, 40–50 (1961)

Bartoshinsky, Z.V.: Mineralogical classification of natural diamonds. Mineral. J. **5**(5), 84–93 (1983)

Bartoshinsky, Z.V.: Parasteric associations of diamonds of Yakutia. Mineral. Coll. Lviv University, **38**(1), 3–6 (1984)

Belov, N.V., Godovikov, A.A., Bakakin, V.V.: Essays on theoretical mineralogy, p. 206. Nauka, Moscow (1982)

Berry, L., Mason, B., Dietrich, R.: Mineralogy. Theoretical foundations, description of minerals, diagnostic tables. Moscow, Mir, p. 592 (1987)

Beskrovanov, V.V.: Evolution of natural diamond formation conditions. In the sat. Geological aspects of the mineral resource base of ALROSA, Current state, prospects, solutions. Mirny, pp. 242–250 (2003)

Beskrovanov, V.V.: Ontogeny of diamond, p. 165. Nauka, Moscow (1992)

Bobrievich, A.P., Bondarenko, M.N., Gnevushev, M.A., Krasov, L.M., Smirnov, G.I., Yurkevich, R.K.: Diamond deposits of Yakutia, p. 527. State publishing house on Geology and protection of the subsoil, Moscow (1959)

Bogush, I.A., Kaftanatiy, A.B., Chernenko, MYu.: Crystallomorphological analysis of pyrite from the ores of the Komsomolsk copper-pyrites Deposit. Zapiski WMO **120**(3), 43–49 (1991)

Bonev, I.R., Radulova, A.: Chalcopyrite and sphalerite "diseases" crystal growth by solid-state diffusion. Abstr. IMA Congr., Pisa (1994)

Boyd, S.R., Kiflawi, I., Woods, G.S.: The relationship between infrared absorption and the A defect concentration in diamond. Phil Mag. b. **69**, 1149–1153 (1994)

Boyd, S.R., Kiflawi, I., Woods, G.S.: Infrared absorption by the B nitrogen aggregate in diamond. Phil Mag. b. **72**, 351–361 (1995)

Brodskaya, R.L., Marina, E. Yu., Shnai, G.K., Saminina, I.A.: Restoration of conditions and kinetics of formation of granites of rare-metal formations by the crystallomorphology of accessory zircon. Notes. WMO, part 115, issue 1 (1986)

Buckley, G.: Growth of crystals. Moscow, Foreign lit. Publishing house, p. 406 (1954)

Bulakh, A.G., Bulakh, K. G.: Physical and chemical properties of minerals and components of hydrothermal solutions. L.: Nedra, p. 167 (1978)

Bulanova, G.R., Baranov, Yu. P., Talnikova, S.B., Smelova, G.B.: Natural diamond-genetic aspects. Novosibirsk, Nauka, p. 176 (1993)

Bulanova, G.P., Novgorodov, P.G., Pavlova, L.A.: The first finding of a molten inclusion in a diamond from the Mir tube. Geochemistry. **5**, 756–765 (1988)

Burns, R.G.: Mineralogical applications of crystal field theory. Cambridge University Press (1993)

Chernov, A.N.: Processes of crystallization. In the book, Modern crystallography. vol. 3. Formation of Crystals. Moscow, pp. 7–233 (1980)

Chervinskaya, A.D.: On the morphology (cutting) of zircon granitoids of Pluto Akzhailau and Flask. In the book. Accessory minerals of igneous and metamorphic rocks. Moscow, Nauka, pp. 176–180 (1982)

Chesnokov, B.V.: Relative age of mineral individuals and aggregates, p. 105. Nedra, Moscow (1974)

Chukhrov, F.V., Petrovskaya, N.V., Zvyagin, B.B.: On some basic concepts of Mineralogy. Mineral. J. **5**(2), 8–18 (1983)

Chrenko, R.M., McDonald, R.S., Darrow, K.A.: Infrared spectrum of diamond coat. Nature **214**, 474–476 (1967)

Cullen, W.H., Jr. Marankowski, M.J., Das, E., S.,P.: A study of the interphase boundary between ilmenite and hematite. Surf. Sci. **36**(2), 395–413 (1973)

Deicha Geores, M.E.: Mineralogie et cristallogenese: Cristallogenese minerale et cavites des materiaux de la lithospere. Bul. Ses. Sci CTHS, **7**, 125 (1984)

Dischler, B.: Handbook of Spectral Lines in Diamond, p. 467. Springer, Berlin (2013)

Dorokhova, G.I., Kaplunnik, L.N.: Morphometry of crystals. Moscow University Publishing house, p. 119 (1986)

Evzikova, N.Z.: Search Crystallomorphology, p. 143. Nedra, Moscow (1984)

Feklichev, V.G.: Microcrystallomorphological Research, p. 177. Nauka, Moscow (1970)

Fersman, A.E.: Diamond Crystallography. Publishing house of the USSR Academy of Sciences, Moscow (1955)

Galimov, E.M. et al.: Isotopic composition of carbon of diamonds containing mineral inclusions from placers of the Northern Urals. Geochemistry, **9**, 1363–1370 (1989)

Galimov, E.M.: Variations of the isotopic composition of diamonds and their connections with the conditions of diamond formation. Geochemistry, **8**, 1091–1118 (1984)

Garanin, V.K., Kudryavtseva, G.P., Marfunin, A.S., Mikhailichenko, O.A.: Inclusions in Diamond and Diamond-Bearing Rocks, p. 256. MSU Publishing house, Moscow (1991)

Garrels, R.M., Christ, C.H.L.: Solutions, Minerals, equillbria, p. 450. Mir, Moscow (1968)

Gbelsky, J.: Composition of main morphological types of garnets from pegmatites of the Male Karpaty, Mts. (the West Carpathians). Geol. Zb., **31**(1–2), 185–199 (1980)

Geguzin, Ya. E.: Problems of modern crystallography. Moscow, Nauka, pp. 110–127 (1975)

Gimeno, D., Pijolrin, L.: Textural evidence of supercooling in intrusive pillows of the guardia Marina Beach (Sardinia island, Western Mediterranean, Italy). Abstr. IMA Congr., Pisa (1994)

Glazov, A.I.: Methods of crystal morphometry. L.: Nedra, p. 147 (1981)

Glikin, A.E.: Method of accounting for supersaturation of the solution in the analysis of crystal typomorphism. Crystallogr. Crystal Chem. 4, 135–139 (1982)

Gnevushev, M.A., Bartoshinsky, Z.V., Zinkov, A.P. et al.: Diamonds In Yakutia. Moscow, Nedra (1961)

Gnevushev, A.M., Shemanina, E.I.: The experience of classification of diamond crystals. Mineralogical Coll. Lviv University. **24**(1), 45–52 (1970)

Goshka, L.L., Kolosov, S.I., Ruzov, V.L.: Periodic precipitation in gel: the rule of the crystal supersaturation. Abstr. IMA Congr., Pisa (1994)

Gravenor, C.P., Leavitt, R.K.: Experimental formation and significance of etch patterns on detrital garnets. Can. J. Scintith, **18**(4), t. 276, 765–775 (1981)

Greer, R.T.: Distribution of pyrite in sheets. SEM, vol. 1, AMF O'hare, pp. 621–626, Discuss. 610 (1978)

Greg, S., Sing, K.: Adsorption, Specific Surface Area, Porosity, p. 407. Mir, Moscow (1970)

Grigoriev, D.P.: Ontogeny of minerals. Lviv: publishing house of Lviv University, p. 284 (1961)

Grigoriev, D.P.: Natural objects of mineralogy: mineral individuals and mineral species. Notes. WMO, 4, 513–514 (1975)

Grigoriev, D.P., Zhabin, A.G.: Ontogeny of minerals. Individuals. Moscow: Nauka, p. 337 (1975)

Harris, J.W., Hawthorue, J.B., Oosterveld, M.M., Wehmeyer, E.: A classification scheme for diamond and a comparative study of South African diamond characteristics. Phis Chem. Eath **9**, 765–783 (1975)

Hirsch, P., Howie, A., Nicholson, R., et al.: Electron Microscopy of Thin Crystals, p. 574. Mir, Moscow (1968)

Honigman, B.: Growth and Shape of Crystals. Moscow: publishing house of foreign Literature, p. 209 (1961)

Izraeli, E.C., Harris, J.W., Navon, O.: Brine inclusions in diamonds: a new upper mantle fluid. Earth Planet. Sci. Lett. **187**, 323–332 (2001)

Izraeli, E.S., Harris, J.W., Navon, O.: Fluid and mineral inclusions in cloudy diamonds from Koffiefontein South Africa. Geochim. Cosmochim. Acta **68**, 2561–2575 (2004)

Jafarov, CH. D.: Crystallomorphology of pyrite and its mineralogical significance. Baku.: ELM publishing house, p. 118 (1970)

Kan, J.: Theory of crystal growth and phase boundary movement in crystalline materials. Success Phys. Sci. **91**(4) (1767)

Khachatryan, G.K.: Improved methodology for assessing nitrogen concentrations in diamond and its practical application. Geological aspects of the mineral resource base of ALROSA: Current state, prospects, solutions. Mirny, pp. 319–322 (2003)

Khachatryan, G.K.: Typification of diamonds from kimberlites and lamproites by distribution of nitrogen centers in crystals. Ores Miner. 2, 46–60 (2010)

Kharkiv, A.D.: Signs of similarity and differences between kimberlite rocks of the North of the Russian platform and other regions, pp. 91–99 (1992)

Kharkiv, A.D., Zinchuk, N.N., Kryuchkov, A.I.: Geological-Genetic Basis Sligo-mineralogical Method of Prospecting of Diamond Deposits, p. 348. Nedra, Moscow (1995)

Khvorova, I.V., Dmitrik, A.L.: Microstructures of Siliceous Rocks. Nauka, Moscow (1972)

Klein-Ben David, O., Izraeli, E.S., Hauri, E., Navon, O.: Mantle fluid evolution—a tale of one diamond. Lithos. **77**, 243–253 (2004)

Klein-Ben, D.J., Wirth, R., Navon, O.: TEM imaging and analysis of microinclusions in diamonds: a close look at diamond-growing fluids. Am. Miner. **91**, 353–365 (2006)

Klyuev, Y.A.: Intensity of bands in the IR absorption spectrum of natural diamonds. Diamonds, 6, 9–12 (1971)

Klyuev, Y.A., Kaminsky, F.V., Smirnov, V.I., et al.: Diamonds of the Northern Timan. " Minerals and paragenesis of minerals, Gorn. rocks and ores". L., pp. 96–100 (1979)

Kovalsky, V.V., Nikishov, K.N., Varshavsky, A.V., Grigoriev, A.P.: The problem of establishing minerals of kimberlite paragenesis and some questions of the genesis of diamonds and minerals of kimberlite rocks. In the collection Paragenesis of minerals of kimberlite rocks, Yakutsk, (1981)

Kreiter, V.M.: The size of gold particles in sulfide deposits as a sign of post-ore metamorphism. WPI. An USSR, ser. GEOL., No 1 (1948)

Kriulina, GYu.: Constitutional Characteristics of Diamond from Deposits of the Arkhangelsk and Yakut Diamond-bearing Provinces, p. 24. Abstract. PhD Diss, Moscow (2012)

Kriulina, G.Y., Garanin, V.K., Samosorov, G.G.: Forecasting the quality of diamond raw materials in deposits of various petrochemical types. In the journal News of higher educational institutions. Geol. Explor. 6 (2013)

Kukharenko, A.A.: Diamonds of the Urals, p. 427. Gosgeolizdat, Moscow (1955)

Kukharenko, A.A.: Mineralogy of Placers. Moscow: Gosgeoltehizdat, p. 247 (1961)

Kuroda, T., Jrisawa, T., Ookawa, A.: Growth of a polyhedral crystal from solution and its morphological stability. J. Cryst. Growth **42**, 41–46 (1977)

Kvasnitsa, N., Krochukh, V.M., Afanasiev, V.P., Tsymbal, Y.S.: Crystallomorphology of chrom-spinelids from kimberlites. Miner. J. **10**(3), 45–51 (1988)

Lavrova, L.D.: A new type of diamond deposits. Nature. 12, pp. 62–68 (1991)

Lemmlein, G.G.: The process of geometric selection in a growing aggregate of crystals. Docl. An USSR. **48**(3) (1945)

Litvin, Y.A.: Experimental studies of the physical and chemical conditions of diamond formation in the mantle substance. Geol. Geophys. **50**(12), 1530–1546 (2009)

Logvinova, A.M., Wirth, R., Fedorova, E.N., Sobolev, N.V.: Nanometre-sized mineral and fluid inclusions in cloudy Siberian diamonds: new insights on diamond formation. Eur J. Mineral. **20**, 317–331 (2008)

McMillan, P.F., Hofmeister, A.M.: Infrared and Raman spectroscopy. In: Hawthorne F.C.(Ed.) Spectroscopic methods in mineralogy and geology, Reviews in Mineralogy. Min. Soc. America, 18, pp. 99–160 (1988)

Martovitsky, V.N., Nadezhdina, E.D., Ekimova, T.E.: Internal structure and morphology of small non-kimberlite diamonds. Miner. J. **9**(2), 26–27 (1987)

Masaitis, V.L., Futergendler, S.I., Gnevushev, M.A.: Diamonds in impactites of the popigai meteorite crater. Notes of the WMO **101**(1), 108–113 (1972)

Melchorre, G., Scandale, E., Lucehesi, S.: Growth history and crystal chemistry of members of the elbaite-tsilaisite series. Abstr. IMA Congr., Pisa (1994)

Meyer, H.O.: Inclusions in diamonds. Mantle xenoliths: Gogn Wiley & Sons / P.H. Nixon (ed.), Chichester (England), pp. 501–521 (1987)

Mokievsky, V.A.: Morphology of crystals. Methodological guide L.: Nedra, p. 285 (1983)

Mosing, R.W.: Morphology of indicator minerals as guide to proximity of source. Publs Geol. Dep. & Extention Service, Univ. West. Aust., 5, pp. 10–20 (1980)

Murowchick, J.B., Barnes, H.L.: Effects of temperature and degree of supersaturation on pyrite morphology. Amer Miner. **72**(11–12), 1241–1250 (1987)

Nadezhdina, E.D., Posukhova, T.V.: Morphology of diamond crystals from metamorphic rocks. Mineral. J. **12**(2), 3–15 (1990)

Nadezhdina, E.D., Shalashilina, T.Y., Baevskaya, G.M.: Typomorphism of non-kimberlite diamonds. Miner. J. **15**(1), 9–19 (1993)

Nakamuta, Y., Aoki, Y.: Mineralogical evidence for the origin of diamond in ureilites. Meteorit. Planet. Sci. **35**(3), 487–493 (2000)

Kvaskov, V.B. (ed.).: Natural Diamonds of Russia. Moscow: Polyaron, p. 303 (1997)

Navon, O., Hutcheon, I.D., Rossman, G.R., Wasserburg, G.J.: Mantle-derived fluids in diamond micro-inclusions. Nature **335**, 784–789 (1988)

Nishida, T., et al.: Geometry of the surface of microcrystals using a raster electron microscope. Kobutsugaku zasen, Issue 11, pp. 150–159 (1973) (in Japanese)

Novgorodov, P.G., Bulanova, G.P., Pavlova, L.A., Mikhailov, V.N., Ugarov, V.V., Shebanin, A.P., Argunov, K.P.: Inclusions of potassium phases, coesite and omphacite in a diamond crystal with a shell from the Mir tube. Dokl RAS. **310**(2), 439–443 (1990)

Orlov, Yu.L.: Mineralogy of diamond, p. 263. Nauka, Moscow (1984)

Orlov, Y.L.: Morphology of Diamond. Moscow: Publishing house of the USSR Academy of Sciences, p. 235 (1963)

Pavlishin, V.I., Yushkin, N.P., Popov, V.A.: Ontogenic Method in Mineralogy, p. 120. Naukova Dumka, Kiev (1988)

Perchuk, L.L.: Fluids in the lower crust and upper mantle of the Earth. Bull. MSU Ser. Geol. **4**, 25–36 (2000)

Popenko, G.S., Posukhova, T.V., Kudryavtseva, G.P. Fedorov, V.N.: Application of SEM methods for determining the age of native gold from placers. Izvestiya Vuzov, ser. "Geology and exploration", No 2 (1993)

Popov, V.A.: Morphological laws of recrystallization of mineral aggregates. New Ideas Genet. Mineral. LSU, pp. 31–38 (1983)

Putnis, A., McConnell, J.: The Main Features of the Behavior of Minerals, p. 304. Mir, Moscow (1983)

Safonov, O.G., Perchuk, L.L., Litvin, Y.A.: Melting relations in the chloride-carbonate-silicate systems at high-pressure and the model for formation of alkalic diamond-forming liquids in the upper mantle. Earth Planet. Sci. Lett. **253**, 112–128 (2007)

Sarasadskikh, N.N., Popugaeva, L.A.: New data on the manifestation of ultrabasic magmatism on the Siberian platform. Explor. Prot. Min. Res. **6**, 11–20 (1955)

Schrauder, M., Navon, O.: Hydrous and carbonatitic mantle fluids in fibrous diamonds from Jwaneng Botswana. Geochim. Cosmochim. Acta. **58**(2), 761–771 (1994)

Schuize, D.J., Weise, D., Steude, J.: Abundabce and distribution of diamonds in eclogite revealed by volume visualization of CT X-ray scans. J. Geol. **104**(1), 109–113 (1996)

Sergeeva, N.E., Eremin, N.I., Kozlova, O.G., Ponomareva, I.A.: Deformation microstructures of sphalerite in Pyrrhic-polymetallic deposits of the Ore Altai. Vestn. MSU, ser. GEOL., No 3, pp. 33–42 (1986)

Shafranovsky, I.I.: Essays on mineralogical crystallography. L.: Nauka, p. 152 (1974)

Shafranovsky, I.I.: Mineral Crystals: Curved, Skeletal and Granular Forms, p. 332. Gosgeoltekhizdat, Moscow (1961)

Sheftal, N.I.: Accessories of crystal growth. Works of IKAN SSSR, Issue 3 (1947)

Sheftal, N.N.: Normal growth mechanism. In the book. Processes of real crystal formation, Moscow, Nauka, pp. 22–30 (1977)

Shiryaev, A.A., Izraeli, E.S., Khauri, E.G., Zakharchenko, O.D., Navon, O.: Chemical, optical and isotopic features of fibrous diamonds from Brazil. Geol. Geophys. **46**(12), 1207–1222 (2005)

Shubnikov, A.V.: On the Relationship Between the Crystalline Individual and the Crystalline Medium. In sat Nauka, Selected works on crystallography. Moscow (1975)

Sobolev, N.V., Efimova, E.S., Reimers, L.F., Zakharchenko, O.D., Makhin, A.I., Usova, L.V.: Mineral inclusions in diamonds of the Arkhangelsk Kimberlite Province. Geol. Geophys. **38**(2), 358–370 (1997)

Sobolev, E.V., Lisovain, V.I., Lenskaya, S.V.: About plate formations in the structure of natural diamonds. J. Struct. Chem. **6**, 1029–1033 (1968)

Soos, M., Dodony, J.: TEM stydy of "orthopyroxene" from a lherzolite nodul. Z. Kristallogr., suppl. No 2, p. 157 (1990)

Strickland-Constable, R.F.: Kinetics and mechanism of crystallization. L.: Nauka, p. 310 (1971)

Sunagava, I.: Effect of impurity components upon growth morphology of crystals. Abstr. IMA Congr., Pisa (1994)

Sunagawa, J.: Morphology of diamonds. In: Morphology and Phase Equilibria of minerals. Sofia, pp.195–207 (1986)

Takaoka, N., Ozima, M.: Rage gas elemental abundance and isotope compositions in diamonds. Geol. Surv. Open.-Fil Rep, **701**, 418–420 (1978)

Talanov, V.M., Vorob'ov, Yu.P., Men, A.N.: Interpretation of the concentration dependent prorerties of garnets and spinels by the cluster components method. J. Phys. Chem. Solids **36**, 641–653 (1975)

Tarbaev, M.B., Ovchinnikova L.A.: Non-traditional method of morphological analysis of gold—a new possibility of genetic information. Abstr. of II Komi resp. young. scientific. conf., Syktyvkar (1990)

Taylor, W.R., Bulanova, G.P., Milledge, H.J.: Quantitative nitrogen aggregation study of some Yakutian diamond crystals: constraints on the growth, thermal, and deformation history of peridotitic

and eclogitic diamond crystals. In Sixth Int. Kimberlite Conf. (Novosibirsk). Extended Abstr., pp. 608–610 (1995)
Tiller, U.: Formation of dislocations during the growth of a crystal from a melt. In sat. Elementary processes of crystal growth. Moscow, IL, pp. 273–292 (1959)
Titkov, S.V.: Isomorphic impurities in natural diamonds and their genetic significance. Abstract of the Ph.D. in geology and mineralogy. IGEM RAS, Moscow (2018)
Tomlinson, E.L., Jones, A.P., Harris, J.W.: Co-existing fluid and silicate inclusions in mantle diamond. Earth Planet Sci. Lett. **250**, 581–595 (2006)
Vasiliev, E.A.: Planar optically active centers of diamonds as indicators of diamond formation conditions. Abstract of the Ph.D. in geology and mineralogy, Saint Petersburg state mining Institute, 2007
Vasilyev, E.A., Kozlov, A.V., Nefedov, Y.V., Petrovskiy, V.A.: Comparative analysis of Uralian, Anabar and Brasilian diamonds by infrared spectrometry method. Proc. Saint-Petersburg Min. Univ. **200**, 167–171 (2013)
Vince, V.G., Eliseev, A.P., Surin, V.A.: Physical bases of modern methods of ennobling natural diamonds and diamonds. Precious metal. Precious stones: 2008, No. 12 (180), pp. 155–163; 2009, No 2 (182), pp. 132–145; 2009, No 3 (183), pp. 127–148
Voitsekhovsky, V.G., Mokievsky, V.A.: Some questions of interrelation of growth and dissolution of crystals. WMO, pV. **94**(1), 71–82 (1971)
Walter, A.A., Kvasnitsa, V.N., Eremenko, G.K.: Structure, composition and optical properties of diamond paramorphoses on graphite // Mineral. Journal **12**(3), 3–16 (1990)
Shi, X., Okay, A.I., Shi, J. et al.: Diamond from the Dabie Shan metamorphic rocks and its implication for tectonic setting. Science. **256**(5053), pp. 80–82
Yushkin, N.P.: Mechanodiffusion in crystals and its geological consequences. Works of the IG Komi branch of the USSR Academy of Sciences, Issue 20, pp. 17–28 (1976)
Yushkin, N.P.: Theory and Methods of Mineralogy. Selected issues. L.: Science. p. 291 (1977)
Zagalskaya, Y.G., Litvinskaya, G.P.: Geometric Microcrystallography. Moscow, p 164 (1976)
Zhabin, A.G.: Ontogeny of Minerals. Nedra, Units. Moscow (1979)
Zinchuk, N.N., Koptil, V.I.: Typomorphism of Diamonds of the Siberian Platform. Moscow, Nedra, p. 603 (2003)

Afterword

Now we completed this story about diamonds from deposits of the European part of Russia. On today their known two: named after M. V. Lomonosov and V. Grib. As can be seen from the above materials, only two classical kimberlite-type diamond deposits have been discovered in the European part to date. At the same time, we note that the V. Grib deposit in an amazing way like known deposit of Yakutia—pipe Udachnaya. Moreover, a large diamond weighing 188 carats was recently extracted from this deposit. Another field is similar in many respects to the deposits of the Nakyn field of Yakutia. The main difference of this type of diamond deposits is the absence or negligible content of picroilmenite. Thus, bright representatives of two contrasting types of diamond deposits were found in ADR.

Of course, the main mineral of these deposits is diamond, which is the main focus of the book. Since 1980 of the last century the extensive material about diamonds and their deposits in ADR is saved up. The first data were published in the book "Arkhangelsk diamondiferous province" (1999). There were no quarries and factories, and the book was the first data about this mineral, which were obtained from the materials of exploration, not production. To production was still far away. But even at that time it was obvious that there are good quality diamonds in the fields of ADR, from which you can get polished diamonds of "pure water". An unusually wide range of colors of diamonds, and in the modern world it attracts the attention of jewelers.

10 years after the book "Arkhangelsk diamondiferous province" in 2005 the next book "Atlas: morphogenesis of diamond and minerals-satellites in kimberlites and related rocks of the Arkhangelsk kimberlite province" was published. By this time, beginning the activities of the diamond company Severalmaz, a subsidiary of ALROSA. Development of the deposit M. V. Lomonosov began with the Arkhangelsk pipe, a feasibility study for the development of the field was compiled. In fact, the active development of diamond deposits began, and by that time new materials about diamond deposits were obtained. This made it possible to pay special attention to the diamond in terms of its application in the search for diamond deposits, as narrated in the new book "Atlas...".

V. Garanin et al., *Diamonds from the Arkhangelsk Province, NW Russia*, Springer Mineralogy, https://doi.org/10.1007/978-3-030-35717-7

A new stage in the research of deposits has come—the stage of intensive study of diamond and substance of kimberlite pipes. By this time, large quantities of diamonds had already been mined, their value was estimated, and the production of polished diamonds began. It was necessary to study diamonds in all directions: the content, quantity and quality of diamond raw materials, parameters of its cost, and it was also necessary to consider the issues of its origin. Since 2006 the Laboratory of diamond deposits of the Geological faculty of Lomonosov Moscow state University has been dealing with these issues detailed. By 2010, a feasibility study was prepared for the development of the V. Grib deposit and began its development.

The laboratory of diamond deposits has got a unique opportunity to study diamond and substance (indicator minerals, kimberlite rocks, diamonds) sequentially by phases of intrusion, laterals and depth of quarry. As a result of such an integrated approach, extensive new material was obtained, which had to be generalized. This today's generalization predetermines the following generalization after 10 years, when the quarries will go to depth and new unique materials will be obtained.

We have tried in this book to focus on spectroscopic studies, the application of new local methods of research of solids to the study of diamond. It was necessary to show all these possibilities through the prism of new data. All these are new achievements, new results. But in this book they relate to the diamond of the Arkhangelsk diamondiferous region. The next step is to prepare a book about the diamond from the bodies of the Yakut diamondiferous province, which is already being prepared and will, we hope, next year.

The accumulation of new data leads to a qualitative leap in the understanding of the origin and development of the diamond. And this in turn leads researchers to a new understanding of the nature of diamond, its polygenicity and discreteness, its place in the hierarchy in the system of minerals. These are all very important issues.

The diamond is an important mineral of natural mineral system, it is found in celestial bodies, and in the mantle and crustal rocks. But diamond deposits today are associated only with kimberlites and lamproites.

We have no doubt that in the coming years we will get new data about the nature of the diamond, its physical characteristics. All of them should be memorized, generalized, and on the basis of such generalizations move on. That's what we do and will continue to do until there is a desire, possibility and the power to enforce.

Completing this work, we once again pay tribute to our teachers who have invested in us knowledge. And then we pass this knowledge along the chain to our students, hoping and firmly believing that they will open new knowledge to the world for the benefit of mankind.

List of Terms

Amphibolitization
Transformation of igneous, metasomatic rocks into amphibolites under the influence of mantle metasomatism processes.

Anti-skeleton, Skeleton crystals
Distortion of the shape of crystals grown under conditions of strong supersaturation or increased viscosity of the environment. In this case, there is an advanced growth of the edges and vertices of the crystal and the formation of cavities in the central parts of the faces. The opposite is anti-skeleton form crystals, at which the maximum growth occurs in the central parts of faces. The anti-skeleton crystal in form approaches a sphere with block and mosaic surfaces of faces.

Asthenosphere
Layer of reduced viscosity in the upper mantle of the Earth, located under a harder and colder lithosphere. Allows more rigid blocks of the lithosphere to move relative to each other and provides isostatic equilibrium of these blocks. The thickness of the asthenosphere is about 100 km, but this is a conditional value, since the stiffness increases with depth smoothly.

Aulacogen
Linear moving zone inside the platform. In a simple case it is a deep (with basement lowering up to 5–10 km), narrow (from several tens to the first hundred km) and elongated for hundreds or first thousands of kilometers deflections limited by long-term developing faults.

Autolith
Inclusion of fragments of rocks of the same composition, but earlier generations.

Breccia
Rock consisting of cemented large (>1 cm) debris.

V. Garanin et al., *Diamonds from the Arkhangelsk Province, NW Russia*, Springer Mineralogy, https://doi.org/10.1007/978-3-030-35717-7

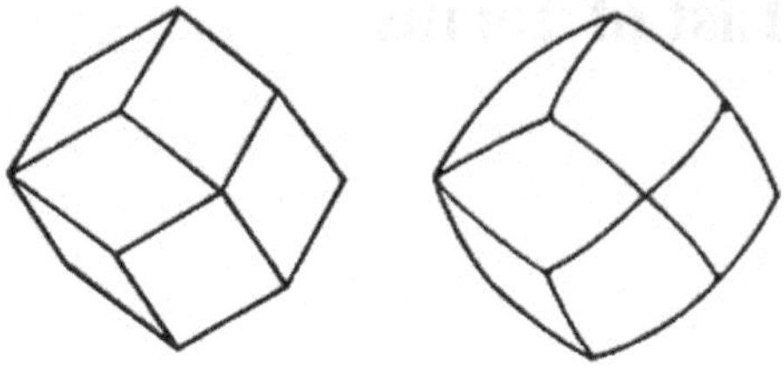

Fig. A.1 Rhombic dodecahedron (left) and dodecahedroid (right)

Chips
Debris and crystals of highly distorted shape (The term is used in trade classifiers).

Chromespinelide, Spinelide
Minerals of the group of complex oxides, solid solutions of chromite $FeCr_2O_4$, spinel $MgAl_2O_4$, hercynite $FeAl_2O_4$, magnetite $FeFe_2O_4$ and ulvospinel Fe_2TiO_4 with the general formula (Mg, Fe^{2+})(Cr, Al, Fe^{3+} Ti)$_2O_4$. The content of Cr_2O_3 in the chromespinelide can reach 75%.

Cleavage
Tendency of crystalline materials to split along definite crystallographic structural planes.

Cliv. Mb
The term of trade classifiers, combines fragments with intersecting cracks.

Coated diamonds
Diamond crystals consisting of a core and a shell, formed under different conditions. In this case, the shell usually has a fibrous structure due to the normal growth mechanism.

Dodecahedroid
Curved-faced crystal formed as a result of partial dissolution processes affecting mainly vertices and edges. Simple primary form is a rhombic dodecahedron (Fig. A.1).

Dunite
Peridotite, containing >90% olivine. Accessory chromite is almost always present.

Eclogite
High metamorphism rock consisting of garnet (pyrope or Mg-rich almandine) and omphacite (Na-Ca-Al-Mg clinopyroxene). As secondary minerals are usually present kyanite, quartz, rutile.

Epigenetically
Formed as a result of any subsequent (secondary) processes.

Fibrous growth mechanisms

This mechanism is based on the rapid growth of spiral structures in one direction. Such growth is possible only in highly non-equilibrium conditions. The complex internal structure of the diamond individuals formed by this mechanism causes their differences from the crystals grown by the mechanism of layer-by-layer growth.

Harzburgite

Peridotite containing <5% hornblende >5% orthopyroxene and <5% clinopyroxene.

Knorringite minal

Theoretical end member of the isomorphous series of garnets—$Mg_3Cr_2[SiO_4]_3$.

Kimberlite

A series of ultrabasic igneous rocks, forming the explosion pipes, dykes and sills. Kimberlite is massive brecciated rock that consists of fine-crystalline (or partly glassy) mass of cement, and plunged into it dissimilar inclusions—more coarse-grained minerals and rocks fragments. Primary magmatic, modified cement is composed of fine-grained aggregate olivine, pyroxene, phlogopite, perovskite, magnetite, serpentine, calcite, etc., as well as serpentinized and carbonated vitreous matrix. Phenocrysts are represented by grains of olivine, pomegranate, pyroxene, ilmenite, phlogopite, etc. In addition, there may be kimberlite autoliths of earlier generations and xenoliths of host rocks. The structure is porphyry; the main mineral of phenocrysts is olivine. However, it should also be noted the existence of aphanitic kimberlites, which are characterized by the absence of inclusions. Such kimberlites can be found, for example, in the form of sills and vein bodies.

Lamproite

A common name for a group of alkaline igneous rocks from ultrabasic to intermediate composition, the main minerals of which are magnesian olivine (forsterite), diopside, leucite, sanidine, phlogopite, richterite and typomorphic minerals: priderite and wadeite. Lamproites differ from kimberlites by a high concentration of Ti, K, P, but there are no significant differences between the diamonds of these two genetic types.

Layered or tangential growth mechanism

The crystal growth occurs under conditions of low supersaturation, carried out by successive growth layers, i.e. the tangential displacements of the stairs. The main stage in this case is the stage of emergence of new stairs. There are two main mechanisms of their formation: spontaneous and on the structural defects. Crystal growth by the mechanism of layer-by-layer growth leads to the formation of flat-faced crystals of mainly octahedral forms.

Lherzolite

Peridotite, containing >40% olivine, <5% hornblende, >5% orthopyroxene and >5% clinopyroxene.

Maccles
The term of trade classifiers, combines spinel twins and crystals with double seams.

Melilitite
Fine-grained extrusion or Intrusive rock containing <10% olivine, consisting of melilite and clinopyroxene (usually titaniferous augite), with minor amounts of feldspar and sometimes plagioclase.

Metasomatite
Rocks formed as a result of the interaction between the primary rock and the fluid, which different from it in chemical composition. The reaction of change occurs when the rock remains solid, but with the active introduction and removal of chemical elements.

Normal growth mechanisms
During normal growth of a crystal atoms are joined to the growing crystal at almost any place of the surface. This is possible in conditions of significant supersaturation and high temperature gradients, when there are a lot of energy-efficient places of fixing atoms on the surface. In this case, the surface in the process of growth moves along the normal to it. Results normal growth: the rounded surfaces of the crystals.

Olivinite
Igneous plutonic rock, of ultramafic composition, the normal range of alkalinity. It consists of more than 90% olivine with a mixture of magnetite; this olivinite differs from dunite, which instead of magnetite is present chromite. A small amount of olivinite may contain impurities of pyroxenes (clinopyroxene, orthopyroxene), plagioclase, etc.

Paragenesis
Regular joint presence of minerals, which associated with common processes of their formation and subsequent changes.

Peridotite
A common name for a group of coarse-grained ultramafic plutonic rocks containing more than 40% olivine (Mg_2SiO_4). This includes dunites, lherzolites, harzburgites and vehrlites. In addition to olivine, these rocks may include pyroxenes, amphiboles and micas, as well as in small amounts of feldspar. Usually there are secondary minerals of the spinel group.

Phlogopitization
Transformation of pyrope and enstatite into phlogopite during mantle metasomatism under the fluid action.

Platelets
Plate formations in the planes of {111} diamond crystals. According to different concepts, they consist of either nitrogen atoms or carbon atoms in interstitial positions.

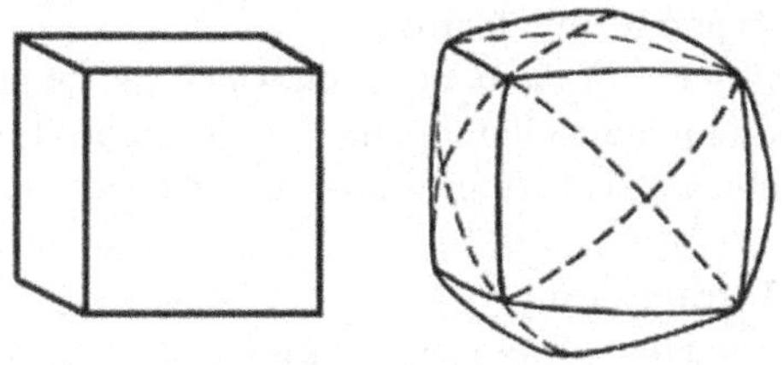

Fig. A.2 Cube (left) and tetrahexahedroid (right)

Policentric growth
Occurs at strong supersaturation, which causes the possibility of mass formation of crystallization centers. Leads to the formation of polycrystalline splices.

Porphyroblastic structures
The structure characterized by the presence of large (0.5–5 cm) round secretions of indicator minerals of kimberlites.

Pseudo-hemimorphic (crystal)
Uneven development of faces of simple forms on different sides of the crystal, and/or their dissolution to varying degrees, leading to apparent differences in the shape of the crystal on opposite sides.

Pyroxenite
Coarse-grained, holocrystalline igneous ultramafic rocks containing >60% pyroxene. It may also contain biotite, hornblende, olivine.

Spinel twin
The splicing of individuals of the diamond in (111) plane, i.e. along the facets of the octahedron, perpendicular to the axis L3. The parts of the double are rotated 180° relative to each other. The diamond is characterized by the flattening of the twins according to the spinel law, resulting in triangular or rhombic forms.

Tetrahexahedroid
A curved-faced crystal formed as a result of partial dissolution processes affecting mainly vertices and edges. Simple primary form is a cube (Fig. A.2).

Tholeiitic basalt
Volcanic rock with SiO_2 content up to 50 wt%; consisting of basic plagioclase, pyroxene (including pigeonite—a variety of diopside-augite with low content of CaO), basaltic hornblende, quartz, and sometimes olivine. In the bulk there are either quartz-feldspar coalescence or glass of acidic composition.

Tuff breccias
Cemented rock in which, along with large (>1 cm) debris present more smaller.

Typomorphic feature

Features of the mineral, on which can be judged on those or other conditions of its origin are called typomorphic features. Typomorphic minerals—minerals that are characteristic (typical) for certain conditions of formation of rocks.

Ultrabasites

The group of igneous rocks, allocated by the content of silica (SiO_2), which varies within 30–45%. In most cases, the this rocks contain a lot of MgO and FeO. Rock-forming minerals are olivine, clinopyroxene (diopside, hedenbergite), orthopyroxene (enstatite, bronzite, hypersthene), chrome spinel, hornblende, melilite, nepheline, leucite, titanomagnetite. Typical accessory minerals are ilmenite, magnetite, apatite, leucoxene, calcite, pyrope, diamond. Characteristic secondary minerals of ultrabasites are serpentine and talc. Characteristic representatives of ultrabasic rocks of plutonic class are dunites and olivinites; volcanic class—picrites. Representatives of a special series of rocks are lamproites and kimberlites.

Urals type (crystal)

The final form of dissolution of the diamond crystal with a rounded shape and aligned relief, close to the dodecahedroid.

Xenolith

Inclusion of fragments of host rocks, including rocks of deep mantle origin.

Xenotuff breccias

Tuff breccias, fragments of which represented the host rocks.

The manufacturer's authorised representative in the EU is Springer Nature Customer Service Centre GmbH, Europaplatz 3, 69115 Heidelberg, Germany. If you have any concerns regarding our products, please contact ProductSafety@springernature.com

Printed and bound by CPI Group (UK) Ltd, Croydon, CR0 4YY
28/09/2026
02224091-0010